Experimental Research and Computational Analysis of Eco- and Bio-Materials

Experimental Research and Computational Analysis of Eco- and Bio-Materials

Tomasz Garbowski
Aleksander Marek

MDPI

Basel • Beijing • Wuhan • Barcelona • Belgrade • Novi Sad • Cluj • Manchester

Tomasz Garbowski
Department of Biosystems Engineering
Poznan University of Life Sciences
Poznan
Poland

Aleksander Marek
Faculty of Engineering and Physical Sciences
University of Southampton
Southampton
United Kingdom

Editorial Office
MDPI AG
Grosspeteranlage 5
4052 Basel, Switzerland

This is a reprint of articles from the Special Issue published online in the open access journal *Materials* (ISSN 1996-1944) (available at: www.mdpi.com/journal/materials/special_issues/Eco_Bio_Materials).

For citation purposes, cite each article independently as indicated on the article page online and using the guide below:

Lastname, A.A.; Lastname, B.B. Article Title. *Journal Name* **Year**, *Volume Number*, Page Range.

ISBN 978-3-7258-2074-0 (Hbk)
ISBN 978-3-7258-2073-3 (PDF)
https://doi.org/10.3390/books978-3-7258-2073-3

Contents

materials

MDPI

Editorial

Experimental Research and Computational Analysis of Eco- and Biomaterials

Tomasz Garbowski

Department of Biosystems Engineering, Poznan University of Life Sciences, Wojska Polskiego 50, 60-627 Poznan, Poland; tomasz.garbowski@up.poznan.pl

This Special Issue of Materials is dedicated to the exploration and analysis of eco- and biomaterials through experimental research and computational methods. These materials are becoming increasingly significant as construction materials and load-bearing elements across various engineering and medical applications. Biomaterials, derived from biological sources such as wood-based products and corrugated cardboard, as well as synthetic and natural materials, are notable for their ability to interact with organic tissues. Ecomaterials, encompassing construction materials and textiles, also play a crucial role. Both categories include composite materials known for their unique properties that address complex challenges where traditional materials are insufficient. This collection provides a platform for scientists and engineers to share the latest advancements in theoretical, experimental, and computational studies related to eco- and biomaterials. Key topics include mechanical properties and strength estimation, numerical and analytical homogenization techniques, laboratory research methods, and linear and nonlinear analyses of structures made from these materials. The focus extends to laminated, corrugated, and fibrous materials, emphasizing experimental validation and empirical evidence. Among submissions that contribute to the understanding and application of bio-, eco-, and composite materials, one can also find comprehensive studies that highlight the mechanical behavior and practical applications of these innovative materials.

In the collection, Kopras et al. [1] analyze the deflection behavior of steel support plates used in temporary excavations, integrating experimental and computational methodologies to evaluate mechanical performance under real-world conditions. This research bridges the gap between empirical data and theoretical models, involving large-scale field experiments to measure plate deflections and comparing them with numerical predictions using the finite difference method. Combining traditional patch measurements and advanced 3D laser scanning under various backfill loads offers a comprehensive performance assessment. The numerical results align closely with experimental data, validating the computational approach. Notably, a new deflection limit criterion ($w_{gr} = L/130$) for temporary excavation support plates is proposed, enhancing design and safety assessments.

Though the study primarily addresses steel plates, the methodologies and insights are broadly applicable to eco- and biomaterials. The integration of experimental validation with computational analysis serves as a model for sustainable materials in construction. Findings on deflection and load-bearing behavior contribute to the broader understanding needed for developing and applying eco-friendly and biological materials. The empirical data and computational models offer valuable benchmarks for future research, enhancing the reliability and safety of construction practices involving diverse materials.

Ma et al. [2] investigate the seismic performance of stiffened corrugated steel plate shear walls (CSPWs) under various conditions, including atmospheric corrosion. This research provides insights into the mechanical behavior and durability of materials, contributing to the field of eco- and biomaterials. The study introduces three types of CSPWs: unstiffened (USW), cross-stiffened (CSW), and asymmetric diagonal-stiffened (ASW). Using a comprehensive numerical model validated against cyclic test data, the research analyzes

Citation: Garbowski, T. Experimental Research and Computational Analysis of Eco- and Biomaterials. *Materials* **2024**, *17*, 4269. https://doi.org/10.3390/ma17174269

Received: 15 July 2024
Revised: 25 July 2024
Accepted: 21 August 2024
Published: 29 August 2024

the lateral and seismic performance of these walls under monotonic and cyclic loading conditions. The results show that stiffeners significantly enhance the elastic critical buckling load, initial stiffness, ultimate shear resistance, energy dissipation capacity, and ductility of CSPWs. Asymmetric diagonal stiffeners outperform cross stiffeners in improving these properties and reducing out-of-plane deformation and stiffness degradation.

The study also examines the impact of atmospheric corrosion on the seismic performance of CSPWs, finding that stiffeners help mitigate corrosion effects, with asymmetric diagonal stiffeners being more effective than cross stiffeners. A fitted formula for predicting the ultimate shear resistance of corroded CSPWs is provided, offering valuable design guidance for engineering applications. Ma et al.'s work significantly enhances the understanding of how structural modifications and environmental factors influence steel material performance, offering a methodological framework adaptable for eco- and biomaterials. The integration of experimental validation with computational analysis sets a strong precedent for future research in developing sustainable construction materials with improved mechanical properties and durability.

Chybiński and Polus [3] present an experimental study on the load-slip behavior of aluminum–timber composite bolted connections reinforced with toothed plates. This research is pivotal for eco- and biomaterials, offering valuable insights into the mechanical performance and reinforcement techniques of sustainable composite structures. The study evaluates the effectiveness of reinforcing aluminum–timber connections through laboratory push-out tests using laminated veneer lumber (LVL) panels, aluminum alloy I-beams, and various bolts. The primary aim is to determine the shear resistance and stiffness of the connections with and without toothed plate reinforcement. Results indicate that while toothed plates reduce timber destruction in bearing zones, they do not significantly protect against splitting of LVL slabs in connections using grade 8.8 bolts. However, for connections with grade 5.8 bolts of 10 mm diameter, toothed plates substantially increase stiffness, though at the cost of reduced strength due to faster bolt shank shearing.

This research significantly enhances the understanding of composite materials' behavior under load, particularly those combining metal and timber. The findings provide a nuanced view of how different reinforcement strategies impact composite connections' performance, directly applicable to the design and development of eco-friendly composite structures. By demonstrating the practical applications and limitations of toothed plate reinforcement, the study contributes to the broader discourse on experimental research and computational analysis of eco- and biomaterials. It underscores the importance of combining experimental validation with advanced modeling techniques to develop innovative, sustainable construction solutions.

Ciesielczyk and Studziński [4] investigate the failure scenarios of various connection types between thin-walled beams and sandwich panels under horizontal loads, simulating bending and lateral–torsional buckling. This research is crucial for understanding the structural behavior and integrity of these connections, which is particularly relevant to this Special Issue. The study examines standard civil engineering connections such as self-drilling fasteners, bolts, blind rivets, and double-sided acrylic tape, both linearly and pointwise. These connections were analyzed under horizontal loads with constant eccentricity to determine their lateral stiffness, initial and secant stiffness, ultimate capacity, and deformation capacity. The research highlights different failure mechanisms: self-drilling fasteners and bolts penetrate all layers, resulting in significant initial stiffness and capacity but also web bow deformation and section rotation under higher loads; blind rivets fail through clamping arm deformation and facing delamination; and double-sided acrylic tapes primarily fail through detachment.

Ciesielczyk et al.'s work provides essential quantitative data on the mechanical response and failure mechanisms of various connection types, offering insights for the design and optimization of sandwich panel connections in eco-friendly and biomaterial structures. This research establishes a comprehensive understanding of how different reinforcement and connection strategies affect the structural integrity and performance of composite materials, laying the groundwork for developing advanced materials and connection methods to enhance the durability and safety of sustainable construction practices.

Sybis and Konował [5] investigate the influence of modified starch admixtures on the rheological properties and compressive strength of cement composites, highlighting the potential of using natural and biodegradable materials to enhance cement performance. The research examines 17 different modified starches, including chemically and physically modified starches and starch hydrolysates, focusing on their effects on the viscosity, tangential stress, yield point, and plastic viscosity of cement slurries. Additionally, the study looks at the flow of fresh cement slurries and the compressive strength of hardened composites.

The results show that modified starches can significantly alter the rheological properties of cement slurries. For example, the retentate LU-1420-0.5%Ac-R increased flow by 82%, and the retentate LU-1412-R improved compressive strength by 25%. However, some starches, like retentate LU-1422-R and retentate OSA-2.5%-R, decreased compressive strength. Sybis et al.'s work demonstrates that natural starch derivatives can effectively modify cement composites, offering potential applications in sustainable construction. The study provides insights into how different starch modifications influence the mechanical behavior and workability of cementitious materials, supporting the development of greener and more efficient building materials. This aligns with the themes of experimental research and computational analysis of eco- and biomaterials, showcasing the practical applications of biodegradable additives in enhancing construction material performance and sustainability.

Qu et al. [6] provide a comprehensive review of stress-fractional plasticity models, which combine fractional calculus with classical plasticity theories to better model the mechanical behavior of materials. This approach is particularly relevant to eco- and biomaterials, offering advanced methodologies for sustainable materials in engineering applications. Fractional plasticity (FP) models are able to capture the non-associated flow behavior of geomaterials like clay, sand, ballast, and rock, which standard plasticity models for metals cannot accurately represent. The review explores the development of FP models; defines the stress length scale (SLS), crucial for fractional differentiation; and discusses two primary branches: past stress and future reference critical states, with a third branch incorporating both for a more holistic approach.

The research highlights the advantages and limitations of FP models and their real-world applications, offering a robust framework for predicting material behavior under various loading conditions and improving simulation accuracy and reliability. Qu et al.'s review contributes significantly to this Special Issue by presenting advanced models that enhance the understanding and performance of sustainable materials. The insights from these models can be applied to develop more resilient and efficient eco- and biomaterials, promoting sustainable practices in construction and engineering. This work underscores the importance of integrating computational approaches with experimental research to innovate and apply eco-friendly materials.

Ksit et al. [7] examine the impact of water vapor and moisture barriers on the energy efficiency and durability of ventilated partitions in buildings, highlighting the importance of moisture analysis in designing sustainable, energy-efficient structures. Excess moisture can degrade building materials, reduce insulation efficiency, and worsen indoor climates, increasing energy demands. The study uses modern digital solutions to model building partitions over an 8-year period under various environmental conditions, focusing on flexible waterproofing materials and their effect on dew point temperature, air temperature, and relative humidity.

The research provides a detailed numerical analysis of different ventilated partition models, assessing the necessity and effectiveness of vapor barriers in preventing moisture damage. By evaluating these barriers over time, the study offers insights into maintaining long-term durability and energy efficiency. Ksit et al.'s work contributes to this collection by demonstrating how advanced computational techniques can enhance sustainable building materials. The findings support the development of resilient construction practices that reduce energy consumption and improve indoor environmental quality, aligning with sustainable architecture goals.

The article by Malewski et al. [8] presents developments in the constitutive material model for architectural soda lime silicate (SLS) glass, updating key modeling parameters. This research is crucial for eco- and biomaterials, providing insights into precise material parameters for accurate numerical modeling of glass in sustainable architectural design. By conducting experimental investigations on the coefficient of thermal expansion (CTE), glass transition temperature, and Young's modulus, the study addresses gaps in contemporary data on SLS glass. These parameters are critical for predicting glass behavior under various conditions, including mechanical and thermal performance.

Malewski et al.'s work significantly enhances the understanding of glass behavior under different environmental conditions. The updated material parameters improve simulations, design, and prototyping of complex glass structures, leading to more efficient production and supporting sustainable building practices. Integrating experimental data with computational analysis, the study bridges the gap between theoretical models and practical applications, ensuring materials used in sustainable construction are reliable and efficient. This research underscores the importance of continuously updating material databases to reflect current technologies and practices, which is essential for the ongoing development of sustainable architecture.

The article by Mrówczyński et al. [9] investigates the compressive strength capacity of open-top corrugated board cartons used for transporting fruits and vegetables. This study is highly relevant to eco- and biomaterials, focusing on optimizing biodegradable and sustainable packaging. Using sensitivity analysis, the research identifies critical geometric parameters affecting load capacity. The authors developed a finite element model to simulate the box compression test (BCT) and analyzed seventeen design parameters. They found that modifications in the length of non-folded sidewall parts and box height significantly influence compressive strength, while dimensions and positions of ventilation holes had negligible effects.

By identifying key parameters affecting structural integrity, the study provides valuable insights for designing eco-friendly packaging solutions. The findings help manufacturers create more efficient and durable corrugated board cartons by focusing on influential geometric factors, thus reducing material usage. Mrówczyński et al.'s work demonstrates how advanced computational techniques and sensitivity analyses optimize biomaterial design. This approach enhances the mechanical performance of materials and supports environmentally friendly packaging solutions, reducing the ecological footprint of the packaging industry. Integrating experimental validation with computational modeling sets a robust precedent for future research in sustainable materials.

The article by Fehér et al. [10] explores the compression strength estimation of corrugated board boxes, focusing on reducing sidewall surface cutouts. This research is relevant to the Special Issue as it optimizes biodegradable packaging materials to enhance mechanical properties while minimizing material usage and waste. The study investigates various cutout configurations on B-flute corrugated cardboard boxes, commonly used in supply chains, and compares experimental observations with the McKee formula for compression strength estimation.

Fehér et al. found that compression strength decreases linearly with increasing cutout size, showing significant discrepancies from the McKee formula, which doesn't account for sidewall cutouts. To address this, the authors propose a modified estimation approach incorporating empirical test data for more accurate predictions. This research demonstrates how experimental and numerical methods can optimize sustainable packaging design, providing valuable insights for packaging engineers to balance material reduction with mechanical performance. The integration of empirical data with computational analysis exemplifies the interdisciplinary approach necessary for advancing sustainable materials in practical applications.

Mrówczyński and Garbowski [11] explore the influence of imperfections on the effective stiffness of multilayer corrugated board, which is crucial for understanding the structural integrity of eco-friendly packaging materials like corrugated cardboard. Widely used for its recyclability, biodegradability, and durability, corrugated cardboard is an increasingly popular choice as companies move away from plastic to improve their environmental impact. The study focuses on how geometric imperfections from the manufacturing process affect the mechanical properties of cardboard.

The authors present a numerical homogenization procedure using the finite element method (FE) to include geometric imperfections in calculating the board's effective stiffness. A 3D model of a representative volumetric element (RVE) is built, incorporating various buckling and distorted shapes from prior analysis. This approach allows for a quick and scalable method to account for imperfections and their impact on material stiffness. Mrówczyński et al.'s work significantly contributes to this collection by providing a framework to understand how imperfections affect the mechanical behavior of sustainable packaging materials. The findings help optimize the design and manufacturing processes of corrugated cardboard, ensuring its integrity and performance despite manufacturing-induced imperfections, supporting the development of resilient and efficient eco-friendly packaging solutions.

Fehér et al. [12] investigate the compressive strength of corrugated paperboard packages with varying cutout rates, which is integral to understanding eco- and biomaterials. The study uses the finite element method (FEM) to model and predict the compression force of corrugated cardboard boxes with different sidewall cutout configurations. Boxes tested had widths and heights of 300 mm, lengths from 200 mm to 600 mm, and cutout rates of 0%, 4%, 16%, 36%, and 64%. The FEM model incorporated a homogenized linear elastic orthotropic material with Hill plasticity.

Results from numerical simulations and experimental box compression tests (BCT) showed that the FEM model accurately predicts the compression strength across various cutout configurations. However, model accuracy slightly decreased with higher cutout rates, highlighting the challenge of maintaining structural integrity. Fehér et al.'s work contributes significantly to sustainable packaging by providing a reliable numerical model for predicting the performance of corrugated cardboard boxes with different cutout designs. This research helps optimize eco-friendly packaging design and manufacturing, ensuring strength and durability while reducing material usage and waste.

Bartkowiak and Słowik [13] focus on developing a predictive model for the stiffness modulus $|E^*|$ of high-modulus asphalt concrete (HMAC) using the four-point bending beam test (4PBB). This study is relevant to eco- and biomaterials, offering insights into optimizing sustainable infrastructure materials through advanced modeling techniques. Building on existing models like the Witczak model, the authors aimed to enhance predictions for HMAC's stiffness modulus. This involved extensive laboratory testing of asphalt mixtures to gather data on stiffness modulus and phase angles under varying temperatures and loading frequencies, including tests on neat and modified bituminous binders using a dynamic shear rheometer (DSR).

The research presents a new model, Model A, which modifies the Witczak model to improve accuracy for the 4PBB method, considering air void content, effective binder content, and aggregate gradation. Results show the model predicts stiffness modulus with high accuracy, as evidenced by low errors. This work contributes to the Special Issue by demonstrating refined computational models that enhance understanding and performance of sustainable construction materials. The findings support developing more resilient and efficient infrastructure using eco-friendly materials like HMAC, aiding engineers in designing and optimizing asphalt mixtures for better durability and performance, thus contributing to sustainable, long-lasting infrastructure solutions.

Acknowledgments: The guest editors would like to extend their heartfelt appreciation to the in-house editor for her unwavering diligence and steadfast support throughout the creation of this Special Issue. We are deeply grateful to all the authors for their invaluable scientific contributions, which were fundamental to this publication. Additionally, we sincerely thank the reviewers for their constructive comments and insightful suggestions, which have greatly enhanced the quality of the presented works.

Conflicts of Interest: The author declares no conflicts of interest.

References

1. Kopras, M.; Buczkowski, W.; Szymczak-Graczyk, A.; Walczak, Z.; Gogolik, S. Experimental Validation of Deflections of Temporary Excavation Support Plates with the Use of 3D Modelling. *Materials* **2022**, *15*, 4856. [CrossRef] [PubMed]
2. Ma, X.; Hu, Y.; Jiang, L.; Jiang, L.; Nie, G.; Zheng, H. Study on the Seismic Performance of Stiffened Corrugated Steel Plate Shear Walls with Atmospheric Corrosion. *Materials* **2022**, *15*, 4920. [CrossRef] [PubMed]
3. Chybiński, M.; Polus, Ł. Experimental Study of Aluminium-Timber Composite Bolted Connections Strengthened with Toothed Plates. *Materials* **2022**, *15*, 5271. [CrossRef] [PubMed]
4. Ciesielczyk, K.; Studziński, R. Experimental Investigation of the Failure Scenario of Various Connection Types between Thin-Walled Beam and Sandwich Panel. *Materials* **2022**, *15*, 6277. [CrossRef] [PubMed]
5. Sybis, M.; Konował, E. Influence of Modified Starch Admixtures on Selected Physicochemical Properties of Cement Composites. *Materials* **2022**, *15*, 7604. [CrossRef] [PubMed]
6. Qu, P.; Sun, Y.; Sumelka, W. Review on Stress-Fractional Plasticity Models. *Materials* **2022**, *15*, 7802. [CrossRef] [PubMed]
7. Ksit, B.; Szymczak-Graczyk, A.; Pilch, R. Numerical Simulation of the Impact of Water Vapour and Moisture Blockers in Energy Diagnostics of Ventilated Partitions. *Materials* **2022**, *15*, 8257. [CrossRef] [PubMed]
8. Malewski, A.; Kozłowski, M.; Podwórny, J.; Środa, M.; Sumelka, W. Developments on Constitutive Material Model for Architectural Soda-Lime Silicate (SLS) Glass and Evaluation of Key Modelling Parameters. *Materials* **2023**, *16*, 397. [CrossRef] [PubMed]
9. Mrówczyński, D.; Gajewski, T.; Garbowski, T. Sensitivity Analysis of Open-Top Cartons in Terms of Compressive Strength Capacity. *Materials* **2023**, *16*, 412. [CrossRef] [PubMed]
10. Fehér, L.; Pidl, R.; Böröcz, P. Compression Strength Estimation of Corrugated Board Boxes for a Reduction in Sidewall Surface Cutouts—Experimental and Numerical Approaches. *Materials* **2023**, *16*, 597. [CrossRef] [PubMed]
11. Mrówczyński, D.; Garbowski, T. Influence of Imperfections on the Effective Stiffness of Multilayer Corrugated Board. *Materials* **2023**, *16*, 1295. [CrossRef] [PubMed]
12. Fehér, L.; Mrówczyński, D.; Pidl, R.; Böröcz, P. Compressive Strength of Corrugated Paperboard Packages with Low and High Cutout Rates: Numerical Modelling and Experimental Validation. *Materials* **2023**, *16*, 2360. [CrossRef] [PubMed]
13. Bartkowiak, M.; Słowik, M. Development and Analysis of High-Modulus Asphalt Concrete Predictive Model. *Materials* **2023**, *16*, 4509. [CrossRef] [PubMed]

materials

MDPI

Article

Experimental Validation of Deflections of Temporary Excavation Support Plates with the Use of 3D Modelling

Marek Kopras [1], Wiesław Buczkowski [2], Anna Szymczak-Graczyk [2,*], Zbigniew Walczak [2] and Sławomir Gogolik [2]

1 Research and Development Department, Kopras Sp. z o. o., Szklarnia 7, 64-510 Wronki, Poland; marek@kopras.pl

2 Department of Construction and Geoengineering, Faculty of Environmental and Mechanical Engineering, Poznan University of Life Sciences, Piątkowska 94 E, 60-649 Poznań, Poland; wabucz@o2.pl (B.W.); zbigniew.walczak@up.poznan.pl (W.Z.); slawomir.gogolik@up.poznan.pl (G.S.)

* Correspondence: anna.szymczak-graczyk@up.poznan.pl

Abstract: Almost every project is accompanied by earthworks, very often involving various types of excavation, and the work of people in the excavations. One of the most important tasks in earthworks is to ensure that the walls of the excavation are protected against sliding and that people working in and around the excavation are safe. Very often, in addition to criteria relating to safety and stability of the excavation, economic considerations are also an important criterion. This issue arises as early as the design stage and is related to the choice of construction and materials of which the shoring is to be made in such a way as to be able to withstand the pressure of the soil, ground loads resulting from stored excavated material and the operation of working machinery. Ongoing monitoring of the excavations and their reinforcement is also very important. The paper describes the unique results of experimental field tests, the purpose of which was to analyse the values of deflections of steel support plates of temporary excavation carried out on the object in 1:1 scale. The course of the experiment is presented for excavation support plates with a total depth of 6 m. Direct tests of the deflection arrow were carried out using two techniques, traditionally with a patch, and with laser scanning. Field tests were carried out for the designed situation without backfill load as well as for backfill load of 3.84, 15.36, 26.88 and 38.4 kN·m^{-2}, respectively, for two measurement stages. Stage-I of the study consisted in collecting the results for soil in intact condition, whereas Stage-II collected results for loosened soil. The research experiment was supported by numerical calculations performed using the finite difference method in variational approach. The measured maximum deflections ranged from 14.40 to 16 mm, and the calculated values were 14.95 and 14.99 mm. The comparison of calculation results and tests proved to be very consistent. The analysis of the values of deflections showed that backfill load does not have a significant effect on the deflection of the lower plate, but it does affect the deflection of the first plate up to a depth of 1.2 m. Based on the obtained results, it is recommended to assume the limit (maximum) deflection arrow for support plates of temporary excavations at least as $w_{gr} = L/130$, where L is the span of the plate. The calculation of deflection values was based on deflection values obtained experimentally and numerically for two steel variants: S235JR and S355JR. The w_{gr} indicator of the maximum deflection arrow proposed by the authors is not stipulated by the industry standards, but it can be very helpful for the designing of excavation reinforcement.

Keywords: 3D scanning; excavation; excavation support; support plate; deflection arrow; finite difference method

check for updates

Citation: Kopras, M.; Buczkowski, W.; Szymczak-Graczyk, A.; Walczak, Z.; Gogolik, S. Experimental Validation of Deflections of Temporary Excavation Support Plates with the Use of 3D Modelling. *Materials* **2022**, *15*, 4856. https://doi.org/10.3390/ma15144856

Academic Editors: Tomasz Garbowski and Aleksander Marek

Received: 14 June 2022
Accepted: 10 July 2022
Published: 12 July 2022

1. Introduction

Sustainable growth of societies and continuous expansion of technical infrastructure result in the emergence of new network, communication, industrial and developmental investments. Systematic increase in the value of land means that more and more structures and installations are located underground. Continuous, conscious, sustainable development is associated with the need to build various underground networks, both technological and the ones that result from the environment requirements related to the protection of soil, water and air against pollution. The construction of water and sewage networks involves performing earthworks and laying pipelines in various types of excavations. In engineering practice, there are several ways (methods) to protect excavation walls against sliding. These include tight walls made of steel sheet piles (Larssen profiles), Berlin walls, cavity walls (concrete) formed in the soil, palisades made of piles (i.e., CFA or micro-piles), column walls made by jet injection as well as reusable system supports [1–3]. In practice, medium-deep and deep rope excavations are not performed without supports, with slopes having inclination corresponding to the angle of internal friction of the soil, as this would require a large strip of land and significant earthworks. Securing excavation walls by making tight walls with the use of various types of steel sheet piles or making Berlin type walls is associated with mechanical drive piling (often impact) of sheet piles or steel profiles (I-sections) [4–6]. The resultant noise and vibrations can be burdensome for residents and pose a threat to the durability of buildings (occurrence of scratches, cracks, subsidence). Therefore, in inhabited areas, works that require a pile driver are impermissible. However, if this is the case, only engines that drive support elements statically into the soil should be used. The above-mentioned methods of securing excavation walls are not suitable for supporting long line excavations. Moreover, cavity walls (concrete) formed in the soil are not suitable for securing walls of temporary line excavations. To protect and secure excavation walls, reusable system supports are used, for instance, those produced by KOPRAS, the Polish company. The components of these supports are poles, struts and plates of a steel or aluminum structure placed between poles, which must safely withstand and transmit the soil pressure acting on them. In general, plates are the most loaded elements of supports—their bearing capacity is determined by the structure, span and depth of excavations and the value of soil pressure. Plates must be constructed in such a way that they can transfer the soil pressure, and sometimes also the backfill load resulting from stored excavated material and operation of working machines (excavators, means of transport, cranes, etc.). The value of soil pressure depends on the condition and type of soil, the depth of excavations, the displacement of excavation supports, the backfill load, as well as the groundwater table. Properly designed support plates must ensure the safety of people's work in any soil and water conditions in connection with the permissible excavation depth for a given type of plate, and they also affect the economic side of investments (production costs, transport, type of excavators used in drive piling due to the weight of plates). In order to be able to properly design excavation support plates, it is necessary to know the effective soil load, its distribution and the amount of pressure, and also the static diagram corresponding to the actual work of the structure should be adopted in calculations [7,8]. Earth pressure is related to the form and size of the displacement of the retaining structures in relation to the soil. The problem of earth pressure has been a frequent subject of scientific studies. Most often they take the form of numerical solutions [9] or model tests [10,11].

The numerical analysis of plates can be successfully performed using the finite difference method in variational approach. The literature on the subject contains numerous fundamental works on the finite difference method [12–25] and publications in which this method was applied and positively verified for calculating plates [26–28], tanks [29–32], shell structures and others [33,34]. Experimental verification of performed numerical calculations plays a key role in contemporary design. Implementation of a research model in 1:1 scale is a unique and undoubtedly the best way to verify numerical calculations, and non-invasive methods of measuring deflections and displacements of structural components are excellent measuring tools.

Validation of the obtained deflection results based on the finite difference method was carried out using a concrete tank model with an innovative measurement tool: a coordinate measuring arm with a touch probe [29]. Validation of calculations with the help of the coordinate measurement arm provided an opportunity to confirm their correctness. The coherence achieved is evidence in favour of the appropriateness of the applied calculation methods, i.e., the finite element method, which provided a basis for calculations using

Autodesk Robot Structural Analysis Professional, as well as the finite difference method in terms of energy used to make traditional calculations [29]. Terrestrial Laser Scanning (TLS) [35] is a non-invasive, non-contact technique that enables fast and, what is very important, precise acquisition of data on the geometry of measured objects in the form of coordinates of points x, y, z. 3D laser scanning is currently used in many practical engineering applications, including inventory [36–43], inspection [44,45], verification of test results [29] and maintenance of buildings [46], particularly historical buildings (constituting the cultural heritage). Inventory can be carried out faster and more accurately than when using traditional solutions. Thanks to the use of TLS techniques, it is possible to obtain more complete data on the geometry of measured objects. It can be assumed that the data are continuous, not point-based—as it is in the case of classical solutions. The accuracy of models created with the use of TLS may vary from a few millimetres [41,47–49] to even decimetres [50]. The accuracy of measurements depends on several factors, including the distance and angle between the scanner and objects being scanned as well as the type of scanned surface, particularly in the case of highly reflective, mirror surfaces [51–53]. The advantages of TLS also include the possibility of obtaining full, three-dimensional information on objects in the form of measurement points, which are their geometric representation, being in fact a dense quasi-continuous cloud of points. The information obtained in this way, the cloud of points, can then be processed, generating i.e., orthophotos, CAD or BIM models, mesh models, models for visualisation, virtual walks, etc. In addition to numerous advantages of TLS, there are also several factors that can make it significantly difficult or even impossible to obtain high-quality data by means of laser scanning. Key issues to mention here are, among others, situations in which some parts of the scanned elements may be invisible, obscured from laser scanning. Moreover, glass elements (windows, mirrors), as well as wet or damp surfaces or surfaces with water on the surface, can cause disturbances. One of the major problems at the stage of further data processing may be the fact that an enormous data set is created, which will require large hardware and computing power for further computing (high requirements for hardware of computer systems).

Problems related to obscuring the visibility of scanned items can be solved via applying techniques combining laser measurements with manual measurements [54], as well as photogrammetry [55–60]. Fawzy [57] indicates that the combination of TLS techniques and short range photogrammetry increases the accuracy of models. Based on the measurements of 20 points, 10 lines and 6 control angles, Fawzy indicates that the maximum improvement in the quality of the model was 80.1, 66.4, and 84.2% for points, lines and angles, respectively [57].

TLS is also widely used for renovation and revitalization of historic buildings for which it is planned to build a BIM model or Heritage or Historic Building Information Modelling (HBIM). As a result, it is possible to build a 3D model to recreate the plans of historic buildings for which they did not exist, use it during planning of modernisation, revitalisation, but also operation or even visualisation of the object [56,61–66]. The use of TLS can also be helpful in assessing the technical condition of facilities [67]. This can significantly facilitate damage assessment and help in proposing an appropriate renovation system [45]. The accuracy of point clouds is sufficient to perform surface regularity checks.

The purpose of the work was to experimentally verify the values of deflections occurring in support plates subjected to soil pressure and backfill load in real conditions reflecting the work of the structure. The results of the measurements obtained based on the natural-scale experiment were compared with the values calculated with the finite difference method. The paper presents the test results of actual deformation of open excavation support plates depending on the acting backfill load. Since neither in the literature nor in the standards is there a deflection limit value for such structures as temporary excavation support plates, the paper includes the recommendation for the deflection arrow value, which can be used to calculate the maximum deflection in this type of construction facilities.

The conducted tests, measurements and calculations can be used to verify existing solutions and improve the structure of excavation support plates in order to ensure greater safety of manufactured and used products, as well as to reduce their cost.

2. Materials and Methods

2.1. Description of Research Location and Objectives

The field tests were conducted from 4 September to 27 November 2019. The values of actual deformations of open excavation support plates were analysed depending on acting backfill load. The test stand was in a place with non-cohesive soils of a possibly homogeneous profile, with the groundwater table below the bottom of planned excavations. The area near Lake Rusałka was selected for the location of the test stand (Figure 1). The terrain where the test stand was installed was situated several meters above the water level. Before the final selection of its location, six boreholes were made in the area in question, from which soil samples were collected (with natural graining NU and natural humidity NW) in order to conduct laboratory tests and sounding with a static CPTU probe. These activities were aimed at checking the soil profile and groundwater level. Based on the measurements, the final location of the test stand was selected. It was a place where the groundwater table was not drilled, with convenient access and sufficient space for all operations related to excavation works and loading the backfill. In the selected place, an open excavation was made with dimensions 4 × 4 m, 6 m deep, reinforced with the system of steel plates (Figure 2).

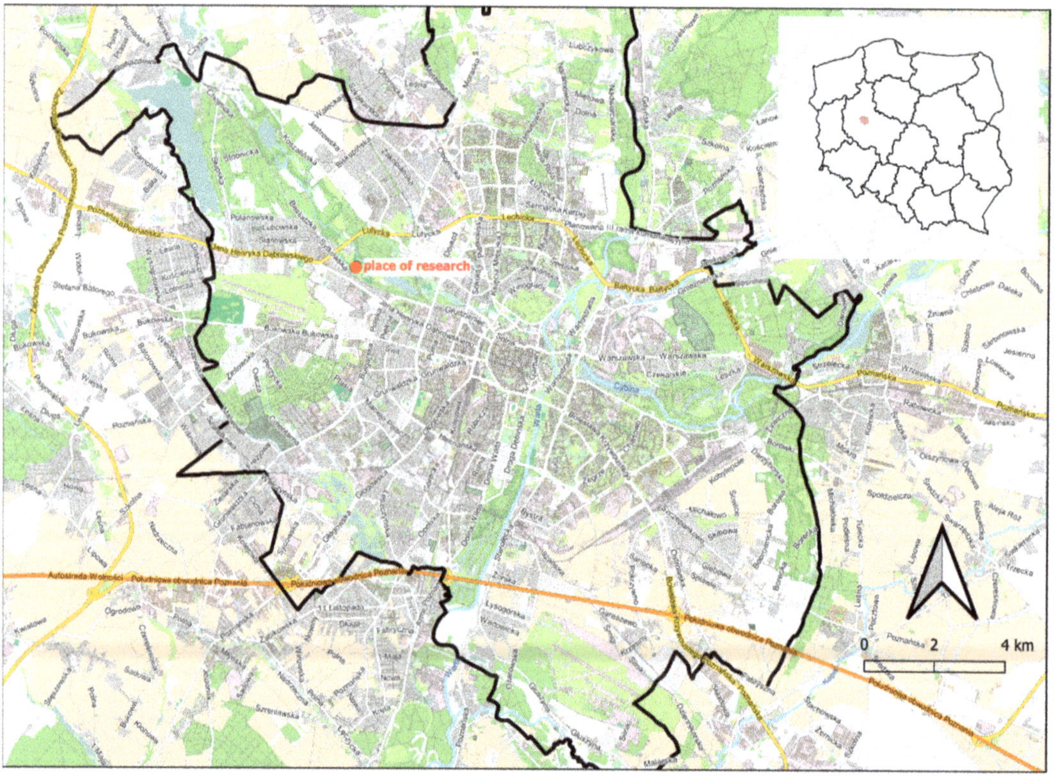

Figure 1. Location of the test stand (Poland, Greater Poland Voivodeship, the city of Poznań).

Figure 2. Test stand—view of the excavation and the excavation support.

The excavation was secured as a point box composed of steel corner posts and a system of fin plates. The fin plate with dimensions 3.92 × 2.40 × 0.12 m was a welded structure made of square and rectangular sections and flat bars. The top horizontal edge, i.e., plate head, was a structure consisting of rectangular and square profiles, additionally reinforced with an overlay made of a 15 mm thick flat bar. The flat bar stiffened the top edge of the plate, preventing it from being dented by an excavator bucket during driving the plate. The bottom horizontal edge was a cutting section, which penetrated the excavation soil under the self-weight of the plate. All profiles on the left and right side of the plate were closed with vertical edges. A special C-profile, pulled on the fronts of horizontal profiles, had welded fins guiding the plate in the guides of the corner posts. Fins transmitted the response of a freely supported beam (which a fin plate is by definition) onto the supports. Each plate was strengthened by corner posts with sliding guides (Figure 3).

2.2. Geotechnical Measurements

During preliminary geotechnical tests, six test boreholes were made to verify the soil profile and groundwater level, which were the basis for selecting the optimal location of the test stand. Detailed geotechnical tests were carried out in the selected location, including drilling to a depth of 6 m, during which samples of soil with natural graining (NU) and natural moisture (NW) were taken for laboratory tests to determine their granulometric composition and natural moisture content. Soundings with a static CPTU probe were also performed. Static sounding tests were performed with a CPTU piezocone in accordance with the TC-16 procedure recommended by the International Society for Soil Mechanics and Geotechnical Engineering [68]. The tests used a standard measuring tip with a diameter of 36 mm, a base area 10 cm^2, an apex angle 60° and a friction sleeve area 150 cm^2. The main test consisted in pressing the measuring tip into the soil with a constant value equal to 2 $cm{\cdot}s^{-1}$. During penetration, three test parameters were recorded every two centimetres of depth increment: resistance of the cone—ci, friction on the friction sleeve area—fs pressure of excess water in pores—u_2. It allowed for determining soil characteristics in the following respects: separation of geotechnical layers with the determination of soil types; identification of the stress level in the soil; determination of the compaction degree (ID); determination of the shear strength parameter—the effective angle of internal friction—(ϕ'); determination of the soil deformability parameter—edometric compressibility modulus (M_o).

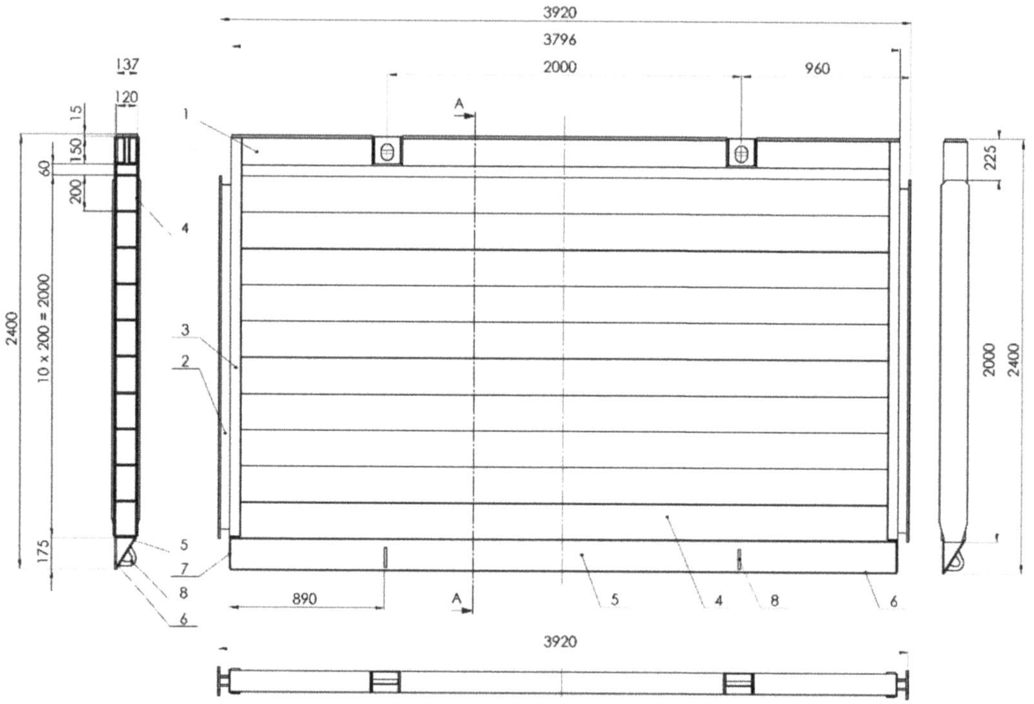

Figure 3. View and cross-sections of the support plate. Denominations: 1—plate head, 2—plate fin, 3—closing C-profile, 4—rectangular profile 200 × 10 × 4, 3760 mm long, 5—steel cutting edge g = 6 mm, 6, 7—closing sheet, 8—transport eye. Units [mm].

2.3. Tests of Stresses in the Soil Using a Hydraulic Probe

Several days before the excavation was made and the support was lowered, hydraulic sensors for measuring soil pressure were inserted into the soil. The probe was inserted into the soil using the same technology as when performing CPTU probing and with the same device. The lowest sensor was at 6.0 m below ground level, and the remaining sensors at 4.5, 3.5, 2.5, 1.5, and 0.5 m below ground level. The measurements of soil pressure taken on 4 and 16 September 2019 corresponded to the resting pressure, as assembly of the support and the excavation had not started yet.

2.4. Soil Stress Tests Using a Hydraulic Probe

Tests were divided into two stages; the work scheme is presented in Figure 4. In the first stage (Stage-I), after the excavation was made and the walls were secured with plates, the backfill was loaded by adding one layer of 2.95 × 0.8 × 0.16 m road slabs, placed near the excavation. The soil around the excavation remained intact. Then, the backfill was additionally loaded using a 40 t vehicle, simulating the volume of traffic that usually takes place on construction sites. Next, the backfill was weighed down with slabs, placed successively in 4, 7 and 10 layers (Figure 5), resulting in obtaining the values of backfill load of 3.84, 15.36, 26.88 and 38.4 kN m^{-2}, respectively. After each successive loading, the deflection arrow was measured with a patch and with the use of laser scanning.

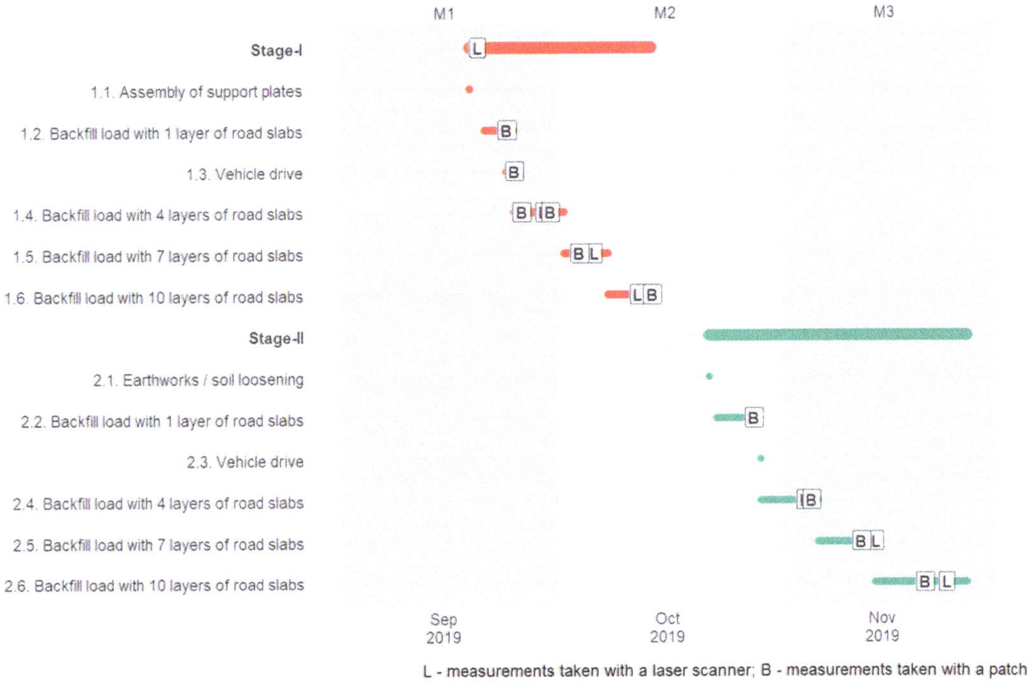

Figure 4. Scheme of tests (L—measurements taken with a laser scanner, B—measurements taken with a patch).

The next stage (Stage-II) involved analysing the loosened soil. To change the condition of soil and its loosening, after removing the backfill material, the plates on the side of the tested wall were excavated within the fragmentation wedge, changing the parameters that affected soil pressure. Then, the excavation was backfilled, and the soil was compacted with layers of max. 50 cm. using a plate vibrator. Sounding with a CPTU static probe was carried out to determine the parameters of the backfill material.

2.5. Measurement of the Deflection Arrow Taken with a Patch

The direct measurement of the deflection arrow was performed using a measurement patch (Figure 6). The patch had previously been calibrated for its straightness in the laboratory.

Figure 5. Loading of the backfill with 10 layers of road slabs (38.4 kN·m^{-2}).

Measurements of the deflection arrow taken with a measurement patch were always made at the same, seven levels (Table 1), each time after changing the value of the backfill load by means of road slabs. They were then compared with the results obtained from laser scanning.

Table 1. Height ordinates of direct measurement points of the deflection arrow taken with a measurement patch.

Measurement No./Support Plate No.	Height Ordinate Below Ground Level
1/plate 1 (upper)	0.64
2/plate 2 (middle)	1.64
3/plate 2 (middle)	2.59
4/plate 2 (middle)	3.51
5/plate 3 (bottom)	3.86
6/plate 3 (bottom)	4.79
7/plate 3 (bottom)	5.72

Figure 6. Direct measurement of the deflection arrow taken with a measurement patch.

2.6. TLS Measurement

Terrestrial Laser Scanning is a technique that uses laser light for measurements. The measured object is illuminated by the scanner with a beam, which then, reflected from the object, returns to the scanner. The measurement of distance consists in measuring the change of the phase between the emitted and reflected beam (phase scanners) or in measuring the time it takes for the beam to travel back and forth (pulse scanners). Phase scanners, more commonly used nowadays, allow the registration of up to about 1,200,000 measurement points per second. By knowing the distance (L) from the scanner and the i-th measurement point, vertical angle (β) and horizontal angle (α) coordinates of a point in a 3D coordinate system in real time can be calculated using the following equation:

$$\begin{cases} X_i = L\ \cos\beta\cos\alpha \\ Y_i = L\ \cos\beta\sin\alpha \\ Z_i = L\ \sin\beta \end{cases} \quad (1)$$

Eight scanning series were performed, four scans for each stage with the use of a 3D scanner Surphaser 100HSX from Surphaser, Redmond, WA, USA. The manufacturer states that the max. accuracy of the distance measurement is ± 0.3 mm and the angular resolution is equal to 1 arcsec. The laser wavelength is 685 nm.

The first scan represents the deformation of the excavation support not subjected to an additional load, which is called the reference scan made to calibrate the device.

Table 2 summarises the dates of scanning measurements for Stages-I and Stage-II with the value of backfill load.

Table 2. Dates of measurements by laser scanning and backfill load.

	Date	Backfill Load [kN·m^{-2}]
Stage-I	20 September 2019	0.00
	29 September 2019	15.36
	06 October 2019	26.88
	12 October 2019	38.40
Stage-II	29 October 2019	3.84
	05 November 2019	15.36
	14 November 2019	26.88
	24 November 2019	38.40

The first scan took place on 20 September 2019, right after the excavation was made and strengthened with steel support plates. Scanning was performed in such a way that the laser beam reached the bottom of the excavation (Figure 7). The initial processing of point clouds related to the registration, filtration and removal of the unnecessary data was carried out in SurphExpress, the manufacturer's software, and then data sets were analysed in the open-source CloudCompare environment in order to estimate deformations.

Figure 7. Location of the scanner.

Subsequent scans were performed each time when the soil was loaded with concrete plates near the excavation (Figure 5). After each scan, a point cloud was created in SurphExpress, which was then exported to CloudCompare, where its further processing and analysis was carried out. Filtered point clouds, after uploading into the program, were aligned to the reference plane, obtaining the distribution of displacements. The deflection arrow value was determined by measuring the basic lengths according to the scheme in Figure 8. Length A-B (half plate width) and angle CAB were read. The value of the deflection arrow was determined from trigonometric relationships for the ordinates corresponding to manual measurements taken with a measurement patch.

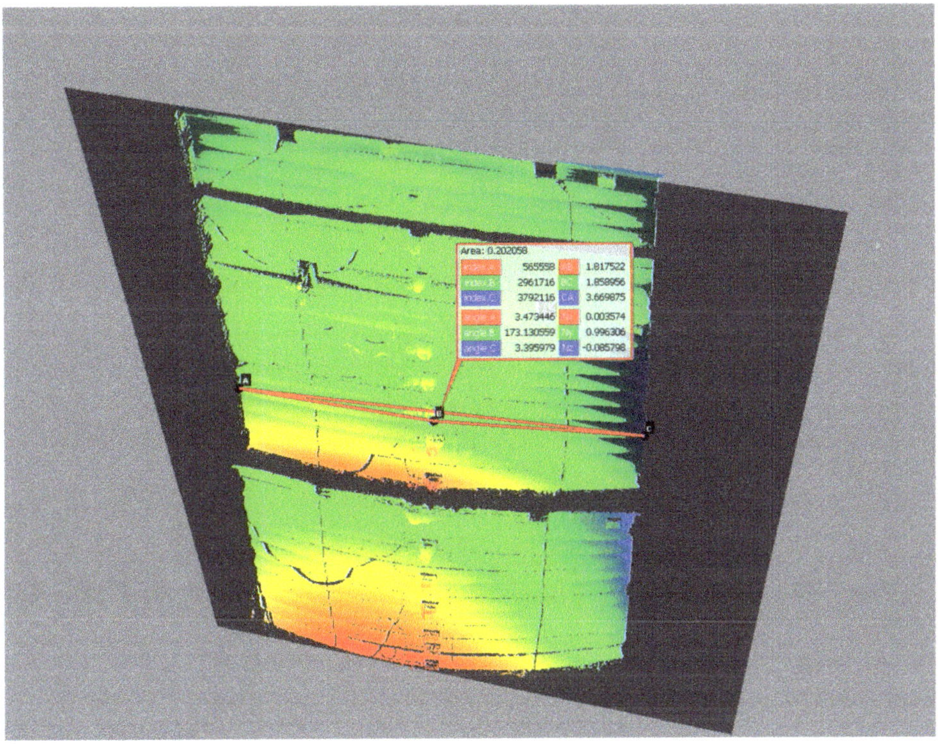

Figure 8. Measurement of the deflection arrow.

Then, the distribution of the differences of the point clouds between the reference model and the analysed point cloud was estimated. The Distance Cloud2Cloud tool of CloudCompare was used. The distances between the two-point clouds were computed as the 'nearest neighbour distance': for each point of the compared cloud, CloudCompare searches the nearest point in the reference cloud and computes their (euclidean) distance. However, because the points in the two point clouds do not exactly correspond to each other, a better estimate of the distance from each point of the compared cloud to its nearest point in the reference cloud can be obtained by replacing the distance from the local mathematical model in the 'nearest' point and several of its surrounding neighbours represented as a local plane [69]. This is statistically more precise and less dependent on the cloud sampling. Quadric local model (Height function) was used. In fact, the corresponding model is a quadratic function (6 parameters: $Z = a.X^2 + b.X + c.XY + d.Y + e.Y^2 + f$).

2.7. Static and Strength Calculations of the Support Plate for the Plate System Calculated with the Finite Difference Method

Calculations were made using the finite difference method in the variational approach.

The linear algebraic equations, from which the desired deflections at the nodes of the subdivision grid will be determined, are obtained in the condition for the minimum of the functional:

$$\frac{\partial V}{\partial w_k} = 0 \tag{2}$$

where for each w_k, where k is the number of meshes of the subdivision mesh. The above condition (2) follows from the theorem that:

"If the system is in steady state equilibrium, its total energy reaches a minimum. Applying this statement to the analysis of plate bending, we must take into account that the total energy in such cases consists of two parts: the elastic energy of bending and the potential energy of the load, located on the plate" [16]. Thus, the problem of plate bending reduces in each particular case to finding such a function in (x, y) that would satisfy the given boundary conditions and give the smallest value of the integral described by the relation (3). The functional describing the total energy of elastic deformation for plate structures, also taking into account the ribs resting on the elastic foundation with potential temperature loads, taken from work [70], has the form:

$$\begin{aligned} V = \frac{D}{2}\iint_A \Bigg\{ \left(\frac{\partial^2 w}{\partial x^2}\right)^2 & +2\left(\frac{\partial^2 w}{\partial x \partial y}\right)^2 + \left(\frac{\partial^2 w}{\partial y^2}\right)^2 + 2v\left[\frac{\partial^2 w}{\partial x^2}\frac{\partial^2 w}{\partial y^2} - \left(\frac{\partial^2 w}{\partial x \partial y}\right)^2\right] \\ & +2(1+v)\frac{\alpha_t \Delta T}{h}\left(\frac{\partial^2 w}{\partial x^2} + \frac{\partial^2 w}{\partial y^2} + \frac{\alpha_t \Delta T}{h}\right)\Bigg\} dA + +\frac{1}{2}\iint_A K w^2 dA - \iint_A q w dA \\ & +\frac{E_z J}{2}\int_S \left(\frac{\partial^2 w}{\partial s^2}\right)^2 dS \end{aligned} \tag{3}$$

where:

w—plate deflection,
ν—Poisson's ratio,
D—flexural rigidity ($D = \frac{Eh^3}{12(1-\nu^2)}$),
h—plate thickness,
E—elasticity modulus for the plate material,
E_z—modulus of elasticity for the rib material,
S—rib area,
s—size of the mesh division used in calculation, A—plate area,
q—load perpendicular to the median plane of the plate,
K—subgrade reaction modulus,
α_t—coefficient of linear thermal expansion of the plate material,
$\Delta T = t_d - t_g$—temperature difference between the plate planes,
J—moment of inertia of the rib cross-section.

When starting the search for solutions for specific plate structures, the area A of the plate was divided by a discrete grid with meshes $s \times s$ into elementary sub-areas.

Further analysis was based on the adopted designations of the second derivative of the deflection function (4):

$$w_{xx}^2 = \left(\frac{\partial^2 w}{\partial x^2}\right)^2 \; w_{xy}^2 = \left(\frac{\partial^2 w}{\partial x \partial y^2}\right)^2 \; w_{yy}^2 = \left(\frac{\partial^2 w}{\partial y^2}\right)^2 \; w_{ss}^2 = \left(\frac{\partial^2 \mathrm{w}}{\partial s^2}\right)^2 \; w_{xx} = \frac{\partial^2 w}{\partial x^2} \; w_{yy} = \frac{\partial^2 w}{\partial y^2} \tag{4}$$

Expressing the partial derivatives of the deflection surface occurring in the energy functional through difference quotients, replacing the integration over the surface with summation over elementary sub-areas and assuming the Poisson's ratio $\nu = 0$ and not considering the temperature load, the functional describing the elastic deformation energy of the plate took the form:

$$V = \frac{D}{2}\iint_A \left(w_{xx}^2 + 2w_{xy}^2 + w_{yy}^2\right) + \frac{1}{2}\iint_A K w^2 dA - \iint_A q w dA + \frac{E_z J}{2}\int_S w_{ss}^2 dS \tag{5}$$

3. Results

3.1. Geotechnical Tests

The conducted geotechnical tests, both laboratory tests on NU and NW samples, and CPTU tests indicated that under the layer of soil, from about 0.2 m to about 1.8 m below ground level, there were mainly wet, medium-compacted medium sands (Medium Sand MSa), while below there were moist, compacted sands with admixtures of silty sand or dust fine sand (Fine Sand FSa). Table 3 shows the development of the geotechnical profile for Stage-I of the study.

Table 3. Geotechnical parameters of the soil in the area of the test stand acc. to PN-81/B-03020 [71].

Layer Gap		Soil Type	Compaction Degree	Volumetric Weight	Angle of Internal Friction
from	to		I_D	γ	φ_u
[m]	[m]	[-]	[-]	[kN·m^{-3}]	[°]
0.0	0.2	Soil	-	17.00	-
0.2	0.7	MSa	0.74	18.86	34.61
0.7	1.4	MSa	0.45	18.43	32.67
1.4	1.8	MSa	0.57	18.61	33.47
1.8	2.2	Saπ	0.43	17.51	30.15
2.2	2.8	MSa	0.79	18.94	34.94
2.8	3.8	FSa	0.74	18.21	31.70
3.8	5.1	FSa	0.88	18.52	32.40
5.1	5.7	Saπ	0.47	17.60	30.35
5.7	6.0	MSa	0.94	19.16	35.94

Figure 9 shows the results of characteristic soil pressure in the vicinity of the test stand, calculated in accordance with recommendations provided in PN-81/B-03010 [72], assuming the geotechnical parameters provided in Table 3. Calculations were made without taking into account backfill load, but in compliance with recommendations of PN-83/B-03010 [72] for individual soil layers, the weight of the higher layers was treated as a substitute load of the backfill.

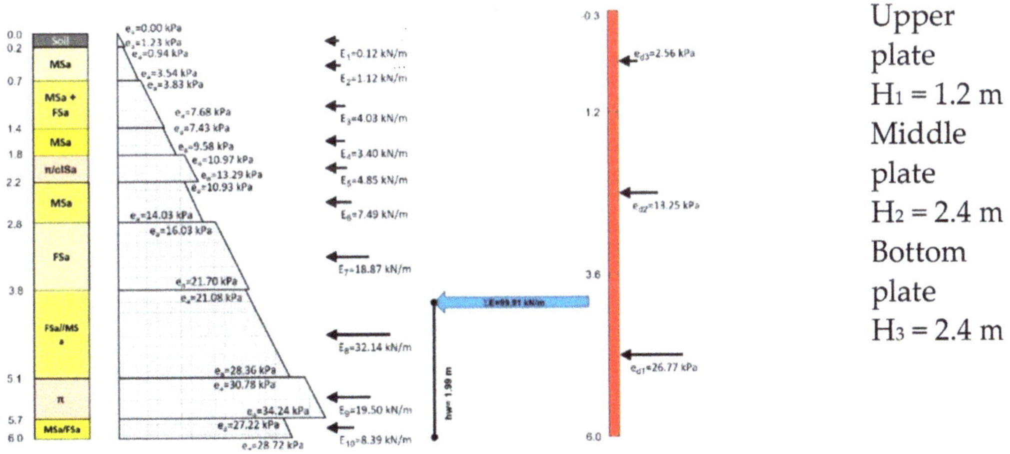

Figure 9. Diagram of computational soil pressure in the vicinity of the test stand for Stage-I.

The assumed value of soil pressure corresponded to the total sum of pressures occurring in individual soil layers. For the unloaded backfill, the total value of soil pressure E_a

was 99.91 kN m^{-1} (Figure 8), thus the resultant ordinate of soil pressure e_a located in the lower part of the plate was equal to 33.30 kN m^{-2}

$$0.5 \cdot e_a \cdot H = 99.91\ \text{kN} \cdot \rightarrow e_a = \frac{99.91}{0.5 \cdot 6} = 33.30\ \text{kN} \cdot \text{m}^{-2}$$

3.2. Measurements of the Deflection Arrow Taken with a Patch

Manual measurement of the deflection arrow was performed each time in the morning hours before the work related to changing the load conditions started. The first measurement of the deflection arrow was taken on 24 September 2019. Table 4 summarizes the deflection arrow values taken with a measurement patch for Stage-I of the study, whereas Table 5 summarizes the deflection arrow values after changing the soil conditions (Stage-II).

Table 4. Results of deflection measurements of support plates taken with a patch (Stage-I).

Date		I 24 September 2019	II 25 September 2019	III 26 September 2019	IV 30 September 2019	V 04 October 2019	VI 14 October 2019
Plate	Ordinate of the measurement m below ground level	Backfill load [kN·m^{-2}]					
		0.00	3.84	15.36	15.36	26.88	38.4
		[mm]	[mm]	[mm]	[mm]	[mm]	[mm]
upper plate	0.64	-	-	-	2.00	2.00	2.00
middle plate	1.64	-	5.50	5.75	5.00	5.50	5.00
	2.59	-	5.00	7.00	5.50	5.25	5.25
	3.51	9.00	9.25	9.75	8.00	8.50	8.00
bottom plate	3.86	7.00	6.75	7.25	7.25	7.50	7.50
	4.79	12.50	12.50	12.00	12.00	12.00	12.00
	5.72	15.50	15.50	15.25	14.75	14.75	14.75

Table 5. Results of deflection measurements of plates in mm—manual measurement taken with a patch (Stage-II).

Date		29 October 2019	05 November 2019	12 November 2019	21 November 2019
Plate	m below ground level	Backfill load [kN·m^{-2}]			
		3.84	15.36	26.88	38.40
upper plate	0.64	1.00	4.50	4.50	5.50
middle plate	1.64	8.50	7.50	7.50	8.00
	2.59	6.00	6.25	7.00	6.75
	3.51	11.75	12.00	12.00	12.50
bottom plate	3.86	5.00	4.00	4.50	4.00
	4.79	9.00	10.00	10.00	10.00
	5.72	10.00	10.50	10.50	11.00

The maximum value of the deflection arrow for Stage-I (15.5 mm) was read for the bottom plate of the excavation support. The measurement was taken at 5.72 m below ground level. After changing the soil conditions (Stage-II), as before, the maximum deflection arrow (11.0 mm) was recorded at 5.72 m below ground level.

3.3. Scanning

The first scans were carried out on 20 September 2019, right after the excavation was made and its walls were strengthened with support plates. The point cloud was assumed as referential, to which the value of the deformation of support plates was then estimated, by comparing the point clouds from the next scan performed after subsequent loading of the soil near the excavation area. Subsequent scans were performed in the same way as the first scan. Deformation distributions of excavation support plates were obtained by comparing two Cloud2Cloud point clouds in CloudCompare. Exemplary deformation distribution of excavation support plates as of 14 November 2019, for Stage-II and the value of the backfill load 26.88 kNm^{-2} is shown in Figure 10. Table 6 presents the measured values of the deflection arrow for Stage-I, whereas Table 7 includes the values for Stage-II. Measurements were taken at levels consistent with direct measurements using a measurement patch (Tables 4 and 5).

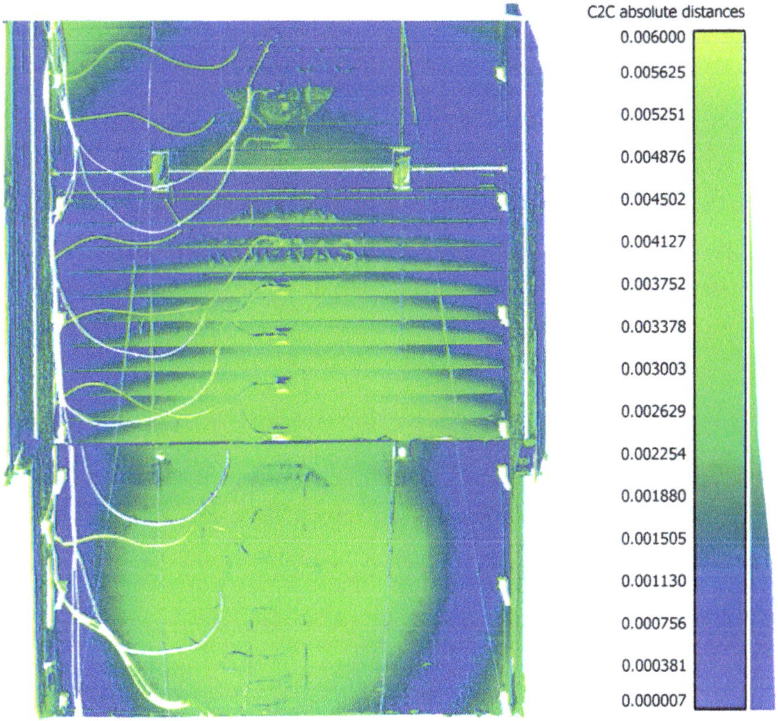

Figure 10. Deformation of excavation support plates on 14 November 2019, Stage-II, backfill load 26.88 kNm^{-2}.

Table 6. Results of deflection measurements of excavation support plates for Stage-I (intact soil) obtained with laser scanning.

	Date	**20 September 2019**	**29 September 2019**	**06 October 2019**	**12 October 2019**
		Backfill load [kN·m^{-2}]			
Plate	m belowground level	0.00 [mm]	15.36 [mm]	26.88 [mm]	38.40 [mm]
upper plate	0.64	0.30	3.50	4.00	9.20
middle plate	1.64	6.80	7.00	7.00	7.20
	2.59	5.50	4.90	5.60	5.40
	3.51	9.30	9.30	9.10	8.80
bottom plate	3.86	8.10	7.60	7.40	7.90
	4.79	12.60	12.50	12.90	12.40
	5.72	16.00	15.70	14.40	15.00

Table 7. Results of deflection measurements of excavation support plates for Stage-II (loosened soil) obtained with laser scanning.

Plate	Date	29 October 2019	05 November 2019	12 November 2019	24 November 2019
	m belowground level	Backfill load [kN·m^{-2}]			
		3.84	15.36	26.88	38.40
upper plate	0.64	4.68	5.67	5.87	6.65
middle plate	1.64	6.98	7.65	7.57	8.68
	2.59	7.62	8.57	7.53	8.44
	3.51	11.90	11.96	12.90	13.21
bottom plate	3.86	4.26	4.18	4.26	4.36
	4.79	7.39	7.52	7.27	7.95
	5.72	11.50	11.23	11.51	11.11

3.4. Static and Strength Calculations for a Support Plate Freely Supported on Two Opposite Edges, Based on Tables

To verify the deflections obtained in situ, numerical calculations were performed using the finite difference method. First, verification calculations were made based on Design Tables [73], and second, a detailed numerical analysis was carried out, the results of which are provided in the next section of the paper.

The structure of the plate consisted of closed rectangular steel profiles with dimensions 200 × 120 × 4 mm (Figure 3), the overall dimensions of the plate were 3920 × 2400 × 120 mm (Figure 11). The plate was subjected to a uniformly distributed load q_1 and soil pressure q_2.

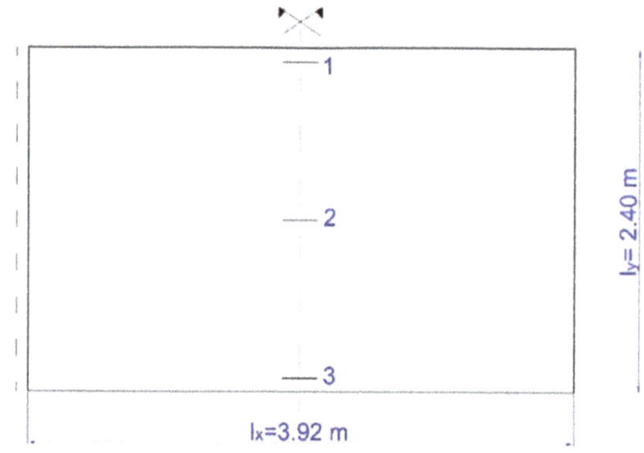

Figure 11. Diagram of the plate with dimensions 3920 × 2400 × 12 mm with marked points for which the deflections were calculated.

By assuming the Young's modulus of steel (also referred to as the modulus of elasticity for steel) as E = 210 GPa and the rigidity of the plate (ribs excluded) as $D = \frac{E\cdot(H^3-h^3)}{12} = \frac{210\cdot10^6\cdot(0.12^3-0.112^3)}{12} \cong 5654$ kNm, based on Tables [73], the following value of hydrostatic thrust q_2 was achieved:

$$\frac{l_y}{l_x} = \frac{2.40}{3.92} = 0.61$$

$$w_1 = 0.00570\frac{q_2\cdot l_x^4}{D} = 0.000238q_2$$

$$w_2 = 0.00652\frac{q_2\cdot l_x^4}{D} = 0.000272q_2$$

$$w_3 = 0.00735\frac{q_2\cdot l_x^4}{D} = 0.000307q_2$$

For an evenly distributed load (q_1), the value of deflection $w_1 = w_2 = w_3 = w$, was:

$$w = \frac{5}{384} \cdot \frac{q_1 \cdot l_x^4}{D} = 0.01302 \cdot \frac{q_1 \cdot l_x^4}{D} = 0.000544 q_1$$

In order to estimate the values of deflections, Figure 12 shows the assumed value of soil pressure calculated for Stage-I (Figure 9).

$$\frac{33.30}{6} = \frac{q_{2.4}}{3.6}$$
$$q_{2.4} = \frac{33.30 \cdot 3.6}{6} = 19.98 \text{ kN} \cdot \text{m}^{-2}$$

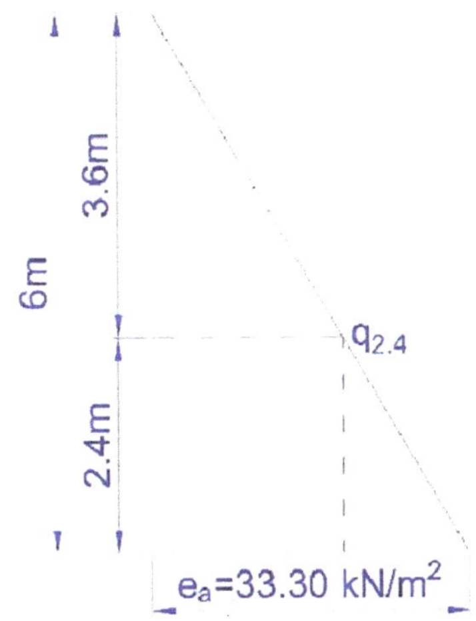

Figure 12. Diagram of the assumed value of soil pressure.

Calculations considered the load acting on the bottom plate as $q_1 = 19.98 \text{ kN} \cdot \text{m}^{-2}$ (evenly distributed load) and $q_2 = 33.30 - 19.98 = 13.32 \text{ kN} \cdot \text{m}^{-2}$ (hydrostatic thrust).

For the above data, it was calculated:

$$w_1 = 0.000544 \cdot q_1 + 0.000238 \cdot q_2 = 14.03 \text{ mm}$$
$$w_2 = 0.000544 \cdot q_1 + 0.000272 \cdot q_2 = 14.49 \text{ mm}$$
$$w_3 = 0.000544 \cdot q_1 + 0.0003070 \cdot q_2 = 14.95 \text{ mm}$$

3.5. Static and Strength Calculations of a Support Plate for the Plate System Using the Finite Difference Method

For detailed calculations using the finite difference method, square meshes of the division grid (s = 20 cm) and a static plate system with two opposite free edges and two other freely supported edges were adopted. The adopted division mesh with the numbering of nodes is shown in Figure 13.

By solving the systems of equations with 130 unknowns for a uniformly distributed load and hydrostatic thrust, the coefficients proportional to deflections at individual nodes of the division mesh were obtained.

To obtain the values of deflection $w(q_1)$, the obtained coefficients should be multiplied by $\frac{q_1 s^4}{D}$, and for the values of deflection $w(q_2)$ by $\frac{q_2 s^4}{D}$.

To compare the values with the calculations given above, according to [73], the exact values of deflections at points 10, 70 and 130 were calculated (Figure 13) and they were: for an evenly distributed load:

$$w_{10} = w_{70} = w_{130} = 0.01305 \frac{q_1 \cdot l_x^4}{D} = 0.000545 q_1$$

and for hydrostatic load:

$$w_{10} = 0.00570\frac{q_2 \cdot l_x^4}{D} = 0.000238q_2$$
$$w_{70} = 0.00653\frac{q_2 \cdot l_x^4}{D} = 0.000273q_2$$
$$w_{130} = 0.00737\frac{q_2 \cdot l_x^4}{D} = 0.000308q_2$$

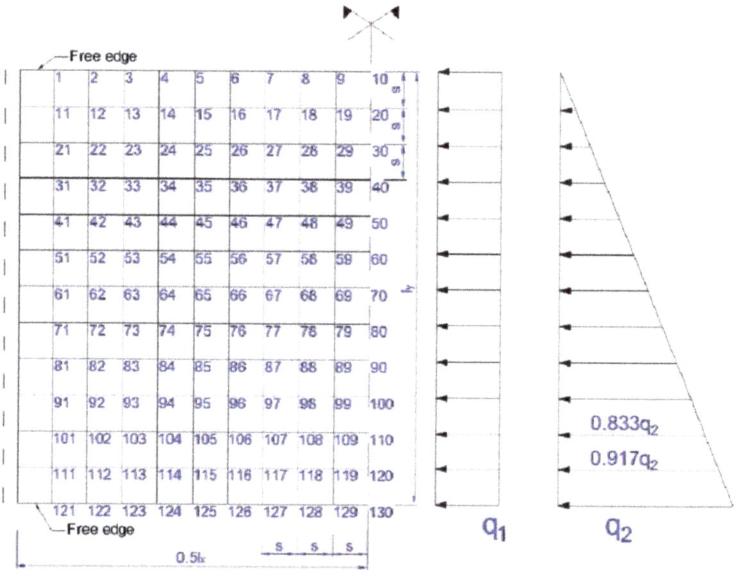

Figure 13. Adopted division mesh and load diagrams: evenly distributed load q_1 and hydrostatic load q_2.

By assuming $q_1 = 19.98\ \frac{kN}{m^2}$ (evenly distributed load) and $q_2 = 13.32\ \frac{kN}{m^2}$ (hydrostatic thrust), it was calculated:

$$w_{10} = 0.000545 \cdot q_1 + 0.000238 \cdot q_2 = 14.06\ \text{mm}$$
$$w_{70} = 0.000545 \cdot q_1 + 0.000273 \cdot q_2 = 14.52\ \text{mm}$$
$$w_{130} = 0.000545 \cdot q_1 + 0.000308 \cdot q_2 = 14.99\ \text{mm}$$

The comparison of results obtained with calculations based on Tables [73] and detailed calculation made with the finite difference method showed a very high consistence.

3.6. Calculations of the Deflection Arrow of the Plate

The serviceability limit state is the state beyond which a plate no longer meets its operational requirements. According to the guidelines of PN-EN 13331-1-1:2002 [74], the maximum deflection of plates and sliding guides should be certified with reference to the load equal to the immediate strength divided by partial factors of γ_F = 1.5 (for soil action) and γ_M = 1.1 (for strength).

The above recommendations correspond to the guidelines of PN-EN 1990:2004 [75], which for the serviceability limit states recommend taking partial factors with the value of γ = 1.

The literature on the subject and applicable standards do not specify the deflection limit for temporary excavation support plates. This is an inconvenience for site managers, as there is no value with which to compare the deflection of support plates, which can be easily measured with a patch at any construction site.

On the basis of the obtained deflection values for the lower edge of the plate for tabular [73] and detailed (FDM) solutions, an attempt was made to determine the permissible (limit) value of the deflection arrow, which was also compared with the values read from measurements taken with a patch and a laser scanner.

Deflection at Point 3 (Figure 11) was 14.95 mm, and at Point 130 (Figure 13) it was 14.99 mm. For the plate with the length l_x = 3.92 m it took the value:

$$\frac{3920}{14.95} = 262 \rightarrow w = \frac{l_x}{262} \text{ and } \frac{3920}{14.99} = 261.5 \rightarrow w = \frac{l_x}{261}. \quad (6)$$

Based on the actual measurements of deflections in situ (Tables 4 and 6), the greatest value of the deflection arrow was 15.5 mm (patch) and 16 mm (scanner); for these values it was then designated:

$$\frac{3920}{15.5} = 253 \rightarrow w = \frac{l_x}{253} \text{ and } \frac{3920}{16} = 245 \rightarrow w = \frac{l_x}{245}.$$

4. Analysis of the Results and Discussion

Analysis of the measured deflections, both taken with a patch and using a laser scanner (Tables 4–9), showed that the backfill load has no significant effect on the deflections of the lower plate. The maximum values of deflections were recorded in the lower edge of the bottom support plate, 5.72 m below ground level and did not exceed 16 mm for measurement taken with a laser scanner and 15.5 mm for direct measurement taken with a patch. Direct comparison of the deflection arrow values measured with a laser scanner and a patch (Tables 8 and 9) indicates a high compliance of measurements. The values of measurements of the deflection arrow taken at the same time, with a difference of one to two days between measurements taken with a patch and a laser scanner, for both Stage-I and Stage-II, differed slightly, on average by 0.67 mm, with the value of standard deviation (SD) 1.45 mm (Figures 14 and 15). Greater values of the deflection arrow in individual support plates were recorded in their lower edges, for the middle plate and for the bottom plate. Lower values of the deflection arrow in the upper parts of support plates resulted from the use of strong rib reinforcement in the upper part of the plate (i.e., plate head). Additionally, the middle plate partially overlapped the bottom plate, thus both plates contributed to the load transmission at their contact point. The greater value of deflection 14.4 to 15.7 mm, measured in the lower edge of the bottom plate, occurred probably because the cutting edge of the bottom support plate was less rigid than the rest of the plate surface.

Table 8. List of deflection measurements of support plates using a patch and a laser scanner taken for Stage-I. Denominations: P—measurement taken with a patch, S—measurement taken with a laser scanner.

Date		30 September 2019 (Scanner: 29 September 2019)		04 October 2019 (Scanner: 06 October 2019)		14 October 2019 (Scanner: 12 October 2019)	
Backfill load [kN·m^{-2}]		15.36		26.88		38.4	
		P	S	P	S	P	S
	m belowground level			[mm]			
upper plate	0.64	2.00	3.50	2.00	4.00	2.00	9.20
middle plate	1.64	5.00	7.00	5.50	7.00	5.00	7.20
	2.59	5.50	4.90	5.25	5.60	5.25	5.40
	3.51	8.00	9.30	8.50	9.10	8.00	8.80
bottom plate	3.86	7.25	7.60	7.50	7.40	7.50	7.90
	4.79	12.00	12.50	12.00	12.90	12.00	12.40
	5.72	14.75	15.70	14.75	14.40	14.75	15.00

Table 9. List of deflection measurements of support plates using a patch and a laser scanner taken for Stage-II. Denominations: P—measurement taken with a patch, S—measurement taken with a laser scanner.

Date		28 October 2019 (Scanner: 29 October 2019)		05 November 2019		12 November 2019 (Scanner: 14 November 2019)		21 November 2019 (Scanner: 24 November 2019)	
Backfill load [kN·m^{-2}]		3.84		15.36		26.88		38.40	
	m below ground level	P	S	P	S	P [mm]	S	P	S
upper plate	0.64	1.00	4.68	4.50	5.67	4.50	5.87	5.50	6.65
middle plate	1.64	8.50	6.98	7.50	7.65	7.50	7.57	8.00	8.68
	2.59	6.00	7.62	6.25	8.57	7.00	7.53	6.75	8.44
	3.51	11.75	11.90	12.00	11.96	12.00	12.90	12.50	13.21
bottom plate	3.86	5.00	4.26	4.00	4.18	4.50	4.26	4.00	4.36
	4.79	9.00	7.39	10.00	7.52	10.00	7.27	10.00	7.95
	5.72	10.00	11.50	10.50	11.23	10.50	11.51	11.00	11.11

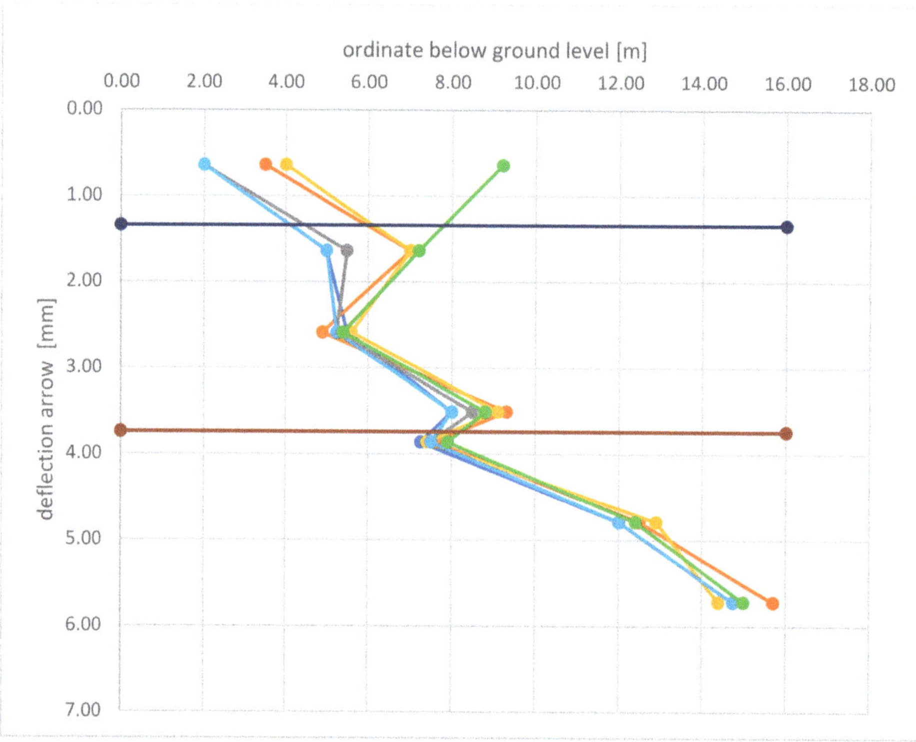

Figure 14. Deflection arrows—Stage-I.

There are few works in the subject literature analysing the results obtained in field experiments as well as analysing measurements taken with one of the most modern technologies, namely laser scanning.

The authors of the paper [76] presented the results of the numerical calculations using the finite element method, which were carried out with the PLAXIS-3D Finite Element computer program. They analysed the value of soil pressure on the steel box cover, acting as an excavation support. These tests included a simulation of the static operation of two steel (or aluminum) excavation covers placed one on the other with a total depth of the excavation equal to 6 m. The finite element method (FEM) was adopted for the analysis and the Mohr-Coulomb constitutive material model (MC) was selected. The results were compared with the empirical diagrams of apparent soil pressure for clay. The comparison showed that the parameters related to the soil and the excavation support structure have a significant impact on the values of received loads and the use of load-bearing capacity of the material from which the support was made. The authors of the work [77] presented the results of experimental field tests, the purpose of which was to examine the value of soil pressure for temporary excavation supports. They provided procedures for instrumentation installation and field testing for two supports of "stacked" excavations, in accordance with the US OSHA regulations, placed in the excavation at a total depth of 6 m. The results of field tests were presented considering the distribution of soil pressure at the excavation depth considering the backfill load and excluding it. The results indicated the possibility of using the tested support system of excavations and ditches in cohesive soils and confirmed the compliance with theoretical results. The authors of the paper [77] found that the results obtained by means of analytical formulas generally underestimated the experimental values for the clay considered in their study.

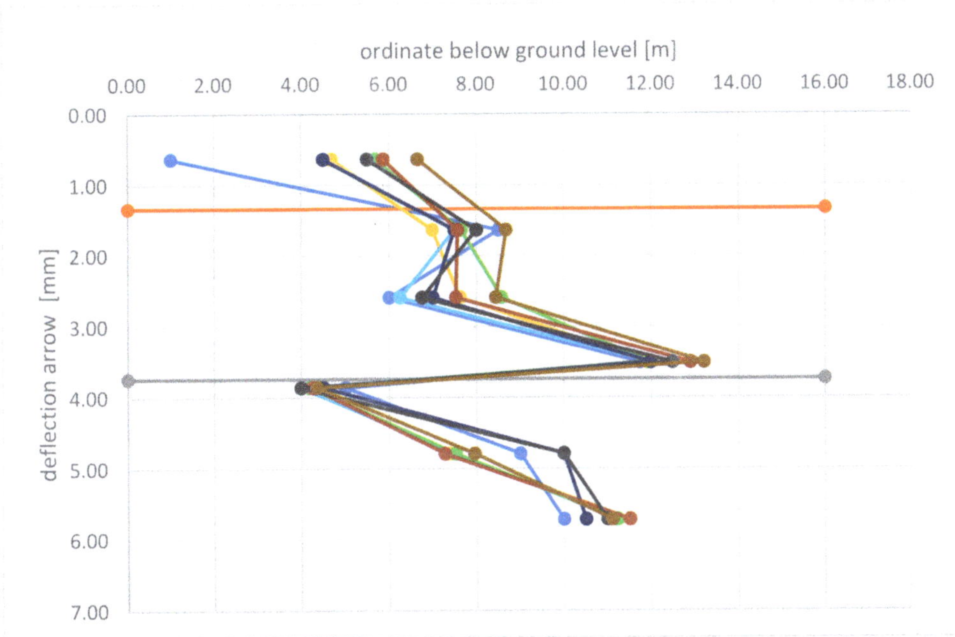

Figure 15. Deflection arrows—Stage-II.

The effectiveness of applying the finite difference method has been reported in numerous scientific papers. The authors of the article [29] verified the values of deflection experimentally by means of a coordinate measuring arm, comparing them with the values calculated with the finite difference method for hydrostatic thrust acting on the tank. The paper [27] presents calculations for a thermally loaded plate, which were experimentally verified.

Table 10 summarizes the values of deflections in the lower edge of the excavation support (at the bottom of the excavation) obtained from tests carried out with a measurement patch, a laser scanner and calculated using the finite difference method based Tables [73] and a detailed solution with no backfill load.

Table 10. Comparison of the values of deflections for support plates using a patch and a laser scanner taken for Stage-I and calculated using Tables and detailed calculations, with the value of backfill load equal to 0.00 kN·m^{-2}.

The Method of Obtaining the Values of Deflections	The Value of Deflection in the Lower Edge of the Plate, at the Bottom of the Excavation, for Backfill Load 0.00 kN·m^{-2}
patch (Table 4)	w = 15.50 mm
scanner (Table 6)	w = 16.00 mm
calculations acc. [73]	w = 14.95 mm
detailed FDM calculations	w = 14.99 mm

It can be clearly stated that the compliance of the obtained results was very good. Thus, both the research experiment and the calculations were carried out correctly.

Calculations of the deflection arrow values, which could be used to recommend the formula for the limit deflection of steel plates of temporary excavation supports, were made based on the values of deflection obtained experimentally and numerically. To compare the obtained results, the deflection arrow values were determined based on the material data of the support plate, which should be treated as a limit (permissible).

Calculations were made in two variants for steel S235JR and S355JR, for the plate shown in Figure 3. It was assumed that the fin plate is a freely supported plate loaded with a uniformly distributed load along its entire length (Figure 16).

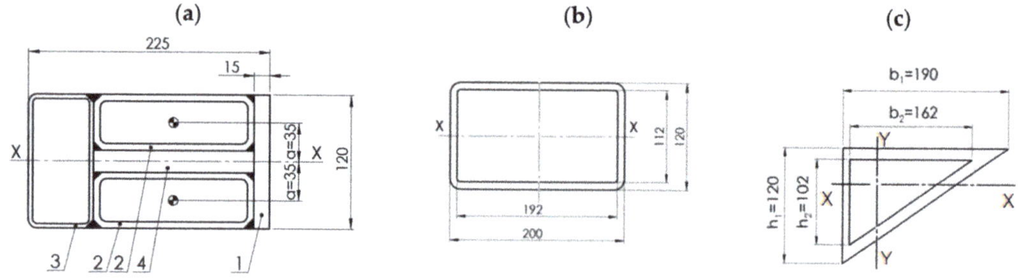

1—Flat bar 120 × 15 mm, 2—Rectangular profile 150 × 50 × 6 mm, 3—Rectangular profile 120 × 60 × 4 mm, 4—Flat bar 30 × 20 × 145 mm

Figure 16. Cross-sections of individual components of the plate, (**a**) head, (**b**) components (rectangular pipe), (**c**) cutting edge.

The moment of inertia for the entire plate ("head" + 10 profiles 200 × 120 × 4 × 4 mm + "cutting edge") was I_x = 7817.01 cm^4.

The assumed load capacity of the cross-section taking into account plastic reserve for S235JR steel was $M_{\text{c,Rd}} = M_{\text{pl,Rd}} = \frac{W_{\text{pl,x}}}{\gamma_{\text{M0}}} \cdot f_y = 33,865.38$ kNcm, and for S355JR steel—$M_{\text{c,Rd}} = M_{\text{pl,Rd}} = \frac{W_{\text{pl,x}}}{\gamma_{\text{M0}}} \cdot f_y = 51,158.34$ kNcm. The maximum bending moment for a freely supported beam, uniformly loaded along its entire length, was assumed: $M_{\text{Ed}} = 0.125 \cdot q \cdot l_x{}^2$.

The maximum plastic load capacity of the cross-section was adopted as the limit value beyond which the steel cross-section plasticization occurs; therefore, the maximum value of the linear load q was: for S235JR steel $q = \frac{M_{Ed}}{0.125 \cdot l_x{}^2} = 1.76$ kN·cm^{-1}, and for S355JR steel—$q = \frac{M_{Ed}}{0.125 \cdot l_x{}^2} = 2.66$ kN·cm^{-1}.

By assuming partial load factors, the continuous linear service load was calculated (q_u):

$$q_u = \frac{q}{\gamma_F \cdot \gamma_M} = \frac{q}{1.5 \cdot 1.1}:$$

for S235JR steel $q_u = \frac{1.76}{1.5 \cdot 1.1} = 1.067\ \text{kN}\cdot\text{cm}^{-1}$,
for S355JR steel $q_u = \frac{2.66}{1.5 \cdot 1.1} = 1.61\ \text{kN}\cdot\text{cm}^{-1}$.

To determine the maximum deflection for the value of plastic resistance of the cross-section, the formula for the deflection arrow was used for a freely supported beam loaded with a load uniformly distributed over its entire length:

$$f = \frac{5 \cdot q_u}{384 \cdot E \cdot I_x} \cdot l_x^4$$

where:

l_x—length of the plate (l_x = 392 cm),
E—modulus of elasticity for steel (I = 210,000 N·mm^{-2}) [78],
I_x—moment of inertia (I_x = 7817.01 cm^4),
q_u—service load.

For S235JR steel:

$f = \frac{5 \cdot 1.067}{384 \cdot 21000 \cdot 7817} \cdot 392^4 = 2.00$ cm

f = 2.00 cm $\rightarrow \frac{392}{2.00} = 196 \rightarrow f = \frac{1}{196} l_x$—the value of deflection was 1/196 of the span.

For S355JR steel:

$f = \frac{5 \cdot 1.61}{384 \cdot 21000 \cdot 7966} \cdot 392^4 = 3.02$ cm

f = 3.02 cm $\rightarrow \frac{392}{3.02} = 130 \rightarrow f = \frac{1}{130} l_x$—the value of deflection was 1/130 of the span.

Based on calculations of the deflection arrow, the following results were obtained and are summarised in Table 11.

Table 11. List of the limit value of the deflection arrow obtained based on experimental tests and calculations.

The Method of Obtaining the Values of Deflection	Calculated Value of the Deflection Arrow (Where l_x Is the Length of the Plate)
Patch (w_{max} = 15.50 mm)	$w_{gr} = \frac{l_x}{253}$
Scanner (w_{max} = 16.00 mm)	$w_{gr} = \frac{l_x}{245}$
Calculations acc. [73] (w_{max} = 14.95 mm)	$w_{gr} = \frac{l_x}{262}$
Detailed FDM calculations (w_{max} = 14.99 mm)	$w_{gr} = \frac{l_x}{261}$
Calculations based on the maximum plastic load capacity for S235JR steel	$w_{gr} = \frac{l_x}{196}$
Calculations based on the maximum plastic load capacity for S355JR steel	$w_{gr} = \frac{l_x}{130}$

5. Conclusions

The paper presents the results of deflection tests for temporary excavation support plates obtained based on measurements taken on the experimental object in 1:1 scale. This is undoubtedly a great cognitive value of the work. In the subject literature, there are few studies in this field, the results of which are based on real-scale experimental tests and on a real research object. The results obtained during the tests were verified by numerical calculations using the finite difference method.

Based on the experimental tests and numerical analysis, it can be concluded that:

- Deflections measured with a patch and obtained using a laser scanner (Tables 8 and 9) showed similar values, corresponding to the values resulting from static calculations made with the finite difference method.
- The measured maximum deflections (Tables 4–7) ranged from 14.40 to 16 mm, and the calculated values were 14.95 and 14.99 mm. The consistency of results is very good.

- Based on the maximum plastic load capacity of the steel fin plate cross-section, the values of permissible deflections were determined. Deflections measured with a patch and a laser scanner were significantly smaller than the values accepted as permissible.
- The analysis of the values of deflections measured both with a patch and a laser scanner (Tables 4–7) showed that the backfill load does not have a significant effect on the deflection of the lower plate, but it does affect the deflection of the first plate up to a depth of 1.2 m. Deflections of the plate without the backfill load are sometimes greater than deflections with the backfill load recorded for the second and third plate.
- According to PN-EN 13331-1-1:2002 [74], there is no obligation to verify the deflection of support plates of temporary excavations; only the value of the maximum deflection should be provided to users. The authors of the paper believe that this reflects an oversight on the part of the legislator. It would be advantageous if the person conducting construction works (site manager) knew the boundary value of the deflection arrow for excavation support plates, which depends only on the span of the plate, so there is no need to perform any numerical calculations.
- It is recommended, based on Table 11, to assume the limit (maximum) deflection arrow for support plates of temporary excavations at least as $w_{gr} = \frac{L}{130}$, where L is the span of the plate.

The conducted tests, measurements and calculations can be used to verify existing solutions in the subject matter and to improve plate structures in order to ensure greater safety of the manufactured and support plates in use as well as to reduce their production costs.

Author Contributions: Conceptualization, K.M., B.W. and S.-G.A.; methodology, K.M., B.W. and S.-G.A.; validation, K.M., B.W., S.-G.A. and W.Z.; formal analysis, K.M., S.-G.A., W.Z. and G.S.; investigation, K.M., B.W., S.-G.A. and W.Z.; writing—original draft preparation, K.M., B.W., S.-G.A., W.Z. and G.S.; writing—review and editing, K.M., B.W., S.-G.A. and W.Z.; visualization, S.-G.A., W.Z. and G.S. All authors have read and agreed to the published version of the manuscript.

Funding: This research received no external funding.

Institutional Review Board Statement: Not applicable.

Informed Consent Statement: Not applicable.

Data Availability Statement: Not applicable.

Acknowledgments: The publication was co-financed within the framework of Ministry of Science and Higher Education programme as "Regional Initiative Excellence" in years 2019–2022, Project No. 005/RID/2018/19.

Conflicts of Interest: The authors declare that they have no conflict of interest.

References

1. Siemińska-Lewandowska, A. *Głębokie Wykopy: Projektowanie i Wykonawstwo*; Wydawnictwa Komunikacji i Łączności: Warsaw, Poland, 2010; ISBN 8320617650. (In Polish)
2. Nazarova, E.V.; Kaloshina, S.V.; Zolotozubov, D.G. The choice of a metal sheet piling for the construction of the foundation pit. *J. Phys. Conf. Ser.* **2021**, *1928*, 12049. [CrossRef]
3. Marcinkowski, A.; Gralewski, J. The comparison of the environmental impact of steel and vinyl sheet piling: Life cycle assessment study. *Int. J. Environ. Sci. Technol.* **2020**, *17*, 4019–4030. [CrossRef]
4. Tatiya, R. *Surface and Underground Excavations. Methods, Techniques and Equipment*, 2nd ed.; CRC Press: Balkema, The Netherlands, 2013; ISBN 9780203440940.
5. Liu, W.; Li, T.; Wan, J. Deformation Characteristic of a Supported Deep Excavation System: A Case Study in Red Sandstone Stratum. *Appl. Sci.* **2021**, *12*, 129. [CrossRef]
6. Saleh, M.; Vanapalli, S.K. Analysis of Excavation Support Systems Considering the Influence of Saturated and Unsaturated Soil Conditions. *J. Geotech. Geoenviron. Eng.* **2022**, *148*, 4022034. [CrossRef]
7. Kopras, M. Weryfikacja Metod Projektowania i Obliczeń Konstrukcji Płyt Obudowy Wykopów Tymczasowych. Ph.D. Thesis, Poznań University of Life Sciences, Poznań, Poland, 2021.
8. Jovanović, M.; Zlatkov, D. Excavation support system with piles—Case study. In Proceedings of the 8th International Conference Contemporary Achievements in Civil Engineering, Subotica, Serbia, 22–23 April 2021.

9. Alam, M.; Chaallal, O.; Galy, B. Flexible temporary shield in soft and sensitive clay: 3D FE modelling of experimental field test. *Model. Simul. Eng.* **2021**, *2021*, 6626750. [CrossRef]
10. Salehi Alamdari, N.; Khosravi, M.H.; Katebi, H. Distribution of lateral active earth pressure on a rigid retaining wall under various motion modes. *Int. J. Min. Geo-Eng.* **2020**, *54*, 15–25.
11. Wang, K.; Liu, G.; Zhang, Y.; Lin, J. Quantification of the active lateral earth pressure changes on retaining walls at the leading edge of steep slopes. *Front. Earth Sci.* **2022**, *223*. [CrossRef]
12. Gołaś, J. *Introduction to the Theory of Plates*; Opole University of Technology Publishing House: Opole, Poland, 1972.
13. Donnell, L.H. *Beams, Plates and Shells*; McGraw-Hill Companies: New York, NY, USA, 1976; ISBN 0070175934.
14. Naghdi, P.M. The theory of shells and plates. In *Linear Theories of Elasticity and Thermoelasticity: Linear and Nonlinear Theories of Rods, Plates, and Shells*; Truesdell, C., Ed.; Springer: Berlin/Heidelberg, Germany, 1973; pp. 425–640, ISBN 978-3-662-39776-3.
15. Panc, V. *Theories of Elastic Plates*; Springer Science & Business Media: Berlin/Heidelberg, Germany, 1975; ISBN 902860104X.
16. Timoshenko, S.; Woinowsky-Krieger, S. *Theory of Plates and Coatings*; Arkady: Warsaw, Poland, 1962.
17. Szilard, R. Theories and applications of plate analysis: Classical, numerical and engineering methods. *Appl. Mech. Rev.* **2004**, *57*, B32–B33. [CrossRef]
18. Ugural, A.C. *Stresses in Plates and Shells*; McGraw-Hill Science, Engineering & Mathematics: New York, NY, USA, 1999; ISBN 0070657696.
19. Wilde, P. Variational approach of finite differences in the theory of plate. In Proceedings of the Materials of XII Scientific Conference of the Committee of Science PZiTB and the Committee of Civil Engineering of Polish Academy of Sciences, Krynica, Poland, 12–17 September 1966; pp. 12–17.
20. Tribiłło, R. Application of the generalized finite difference method for plate calculations. *Arch. Inżynierii Lądowej* **1975**, *2*, 579–586.
21. Son, M.; Jung, H.S.; Yoon, H.H.; Sung, D.; Kim, J.S. Numerical study on scale Effect of repetitive plate-loading test. *Appl. Sci.* **2019**, *9*, 4442. [CrossRef]
22. Nowacki, W. *Z Zastosowań Rachunku Różnic Skończonych W Mechanice Budowli*; Wydawnictwo Zakładu Mechaniki Budowlanej Politechniki Gdańskiej: Gdańsk, Poland, 1952.
23. Rapp, B.E. Chapter 30—Finite Difference Method. In *Microfluidics: Modelling, Mechanics and Mathematics, Micro and Nano Technologies*; Rapp, B.E., Ed.; William Andrew: Norwich, NY, USA, 2017; pp. 623–631.
24. Blazek, J. Principles of solution of the governing equations. In *Computational Fluid Dynamics: Principles and Applications*, 3rd ed.; Blazek, J., Ed.; Butterworth-Heinemann: Woburn, MA, USA, 2015; pp. 121–166.
25. Sadd, M.H. Chapter 5—Formulation and Solution Strategies. In *Elasticity, Theory, Applications, and Numerics*; Sadd, M.H., Ed.; Academic Press: Cambridge, MA, USA, 2005; pp. 83–102.
26. Szymczak-Graczyk, A. Numerical analysis of the impact of thermal spray insulation solutions on floor loading. *Appl. Sci.* **2020**, *10*, 1016. [CrossRef]
27. Szymczak-Graczyk, A. Rectangular plates of a trapezoidal cross-section subjected to thermal load. In *IOP Conference Series: Materials Science and Engineering*; IOP Publishing: Bristol, UK, 2019; p. 32095, ISBN 1757-899X.
28. Numayr, K.S.; Haddad, R.H.; Haddad, M.A. Free vibration of composite plates using the finite difference method. *Thin-Walled Struct.* **2004**, *42*, 399–414. [CrossRef]
29. Buczkowski, W.; Szymczak-Graczyk, A.; Walczak, Z. Experimental validation of numerical static calculations for a monolithic rectangular tank with walls of trapezoidal cross-section. *Bull. Pol. Acad. Sci. Tech. Sci.* **2017**, *65*, 6. [CrossRef]
30. Szymczak-Graczyk, A. Floating platforms made of monolithic closed rectangular tanks. *Bull. Pol. Acad. Sci. Tech. Sci.* **2018**, *66*, 2. [CrossRef]
31. Szymczak-Graczyk, A. Numerical Analysis of the Bottom Thickness of Closed Rectangular Tanks Used as Pontoons. *Appl. Sci.* **2020**, *10*, 8082. [CrossRef]
32. Szymczak-Graczyk, A. The Effect of Subgrade Coefficient on Static Work of a Pontoon Made as a Monolithic Closed Tank. *Appl. Sci.* **2021**, *11*, 4259. [CrossRef]
33. Wu, Y.; Qiao, W.; Li, Y.; Jiao, Y.; Zhang, S.; Zhang, Z.; Liu, H. Application of computer method in solving complex engineering technical problems. *IEEE Access* **2021**, *9*, 60891–60912. [CrossRef]
34. Lee, Y.-Z.; Schubert, W. Determination of the round length for tunnel excavation in weak rock. *Tunn. Undergr. Space Technol.* **2008**, *23*, 221–231. [CrossRef]
35. Lemmens, M. Terrestrial Laser Scanning. In *Geo-Information: Technologies, Applications and the Environment*; Lemmens, M., Ed.; Springer: Dordrecht, The Netherlands, 2011; pp. 101–121, ISBN 978-94-007-1667-4.
36. Cheng, X.J.; Jin, W. Study on reverse engineering of historical architecture based on 3D laser scanner. In *Journal of Physics: Conference Series*; IOP Publishing: Bristol, UK, 2006; p. 160, ISBN 1742-6596.
37. Szolomicki, J. (Ed.) Application of 3D laser scanning to computer model of historic buildings. In Proceedings of the World Congress on Engineering and Computer Science, San Francisco, CA, USA, 21–23 October 2015.
38. Skwirosz, A.; Bojarowski, K. The Inventory and Recording of Historic Buildings Using Laser Scanning and Spatial Systems. In Proceedings of the 2018 Baltic Geodetic Congress (BGC Geomatics), Olsztyn, Poland, 21–23 June 2018; pp. 340–343, ISBN 1538648989.
39. Gleń, P.; Krupa, K. The use of 3D scanning for the inventory of historical buildings on the example of the palace in Snopków. *Teka Kom. Archit. Urban. Studiów Kraj.* **2019**, *15*, 73–78. [CrossRef]

40. Lipecki, T. Geodetic and Architectural Inventory of the Historic Wooden Church of St. Szczepan in Mnichów (Poland) in Terms of Safety Assessment of the Geometric Condition of the Structure 2020. Available online: https://assets.researchsquare.com/files/rs-107843/v1/14998c5c-c307-445d-8462-95fede74b381.pdf?c=1631861438 (accessed on 9 July 2022).
41. Przyborski, M.; Tysiąc, P. As-built inventory of the office building with the use of terrestrial laser scanning. *E3S Web Conf.* **2018**, *26*, 1–4. [CrossRef]
42. Klapa, P.; Mitka, B.; Bożek, P. Inventory of various stages of construction using TLS technology. In Proceedings of the International Multidisciplinary Scientific GeoConference Surveying Geology and Mining Ecology Management, SGEM, Dobrich, Bulgaria, 30 June 2018; Volume 18, pp. 137–144.
43. Gardzińska, A. Application of Terrestrial Laser Scanning for the Inventory of Historical Buildings on the Example of Measuring the Elevations of the Buildings in the Old Market Square in Jarosław. *Civ. Environ. Eng. Rep.* **2021**, *31*, 293–309. [CrossRef]
44. Lercari, N. Monitoring earthen archaeological heritage using multi-temporal terrestrial laser scanning and surface change detection. *J. Cult. Herit.* **2019**, *39*, 152–165. [CrossRef]
45. Nowak, R.; Orłowicz, R.; Rutkowski, R. Use of TLS (LiDAR) for building diagnostics with the example of a historic building in Karlino. *Buildings* **2020**, *10*, 24. [CrossRef]
46. Bernat, M.; Janowski, A.; Rzepa, S.; Sobieraj, A.; Szulwic, J. Studies on the use of terrestrial laser scanning in the maintenance of buildings belonging to the cultural heritage. In Proceedings of the 14th Geoconference on Informatics, Geoinformatics and Remote Sensing, SGEM. ORG, Albena, Bulgaria, 17 June 2014; Volume 3, pp. 307–318.
47. Liu, J.; Zhang, Q.; Wu, J.; Zhao, Y. Dimensional accuracy and structural performance assessment of spatial structure components using 3D laser scanning. *Autom. Constr.* **2018**, *96*, 324–336. [CrossRef]
48. Kwiatkowski, J.; Anigacz, W.; Beben, D. A case study on the noncontact inventory of the oldest european cast-iron bridge using terrestrial laser scanning and photogrammetric techniques. *Remote Sens.* **2020**, *12*, 2745. [CrossRef]
49. Skrzypczak, I.; Oleniacz, G.; Leśniak, A.; Zima, K.; Mrówczyńska, M.; Kazak, J.K. Scan-to-BIM method in construction: Assessment of the 3D buildings model accuracy in terms inventory measurements. *Build. Res. Inf.* **2022**, 1–22. [CrossRef]
50. Borkowski, A.; Jóźków, G. Accuracy assessment of building models created from laser scanning data. *Int. Arch. Photogramm. Remote Sens. Spat. Inf. Sci.* **2012**, *39*, B3. [CrossRef]
51. Klapa, P.; Mitka, B.; Zygmunt, M. Study into point cloud geometric rigidity and accuracy of TLS-based identification of geometric bodies. In *IOP Conference Series: Earth and Environmental Science*; IOP Publishing: Bristol, UK, 2017; p. 32008, ISBN 1755-1315.
52. Zámečníková, M.; Wieser, A.; Woschitz, H.; Ressl, C. Influence of surface reflectivity on reflectorless electronic distance measurement and terrestrial laser scanning. *J. Appl. Geod.* **2014**, *8*, 311–326. [CrossRef]
53. Schmitz, B.; Holst, C.; Medic, T.; Lichti, D.D.; Kuhlmann, H. How to efficiently determine the range precision of 3d terrestrial laser scanners. *Sensors* **2019**, *19*, 1466. [CrossRef]
54. Gleń, P.; Krupa, K. Comparative analysis of the inventory process using manual measurements and laser scanning. *Bud. Archit.* **2019**, *18*, 21–30. [CrossRef]
55. El-Omari, S.; Moselhi, O. Integrating 3D laser scanning and photogrammetry for progress measurement of construction work. *Autom. Constr.* **2008**, *18*, 1–9. [CrossRef]
56. Rocha, G.; Mateus, L.; Fernández, J.; Ferreira, V. A scan-to-BIM methodology applied to heritage buildings. *Heritage* **2020**, *3*, 47–67. [CrossRef]
57. Fawzy, H.E.-D. 3D laser scanning and close-range photogrammetry for buildings documentation: A hybrid technique towards a better accuracy. *Alex. Eng. J.* **2019**, *58*, 1191–1204. [CrossRef]
58. Grussenmeyer, P.; Alby, E.; Landes, T.; Koehl, M.; Guillemin, S.; Hullo, J.-F.; Assali, P.; Smigiel, E. Recording approach of heritage sites based on merging point clouds from high resolution photogrammetry and terrestrial laser scanning. *Int. Arch. Photogramm. Remote Sens. Spat. Inf. Sci.* **2012**, *39*, 553–558. [CrossRef]
59. Remondino, F. Heritage recording and 3D modeling with photogrammetry and 3D scanning. *Remote Sens.* **2011**, *3*, 1104–1138. [CrossRef]
60. Sahin, C.; Alkis, A.; Ergun, B.; Kulur, S.; Batuk, F.; Kilic, A. Producing 3D city model with the combined photogrammetric and laser scanner data in the example of Taksim Cumhuriyet square. *Opt. Lasers Eng.* **2012**, *50*, 1844–1853. [CrossRef]
61. Biagini, C.; Capone, P.; Donato, V.; Facchini, N. Towards the BIM implementation for historical building restoration sites. *Autom. Constr.* **2016**, *71*, 74–86. [CrossRef]
62. Mahdjoubi, L.; Moobela, C.; Laing, R. Providing real-estate services through the integration of 3D laser scanning and building information modelling. *Comput. Ind.* **2013**, *64*, 1272–1281. [CrossRef]
63. Osello, A.; Lucibello, G.; Morgagni, F. HBIM and virtual tools: A new chance to preserve architectural heritage. *Buildings* **2018**, *8*, 12. [CrossRef]
64. López, F.J.; Lerones, P.M.; Llamas, J.; Gómez-García-Bermejo, J.; Zalama, E. A review of heritage building information modeling (H-BIM). *Multimodal Technol. Interact.* **2018**, *2*, 21. [CrossRef]
65. Bruno, S.; de Fino, M.; Fatiguso, F. Historic Building Information Modelling: Performance assessment for diagnosis-aided information modelling and management. *Autom. Constr.* **2018**, *86*, 256–276. [CrossRef]
66. Andriasyan, M.; Moyano, J.; Nieto-Julián, J.E.; Antón, D. From point cloud data to building information modelling: An automatic parametric workflow for heritage. *Remote Sens.* **2020**, *12*, 1094. [CrossRef]

67. Szymczak-Graczyk, A.; Walczak, Z.; Ksit, B.; Szyguła, Z. Multi-criteria diagnostics of historic buildings with the use of 3D laser scanning (a case study). *Bull. Pol. Acad. Sci. Tech. Sci.* **2022**, *70*, e140373. [CrossRef]
68. Lunne, T.; Powell, J.J.M.; Robertson, P.K. *Cone Penetration Testing in Geotechnical Practice*; CRC Press: Boca Raton, FL, USA, 2002; ISBN 0429177801.
69. Girardeau-Montaut, D. CloudCompare, v. 2.6.1. Available online: https://danielgm.net/cc/doc/qCC/CloudCompare%20v2.6.1%20-%20User%20manual.pdf (accessed on 9 July 2022).
70. Kączkowski, Z. *Plates. Static Calculations. Wyd. 3 zm*; Arkady: Warszawa, Poland, 2000; ISBN 83-213-4188-8.
71. *PN-81-B-03020*; Grunty Budowlane. Posadowienie Bezpośrednie Budowli. Obliczenia Statyczne i Projektowanie. Polish Committee for Standardization: Warsaw, Poland, 1981.
72. *PN-83-B-03010*; Ściany Oporowe, Obliczenia Statyczne i Projektowanie. Polish Committee for Standardization: Warsaw, Poland, 1983.
73. Buczkowski, W. *Prostokątne Studnie Opuszczone—Tabela do Obliczeń Statycznych Płyt Ściennych*; Budownictwo Przemysłowe: Poznań, Poland, 1988.
74. *PN-EN 13331-1-1:2002*; Obudowy Ścian Wykopów. Część 1. 2002. Polish Committee for Standardization: Warsaw, Poland, 2002.
75. *PN-EN 1990:2004*; Eurokod. Podstawy projektowania konstrukcji. Polish Committee for Standardization: Warsaw, Poland, 2004.
76. Alam, M.; Chaallal, O.; Galy, B. Soil-structure interaction of flexible temporary trench box: Parametric studies using 3D FE modelling. *Model. Simul. Eng.* **2021**, *2021*, 9949976. [CrossRef]
77. Alam, M.; Chaallal, O.; Galy, B. Field test of temporary excavation wall support in sensitive clay. *ISSMGE Int. J. Geoengin. Case Hist. IJGCH* **2021**, *6*, 18–40.
78. *PN-EN 1993-1-1:2006*; Eurocode 3: Design of Steel Structures. Part 1-1: General Rules and Rules for Buildings. Polish Committee for Standardization: Warsaw, Poland, 2006.

materials

Article

Study on the Seismic Performance of Stiffened Corrugated Steel Plate Shear Walls with Atmospheric Corrosion

Xiaoming Ma [1,2,3], Yi Hu [4,5], Liqiang Jiang [5,*], Lizhong Jiang [5], Guibo Nie [1,2] and Hong Zheng [3]

1 Key Laboratory of Earthquake Engineering and Engineering Vibration, Institute of Engineering Mechanics, China Earthquake Administration, Harbin 150080, China; 2020228041@chd.edu.cn (X.M.); nieguibo0323@163.com (G.N.)
2 Key Laboratory of Earthquake Disaster Mitigation, Ministry of Emergency Management, Harbin 150080, China
3 School of Civil Engineering, Chang'an University, Xi'an 710061, China; cehzheng@chd.edu.cn
4 School of Civil Engineering, Central South University of Forestry and Technology, Changsha 410004, China; hyi_1991@163.com
5 School of Civil Engineering, Central South University, Changsha 410075, China; lzhjiang@csu.edu.cn
* Correspondence: jianglq2019@csu.edu.cn

Abstract: Corrugated steel plate shear walls (CSPWs) with three different stiffening methods are proposed in this paper, including unstiffened CSPWs (USWs), cross stiffened CSPWs (CSWs) and asymmetric diagonal-stiffened CSPWs (ASWs). A numerical model was established by ABAQUS 6.13 based on the validation of an existing cyclic test on a CSPW. This paper presents an investigation of the lateral performance under monotonic loading, seismic performance under cyclic loading and seismic performance under atmospheric corrosion of USW, CSW and ASW. The results show that (1) Stiffeners can improve the elastic critical buckling load, the initial stiffness and the ultimate shear resistance of CSPWs, and the effect of asymmetric diagonal stiffeners is more significant than that of cross stiffeners; (2) Stiffeners can improve the energy dissipation capacity and ductility, delay stiffness degradation and reduce the out-of-plane deformation of CSPWs, and the hysteretic performance of ASWs is obviously better than that of CSWs; and (3) Under atmospheric corrosion, stiffeners are conducive to inhibiting buckling and improving the seismic performance of CSPWs, while the seismic performance of CSWs is significantly affected by corrosion, so asymmetric diagonal stiffeners are better than cross stiffeners in improving the seismic performance of CSPWs. Meanwhile, the formula of ultimate shear resistance of corroded specimens is also fitted in this paper, which can provide design suggestions for practical engineering.

Keywords: corrugated steel plate shear wall (CSPW); finite element method (FEM); lateral performance; hysteretic performance; atmospheric corrosion

check for updates

Citation: Ma, X.; Hu, Y.; Jiang, L.; Jiang, L.; Nie, G.; Zheng, H. Study on the Seismic Performance of Stiffened Corrugated Steel Plate Shear Walls with Atmospheric Corrosion. *Materials* **2022**, *15*, 4920. https://doi.org/10.3390/ma15144920

Academic Editor: Carlos Garcia-Mateo

Received: 27 May 2022
Accepted: 6 July 2022
Published: 14 July 2022

1. Introduction

With lighter weight, higher strength and better ductility, the damage degree of steel structures is much lower than that of reinforced concrete structures after earthquakes un-der the same conditions. To address problems existing in traditional structures, the steel plate shear wall (SPSW) has been proposed. The SPSW is a kind of lateral force resistant structural system that emerged in the 1970s. Due to its high initial stiffness, good ductility and seismic performance, simple connection and reduced structural deformation, the SPSW has been widely used in the lateral force resisting system of middle- and high-rise buildings [1–4]. However, the steel plate is prone to buckling, which can cause a reduction in the shear resistance and lateral stiffness. Meanwhile, the steel plate has poor energy dissipation capacity, and it can also form a tension field, which is disadvantageous to the boundary column [5–8].

To solve the problem that the steel plate is prone to buckling, many experiments and theoretical analyses were conducted. Berman and Bruneau [9–11] proposed the strip mod-el of the flat steel plate, and the formulae for calculating the ultimate shear resistance of the SPSW when the infilled steel plate is hinged and rigidly connected to the boundary frame are proposed, which were verified by experiments. Studies have shown that stiffeners can significantly improve the energy dissipation capacity, ultimate shear resistance, initial stiffness and ductility [12]. Nie et al. [13] studied the influence of 6 kinds of stiffeners on the SPSW, and the results showed that among the six kinds of stiffeners, the asymmetric diagonal stiffened SPSW had the largest critical buckling load, followed by the cross stiffener. According to Cao et al. [14,15], compared with unreinforced SPSWs, the new SPSW with X-shaped restrainers has been improved in some respects, including the shear resistance, peak load, energy dissipation capacity and out-of-plane displacement constrained efficiency, and its hysteretic curve tends to be full. Khaloo et al. [16] conducted numerical investigations on a diagonal stiffened SPSW, and the results showed that the stiffened SPSW showed good crack resistance in both the elastic and plastic stages.

CSPWs have attracted more attention than traditional SPSWs because CSPWs exhibit higher out-of-plane stiffness, shear resistance, ductility and energy dissipation capacity [17,18]. Emami et al. [19,20] conducted numerical investigations and experiments on the cycle performance of CSPWs and SPSWs, and the results showed that compared to SPSWs, CSPWs had higher out-of-plane stiffness and buckling strength and a fuller hysteretic curve. Meanwhile, the beam-only connected SPSW was proposed first to avoid high-flexural and axial demands in the boundary column due to the tension field [21]. Paslar et al. [22] established 57 numerical models to investigate the connection between steel plates and boundary frames. Elastic buckling can be avoided by reasonable design of the corrugated size, but there is a certain relationship between the ultimate shear resistance and the geometric parameters of the CSPW, so geometric parameters should be reasonably selected to avoid a significant decrease in the post-buckling strength [23–28]. To inhibit the out-of-plane buckling of corrugated steel plates, corrugated steel plates with stiffeners have been proposed by researchers. Tong [29] proposed three new types of CSPWs and then conducted experimental, numerical and theoretical investigations. The results showed that stiffeners can effectively restrain the out-of-plane buckling of CSPWs and improve the ultimate shear resistance and energy consumption of CSPWs to some extent. Wang [30] and Zheng et al. [31,32] proposed a new type of cross stiffened CSPW and established FEM models. Then, the lateral resistance, seismic performance and the influence of the dimensions of the CSPW on the mechanical performance were analysed. The results showed that in the middle and late loading stage of the CSPW, the steel plate can have in-elastic buckling, resulting in the decline of the ultimate shear resistance and the final failure. The cross stiffener can play a good role in inhibiting buckling and further improving the mechanical performance of the CSPW.

Meanwhile, corrosion is a problem for a large number of steel structures. In coastal areas, corrosion is an important factor leading to age-related structural degradation of steel structures; it can reduce the strength of steel, cause stress concentration, and finally reduce the bearing capacity of structures [33,34]. Zheng et al. [35–37] conducted accelerated corrosion tests and quasi-static cyclic tests on steel frames and joints, and Xu et al. [38,39] conducted accelerated corrosion tests, quasi-static cyclic tests and finite element analyses on seven H-shaped steel columns. The results showed that with increasing corrosion degree, the forming of the plastic hinge was accelerated, and the ultimate lateral resistance, stiffness and energy dissipation capacity of the specimens decreased significantly.

Although many experimental and theoretical investigations have been conducted on CSPWs with stiffeners, the current research mainly compares the mechanical properties of CSPWs before and after stiffening, and research comparing CSPWs with different stiffening methods is still insufficient. Meanwhile, although many experimental and numerical investigations of steel structures under atmospheric corrosion have been conducted [35–39], there are still few investigations on the seismic performance of corroded CSPWs [40–42].

Therefore, this paper carries out finite element analyses on CSPWs with different stiffening methods by using ABAQUS. The lateral performance, seismic performance and hysteretic performance under atmospheric corrosion are studied, and the analysis results can provide practical suggestions for practical engineering design.

2. Numerical Modelling

2.1. Model Description

This paper proposes three types of CSPWs with different stiffening methods, including USW, CSW and ASW. According to the Chinese standards of GB50017-2017 [43], JGJ/T380-2015 [44] and GB50011-2010 [45], the dimensions of the USW, CSW and ASW are presented in Figure 1. The specimen consists of the boundary frame, infilled corrugated steel plate and the stiffener. The section of the boundary column is H 400 × 300 × 18 × 22. The top-beam section and the bottom-beam section are designed as the same sizes, and they are H 350 × 300 × 14 × 20. The section of the middle beam is H 300 × 250 × 12 × 16. The beam column is rigidly connected, and the stiffeners are welded in the node domain. The dimensions of the infilled corrugated steel plate are presented in Figure 1. Vertical trapezoidal corrugated steel plates are used for these specimens, and the thickness is 6 mm, the wave height is 90 mm, the wavelength is 300 mm, and the CSPW is only welded with the frame beam. The stiffeners are welded to the surface of the corrugated steel plate. The dimensions of the cross stiffener of the CSW are −3000 × 70 × 7. The stiffeners are arranged asymmetrically on the front and back sides of the corrugated steel plate, and the dimensions of the long stiffener and short stiffener are −4240 × 70 × 7 and −2120 × 70 × 7, respectively. Q355 steel is used for the boundary frame and stiffeners, and Q235 steel is used for the corrugated steel plate.

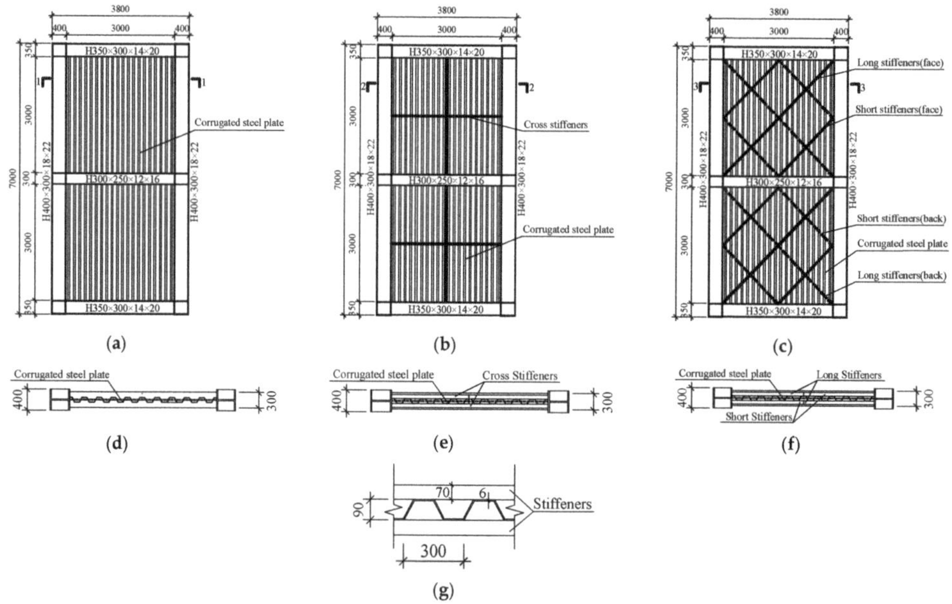

Figure 1. Dimension details of USW, CSW and ASW: **(a)** USW; **(b)** CSW; **(c)** ASW; **(d)** the profile of Section 1-1; **(e)** the profile of Section 2-2; **(f)** the profile of Section 3-3; **(g)** the dimensions of corrugated steel plate.

Numerical models of USW, CSW and ASW are shown in Figure 2. In the numerical models, S4R four-node shell elements are used for modelling the boundary frame, the corrugated steel plates and the stiffeners. The maximum size of the elements is 70 mm for steel, and the mesh sensitivity analysis reveals that the mesh size is dense enough for reasonable accuracy and computation time. The constitutive relation of the steel is a bilinear kinematic model; before the steel reaches the yield strength, the slope is E, and after reaching the yield strength, the slope is 0.03 E. A "Merge" combination command is used to simulate the weld connections in the beams and the columns. A "Tie" command is used to connect the corrugated steel plates to the boundary beams and stiffeners. Initial defects are introduced into the model by modifying the keyword, and the amplitude is taken as 1/1000 [46].

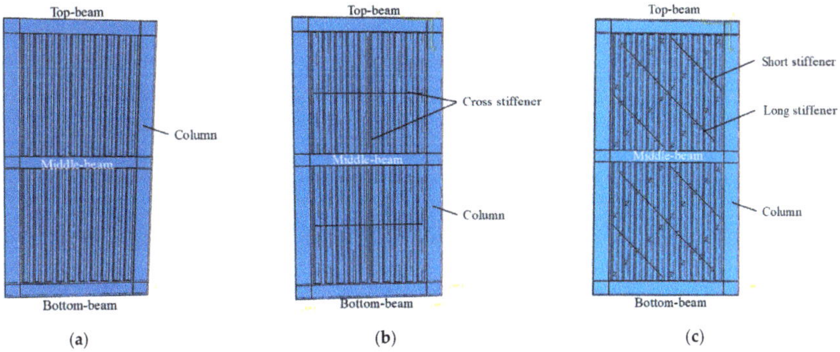

Figure 2. Model diagrams of USW, CSW and ASW: (**a**) USW; (**b**) CSW; (**c**) ASW.

The lateral loads were imposed using the displacement-control scheme, as shown in Figure 3. Loading steps were integer multiples of yield displacement δ_y. The spacing between each control point was δ_y, and it could be known when an inflection appeared in the curve in Figure 3, repeated three times at each control point.

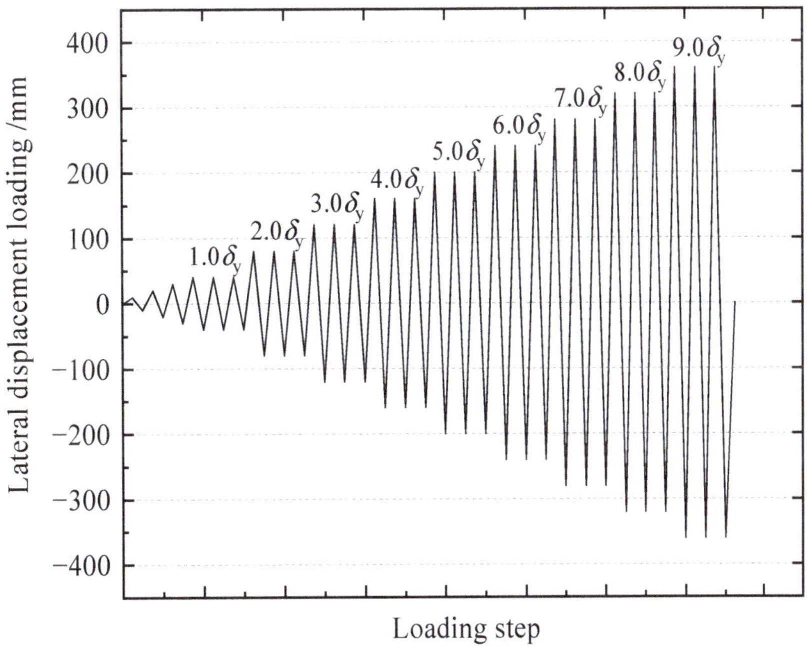

Figure 3. Cyclic loading of the specimens.

2.2. Validation of FEM

Currently, no existing cyclic tests have been performed to validate the FEM model of CSWs and ASWs. This paper selects test results from specimen S-1 from the Ref. [47]. S-1 is shown in Figure 4. The thickness, height and width of the corrugated steel plates were 3 mm, 1 m and 1.75 m, respectively. Q345 steel was used for the boundary frame of S-1, and Q235 steel was used for the infilled corrugated steel plate of S-1. The numerical model of specimen S-1 is then built using the same modelling method presented in Section 2.1, and the material properties of the test specimen are shown in Table 1.

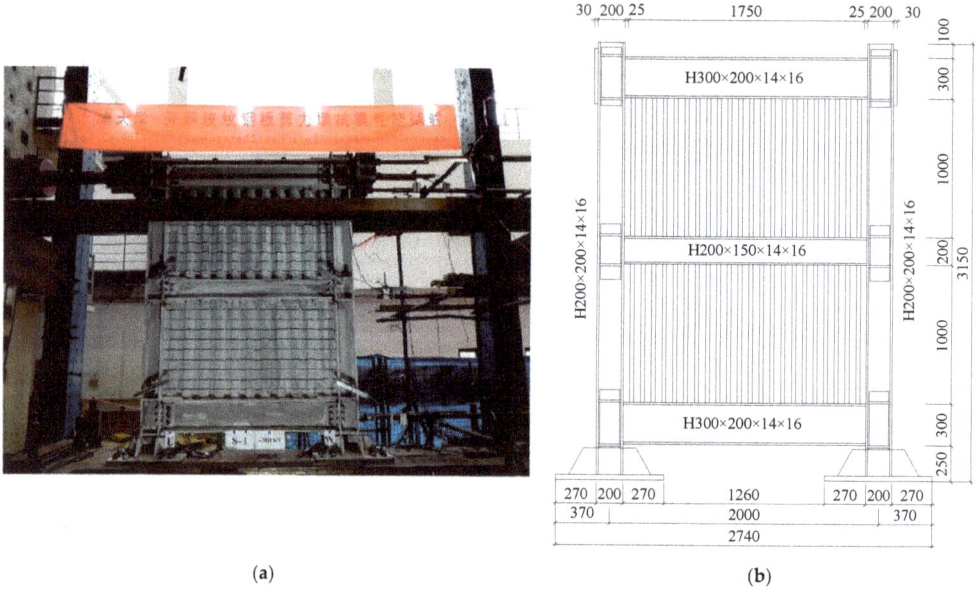

Figure 4. Model and sizes of the test [47]: (**a**) setup of the test; (**b**) size of the test specimen.

Table 1. Material properties of S-1 [47].

Items	Steel Type	Thickness *t*/mm	*E*/MPa	f_u/MPa	f_y/MPa	f_u/f_y	Elongation Ratio
H-web	Q345	14	202,558	513.3	357.3	1.44	34.6%
H-flange	Q345	16.6	183,527	468.9	334.8	1.42	34.1%
Fishplate	Q235	5.7	204,273	460	336	1.37	36.9%
Corrugated steel plate	Q235	3	193,252	451.3	312.3	1.45	39.4%

Note: *E* is the elastic modulus; f_y is the yield strength; f_u is the ultimate strength.

The comparisons of the hysteretic curves and envelope curves obtained from FEM analyses and tests are shown in Figure 5, and the detailed results are shown in Table 2. The hysteretic curves of FEM analyses and tests were full, and the overall trend of envelope curves from FEM analyses and tests was basically consistent. The error of yield strength between FEM analyses and tests is only 0.7%. The yield displacement and peak load of the tests are higher than those of FEM analyses, and the errors are 8.8% and 3.8%, respectively.

The comparisons of the failure modes of S-1 between FEM analyses and tests are shown in Figure 6. The results of the comparisons show that the FEM model accurately simulates the buckling modes and buckling positions of boundary columns and corrugated steel plates. It can be found that the numerical model in Section 2.1 is reasonable, and it can be used for the following study.

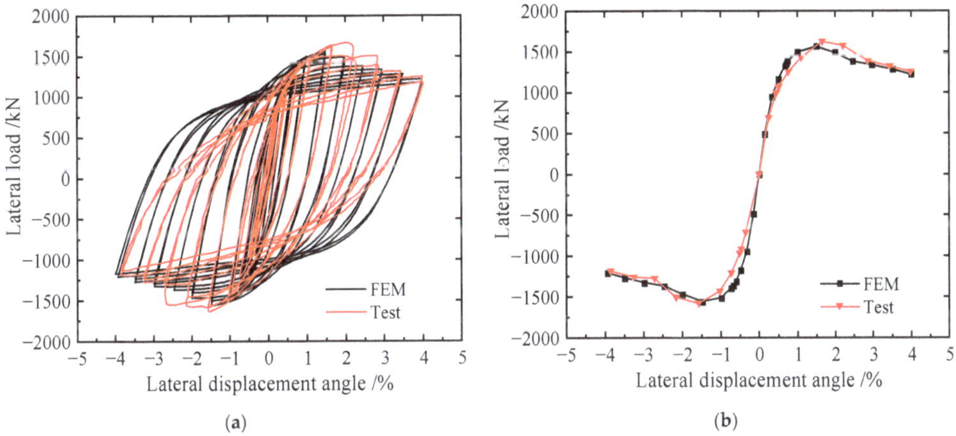

Figure 5. Comparisons of the load–displacement curves of specimen S-1: (**a**) hysteretic curves and (**b**) envelope curves.

Table 2. Detailed results of comparison between FEM and test.

Results	P_y/kN	Δ_y/mm	K_0/(kN/mm)	P_m/kN	Δ_m/mm
Test	1300.08	20.05	99.94	1626.35	46.19
FEM	1309.99	18.28	109.04	1565.01	41.69
Error	0.76%	8.82%	9.10%	3.77%	9.76%

Note: P_y is the yield load, Δ_y is the yield displacement, K_0 is the initial lateral stiffness, P_m is the peak load, and Δ_m is the displacement at P_m.

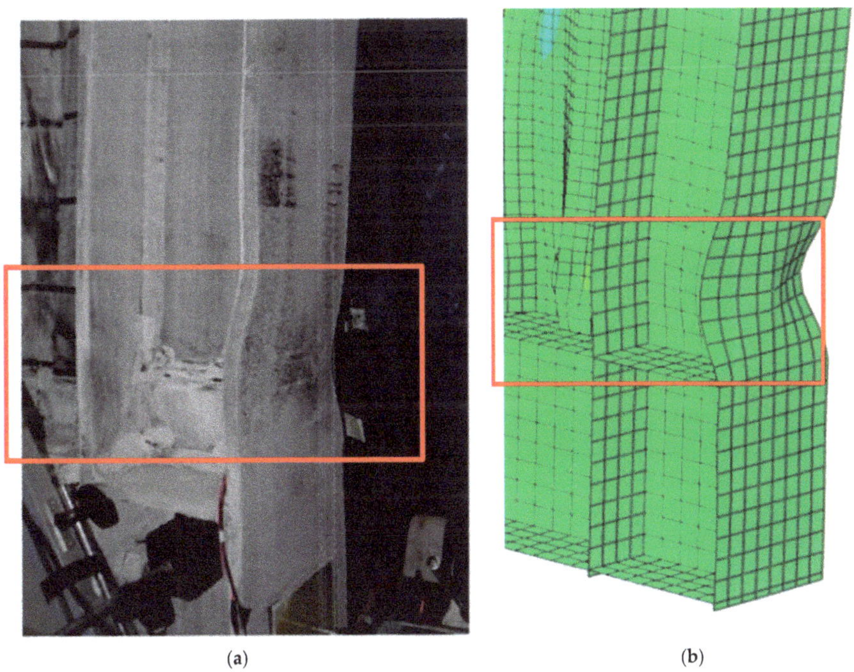

Figure 6. *Cont.*

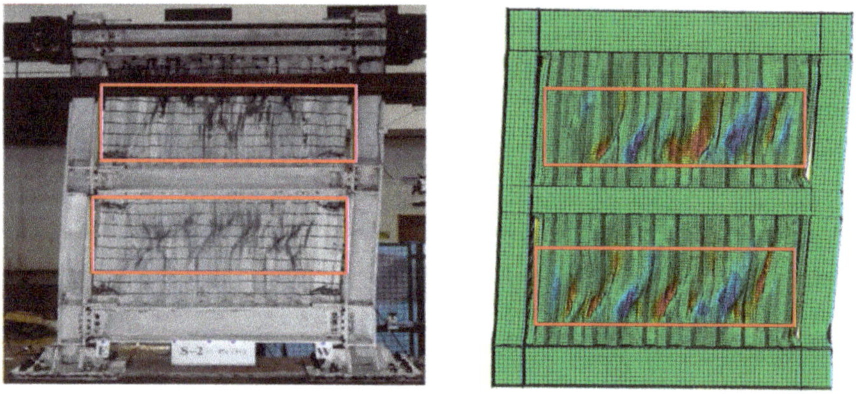

(c) (d)

Figure 6. Comparison of the failure modes of the test observation and FEM analyses: (**a**) local buckling at the bottom of the test column; (**b**) local buckling at the bottom of the FE column; (**c**) global de-formation of the test specimen; and (**d**) global deformation of the FE specimen.

3. Comparative Study of the Lateral Resistance of the CSPW

FEM models of USW, CSW and ASW are built as shown in Figure 7 and their lateral resistance under monotonic loading is compared and analyzed. The modelling method is similar to that in Section 2.1, and a "Tie" combination command is used to simulate the weld connections in the stiffener and the corrugated steel plate. The constitutive relation of the material is linear elasticity. The boundary condition of the corrugated steel plate is that the upper and lower sides are hinged with the boundary frame. Axial forces P_1 and P_2 are applied at two coupling points on the top of the boundary columns to simulate the vertical load, such as gravity, and $P_1 = P_2 = 600$ kN. Then, horizontal load is applied at the coupling point on the height of the centerline of the top beam until the specimen fails.

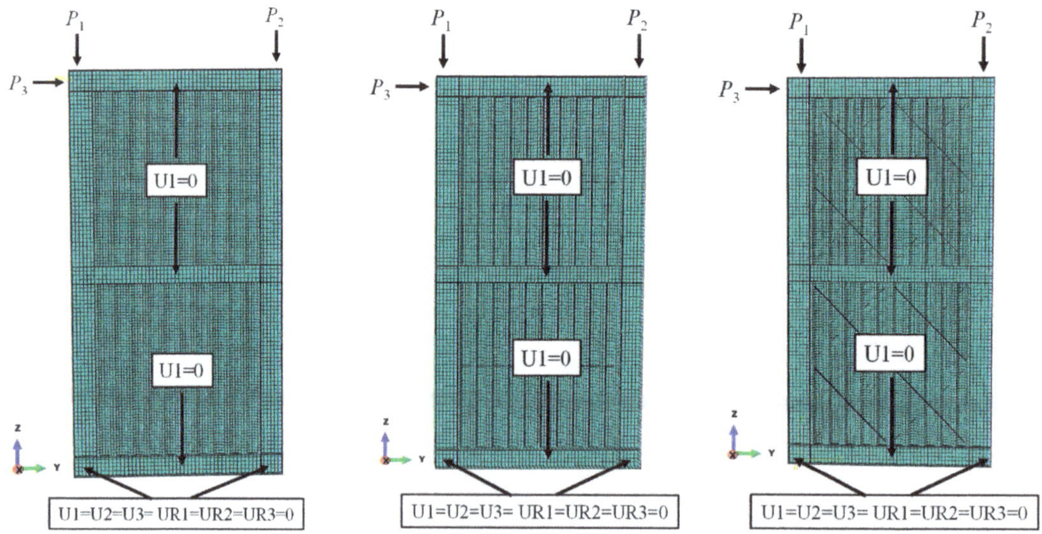

(**a**) (**b**) (**c**)

Figure 7. FE models of USW, CSW and ASW: (**a**) USW; (**b**) CSW; (**c**) ASW.

3.1. Comparison of Monotonic Load–Displacement Curves

The load–displacement curves of the USW, CSW and ASW are shown in Figure 8, and the detailed results are shown in Table 3. Due to stiffeners, the properties of three types of specimens such as the initial stiffness, yield load and ultimate shear resistance are different. The yield load and peak load of ASW were the largest, followed by CSW. With increasing lateral displacement, USW showed the earliest decline period with a large decline range, and ASW showed the last decline period with a gentler decline. This means that ASW has the best ductility and USW has the worst ductility.

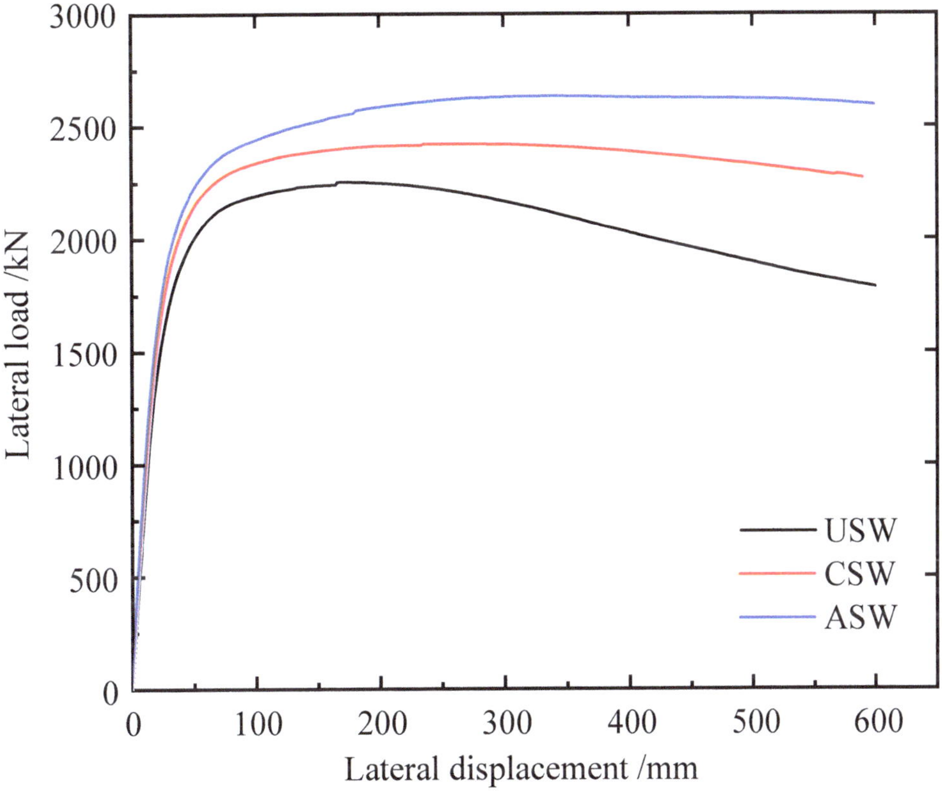

Figure 8. Load–displacement curves of USW, CSW and ASW.

Table 3. Comparisons of the characteristic results of USW, CSW and ASW.

Specimen	K_0/(kN/mm)	P_y/kN	Δ_y/mm	P_u/kN	Δ_u/mm
USW	83.02	1848.81	37.46	2254.15	170.83
CSW	88.04	1983.98	38.00	2422.66	238.89
ASW	92.50	2099.42	40.29	2633.8	334.22

Note: P_u is the maximum load, Δ_u is the displacement at P_u.

Compared to USW, the initial lateral stiffness, yield load, yield displacement, peak load and displacement at the peak load of the CSW increase by 6.05%, 7.31%, 1.44%, 7.48% and 39.84%, respectively. Compared to the CSW, the initial lateral stiffness, yield load, yield displacement, peak load and displacement at the peak load of the ASW increase by 5.1%, 5.8%, 6.02%, 8.7% and 40%, respectively. Therefore, cross stiffeners can effectively improve the lateral resistance of the CSPW, and asymmetric diagonal stiffeners can be more effective than cross stiffeners. Meanwhile, asymmetric diagonal stiffeners can also avoid the stress concentration and complex connection caused by diagonal stiffeners on the same side.

3.2. *Comparison of Out-of-Plane Deformations*

The out-of-deformations of the USW, CSW and ASW at failure under monotonic loading are shown in Figure 9. USW had serious out-of-plane deformation and buckling. Due to the inhibition of cross stiffeners, the out-of-plane deformation of the CSW was lower than that of the USW. The deformation of ASW at failure was mainly local deformation, and the out-of-plane deformation of ASW was the lowest among them. The maximum out-of-plane deformation of USW, CSW and ASW is 231.7 mm, 145.61 mm and 88.01 mm, respectively, so compared to ASW, the maximum out-of-plane deformation of CSW and USW increases 39.56% and 62.02%, respectively. It is revealed that asymmetric diagonal stiffeners can inhibit the out-of-plane deformation of the CSPW more effectively than cross stiffeners.

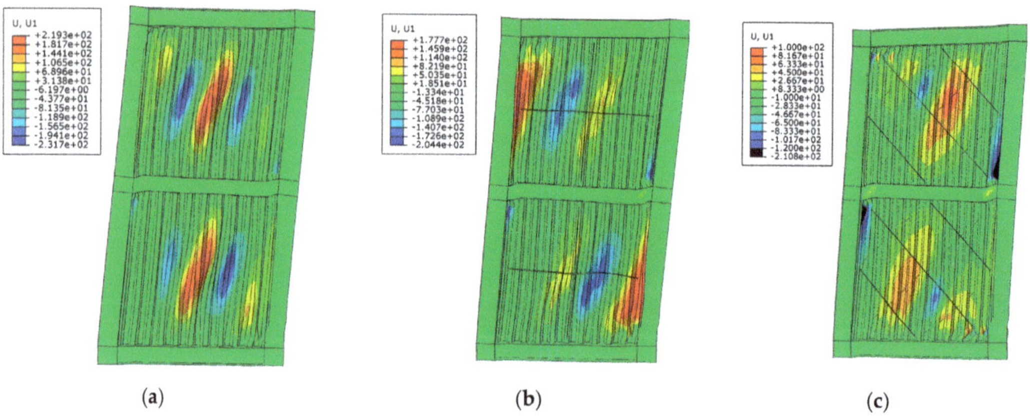

Figure 9. Out-of-plane deformations of USW, CSW and ASW under monotonic load: (**a**) USW; (**b**) CSW; (**c**) ASW.

3.3. *The Comparison of Stress Distributions*

The stress distributions of USW, CSW and ASW at failure are shown in Figure 10a–f. When USW was damaged, the corrugated steel plate buckled seriously, full section yielding occurred at the bottom of the boundary column, and plastic buckling occurred at the end of the middle beam. Due to the inhibition of cross stiffeners, the lateral displacement of the CSW at buckling was larger than that of the USW, the plastic hinge at the bottom of the boundary column was formed later, and the buckling of the corrugated steel plate of the CSW at failure was slighter than that of the USW. Due to the inhibition of asymmetric diagonal stiffeners, the corrugated steel plate of ASW had large in-plane stiffness, the plastic area at the bottom of the boundary column of ASW was smaller than that of CSW, and when ASW was damaged, there was no full section yielding at the bottom of the boundary column.

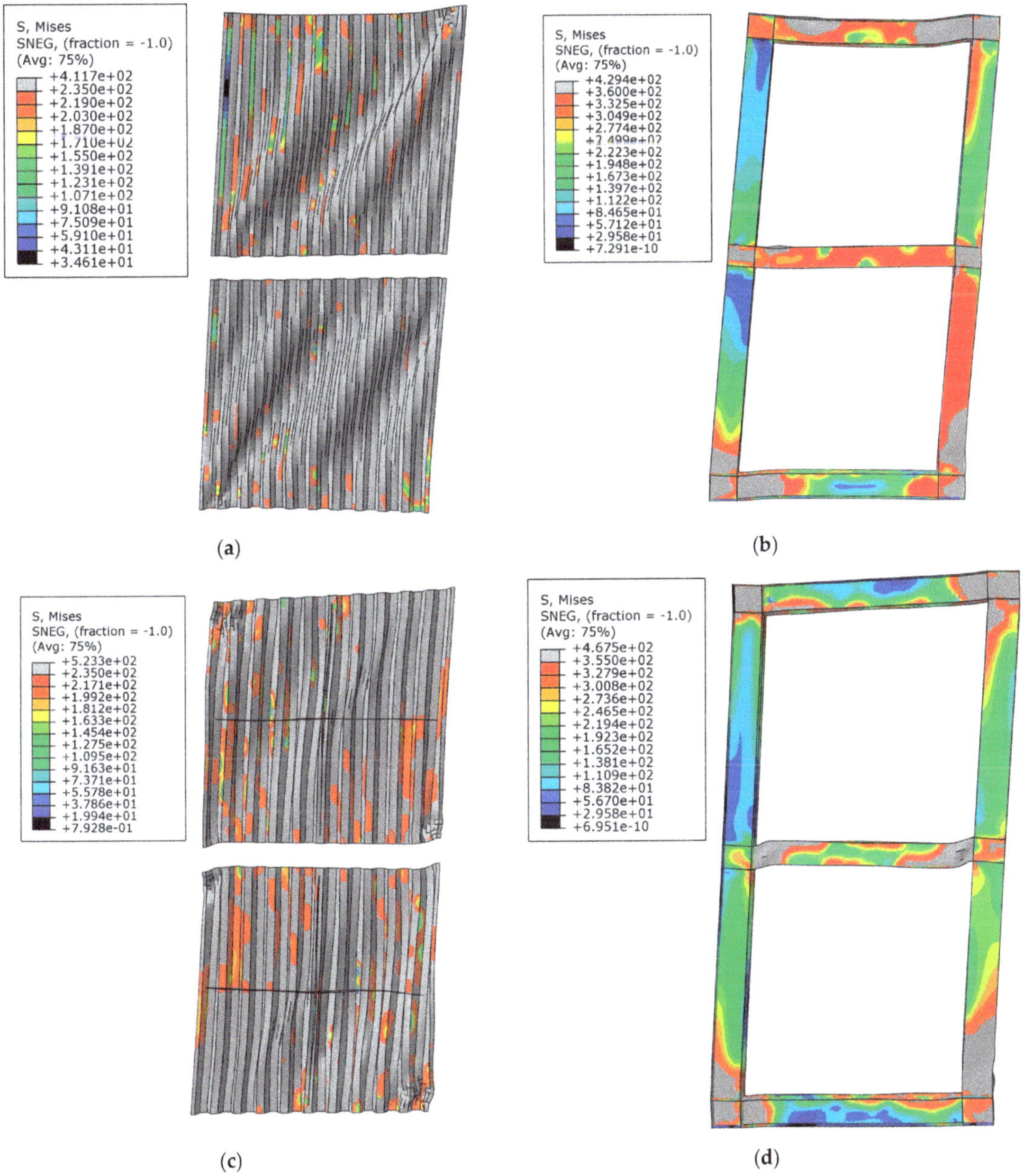

Figure 10. *Cont.*

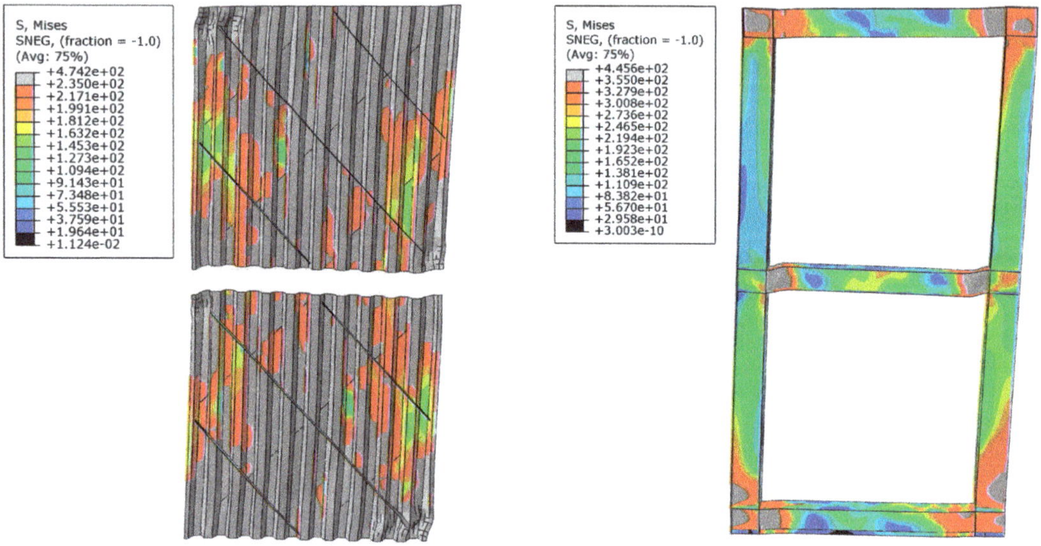

(e) (f)

Figure 10. Stress distribution of damaged specimens under the monotonic load: (**a**) corrugated steel plate of USW; (**b**) boundary frame of USW; (**c**) corrugated steel plate of CSW; (**d**) boundary frame of CSW; (**e**) corrugated steel plate of ASW; (**f**) boundary frame of ASW.

The comparative results show that due to the inhibition of cross stiffeners and asymmetric diagonal stiffeners, the plastic hinge at the bottom of the boundary column is formed later, and the failure of the whole specimen is delayed. Meanwhile, asymmetric diagonal stiffeners are better than cross stiffeners in inhibiting the inelastic buckling in-stability of CSPWs.

4. Comparative Study on Seismic Performance of the CSPW

Finite element analyses of USW, CSW and ASW under cyclic loads are carried out, and the modelling method is the same as that in Section 3. A vertical load is applied at two coupling points on the top of the boundary columns, and then a horizontal load is applied at the coupling point on the beam-column joint.

4.1. Comparison of Cyclic Load–Displacement Curves

The hysteretic curves and the envelope curves of the specimens are shown in Figure 11, and detailed results are shown in Table 4. Compared to USW and CAW, the hysteresis loop of specimen ASW was fuller. With global buckling of the CSPW, the lateral stiffness of the specimens decreased, and the stiffness of the CSW and ASW decreased slightly in the late loading stage. With increasing lateral displacement, the specimen yielded and gradually entered the stage of plastic development, resulting in loss of stiffness. Then, the ultimate shear resistance of the specimen decreased significantly, and the specimen entered the stage of plastic failure. Compared to USW, the initial stiffness, yield load, yield displacement, peak load and ductility of the CSW increase by 4.1%, 6.82%, 11.35, 6.7% and 27.64%, respectively, and the initial stiffness, yield load, yield displacement, peak load and ductility of the ASW increase by 12.15%, 11.66%, 15.32%, 14.4% and 46.4%, respectively. Therefore, asymmetric diagonal stiffeners can inhibit the buckling of CSPWs and improve hysteretic performance more effectively than cross stiffeners.

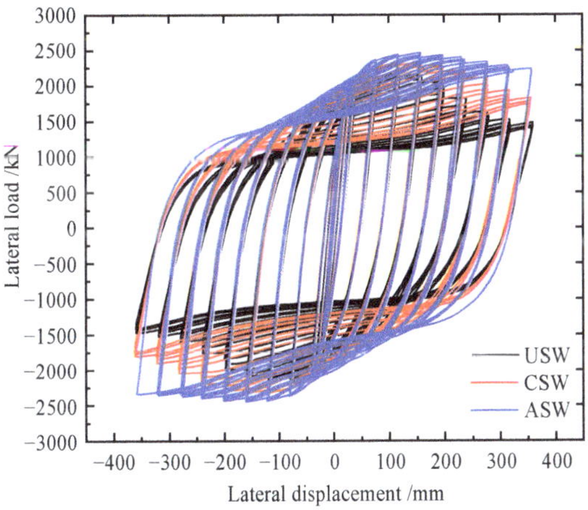

(**a**)

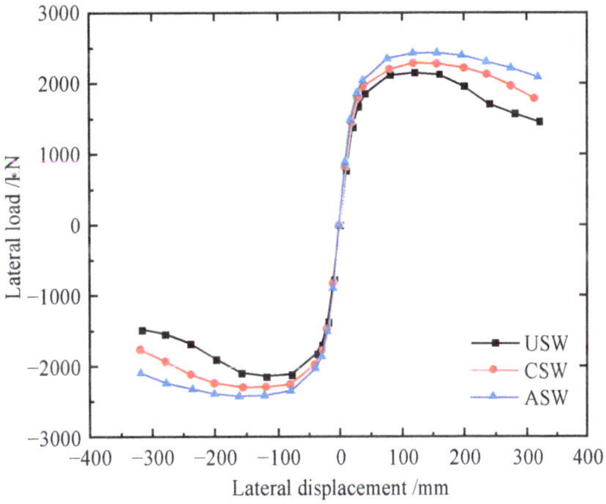

(**b**)

Figure 11. Load–displacement curves of specimens USW, CSW and ASW: (**a**) hysteretic curves; (**b**) envelope curves.

Table 4. Characteristic results of USW, CSW and ASW.

Specimen	K_0/(kN/mm)	P_y/kN	Δ_y/mm	P_m/kN	Δ_m/mm	μ
USW	80.3	1792.36	36.43	2144.82	119.01	5.97
CSW	83.59	1914.53	40.6	2288.62	138.59	7.62
ASW	90.06	2001.27	42.01	2453.59	159.15	8.74

Note: P_m is the peak load, Δ_m is the displacement at P_m, and μ is the ductility factor, $\mu = \Delta_u/\Delta_y$.

4.2. Comparison of Energy Dissipation Capacity

The energy dissipations of the specimens are shown in Figure 12. The energy dissipation of the ASW under each loading step was higher than that of the USW and CSW. After the loading step reached 2δy, the energy dissipation of the ASW increased the fastest, followed by that of the CSW. The maximum values of energy dissipation of USW, CSW and ASW are 1,207,511 kN·mm, 1,476,870 kN·mm and 1,865,640 kN·mm, respectively. Compared to USW and CSW, the energy dissipation of ASW improves 26% and 54%, respectively. Therefore, stiffeners can improve the energy dissipation of CSPWs, and asymmetric diagonal stiffeners can improve the energy dissipation of CSPWs more effectively than cross stiffeners.

4.3. Comparison of the Degradation of the Stiffness and Strength

The curves of stiffness degradation of USW, CSW and ASW are shown in Figure 13. The stiffness degradation of the specimens had the same trend. The stiffness of the ASW at each loading step was higher than that of the USW and CSW. After reaching the peak load, the stiffness degradation of ASW was smaller than that of USW and CSW. Therefore, asymmetric diagonal stiffeners can better inhibit buckling and improve the in-plane stiffness of CSPWs at the late loading stage.

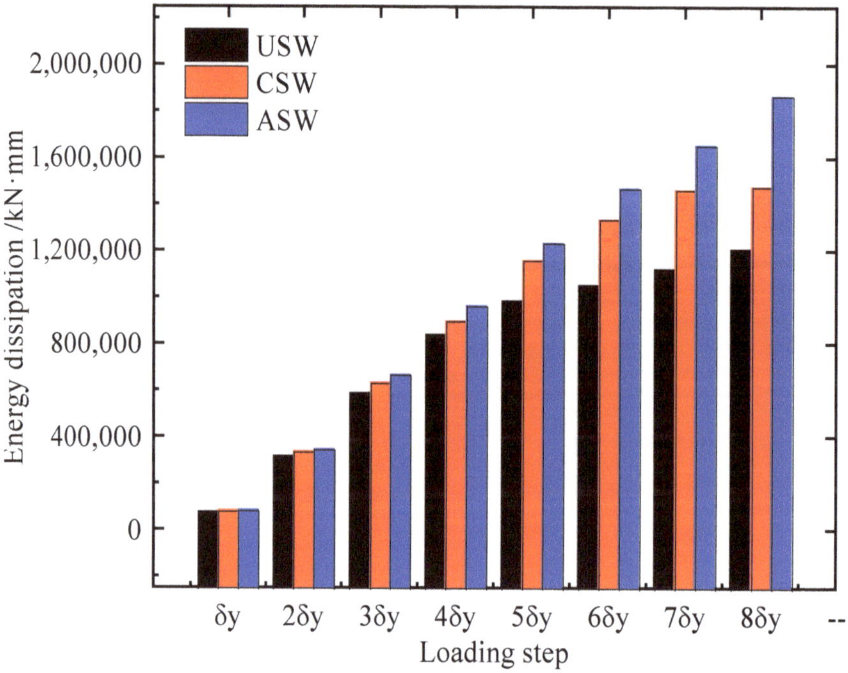

Figure 12. Energy dissipation of USW, CSW and ASW.

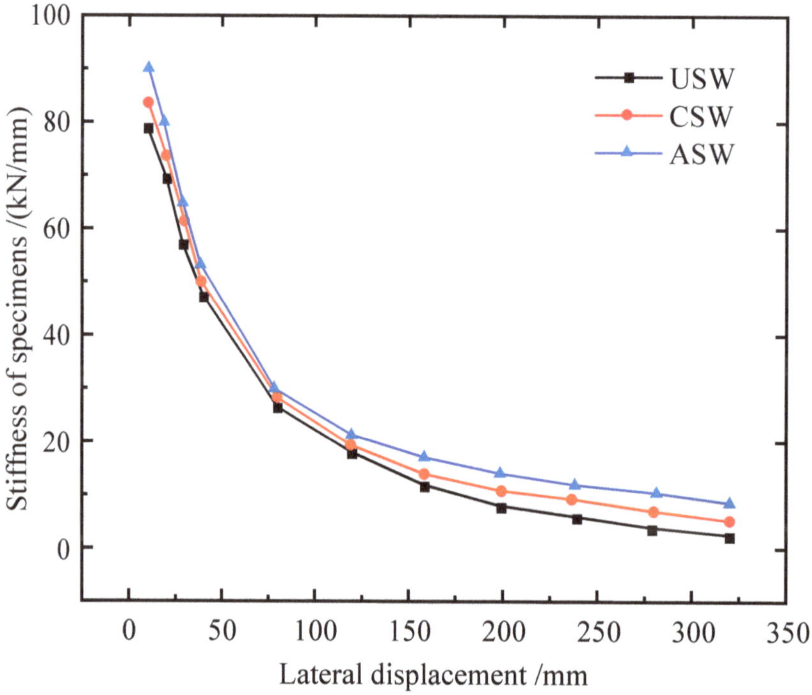

Figure 13. Stiffness degenerations of USW, CSW and ASW.

The strength degradation coefficient (λ_i, $i = 1, 2, \ldots\ldots$) is an index to evaluate the strength degradation of the structure, and the calculation formula is:

$$\lambda_i = \frac{F_j^{i+1}}{F_j^i}, \tag{1}$$

where F_j^{i+1} is the peak load of cycle $i + 1$ at the jth loading step and F_j^i is the peak load of cycle i at the jth loading step.

The variation trend of λ_1 and λ_2 with loading displacement is shown in Figure 14. Before reaching the peak load, λ_i of the USW, CSW and ASW is approximately about 1, and after reaching the peak load, λ_i of the USW, CSW and ASW decreases by 6%, 4.5% and 2%, respectively, and λ_i of the ASW is not less than 0.98. Compared to USW and CSW, λ_i of ASW increases by a maximum of 4.8% and 3.7%, respectively, so stiffeners can improve the stability of ultimate shear resistance to some extent, but the improvement is not as obvious.

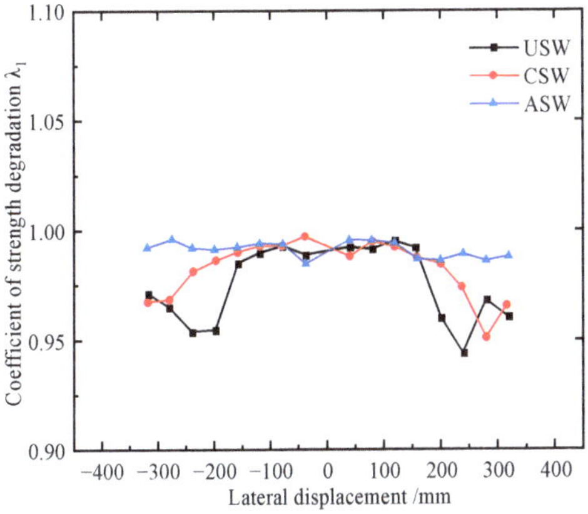

(**a**)

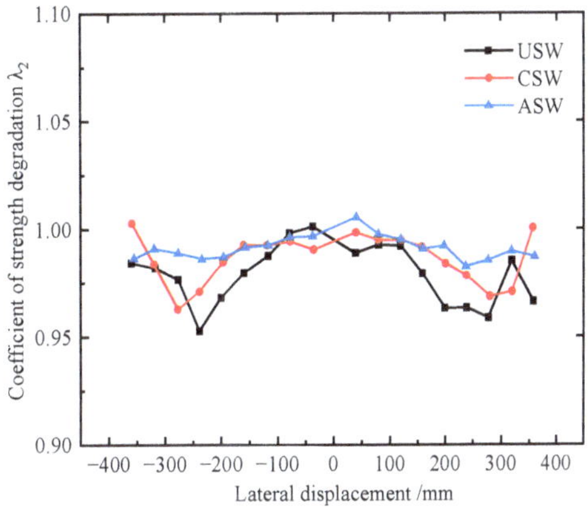

(**b**)

Figure 14. Strength degradation curves of USW, CSW and ASW: (**a**) variation curve of λ_1; (**b**) variation curve of λ_2.

5. Comparative Study of the Seismic Performance of CSPWs under Atmospheric Corrosion

FEM models of USW, CSW and ASW are built under different corrosion levels, and the modelling method is the same as that in Section 3. Then, the hysteretic performance of the specimens is compared and analyzed.

5.1. Material Properties after Corrosion

To achieve the long-term corrosion effect of steel in a chloride environment for a short time, the indoor or outdoor accelerated corrosion testing method is generally adopted. The outdoor accelerated corrosion test and results from the literature [35] were selected in this paper. The NaCl solution concentration was 50 ± 5 g/L, and the pH value was 6.5–7.2. The fitting formula from the literature [35] was selected to calculate the yield strength of corroded steel, as shown in Equation (2).

$$f_y = (1 - 0.902 \times \frac{1 - E/E_0}{0.897}) f_{y0}, \tag{2}$$

where f_{y0} and E_0 are the yield strength and elastic modulus of uncorroded steel, respectively, and f_y and E are the yield strength and elastic modulus of corroded steel, respectively. f_{y0} of steel are 235 MPa and 355 MPa, respectively, and E_0 is 206,000 MPa.

The time of the accelerated corrosion test is converted to the time of actual outdoor corrosion. The rate of accelerated corrosion is 1208 $\mu m \cdot a^{-1}$ in this paper, and the rate of atmospheric corrosion in Beijing is 11.7 $\mu m \cdot a^{-1}$ [48,49]. These two rates are used for conversion, and the results are shown in Table 5. The change in the elastic modulus of steel during the corrosion process is shown in Figure 15. To better reflect the changes in material properties during the corrosion process, the elastic modulus after 20, 60 and 120 days of corrosion [35], which is actual corrosion of 5.657, 16.972 and 33.945 years, are selected for FEM models, as shown in Table 6.

Table 5. Corrosion time of specimens.

Experimental Corrosion Time/d	20	40	60	80	100	120
Actual corrosion time/a	5.657	11.315	16.972	22.63	28.287	33.945

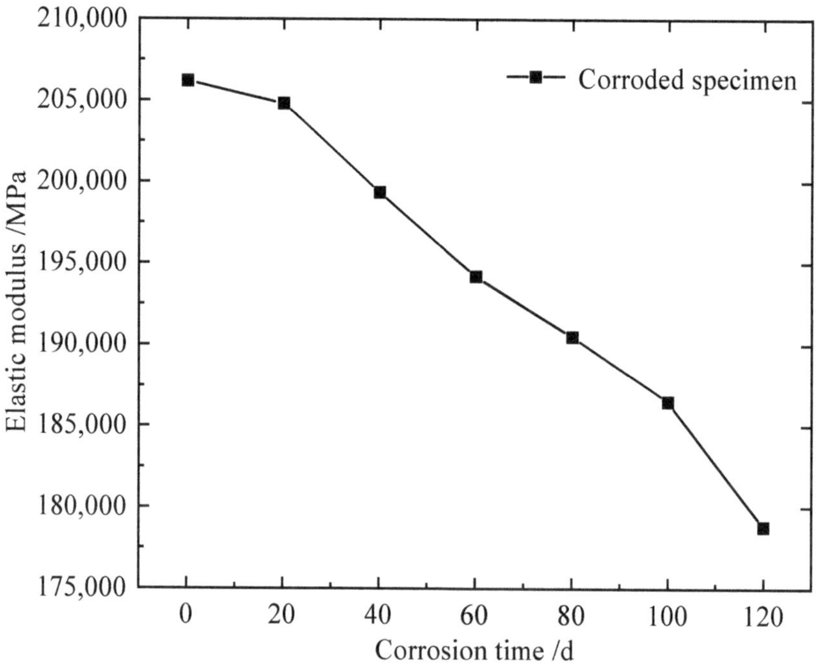

Figure 15. Elastic modulus of corroded specimens [35].

Table 6. The elastic modulus of corroded steel [35].

Corrosion Time/d	20	60	120
E/Mpa	204,798	194,170	178,805

5.2. Comparison of the Hysteretic Performance of Corroded CSPWs

Load–displacement curves of corroded specimens for 20, 60 and 120 days are shown in Figure 16a–f, and the detailed results are shown in Table 7. Under different corrosion levels, the hysteretic loop of ASW was the fullest among them, and when corroded for 20 and 60 days, the stiffness and ultimate shear resistance of CSW and ASW decreased more gently than that of USW, but when corroded for 120 days, the trends of ultimate shear resistance degradation and stiffness degradation of USW and CSW were basically consistent, and the stiffness and ultimate shear resistance of ASW decreased more gently than that of USW and CSW.

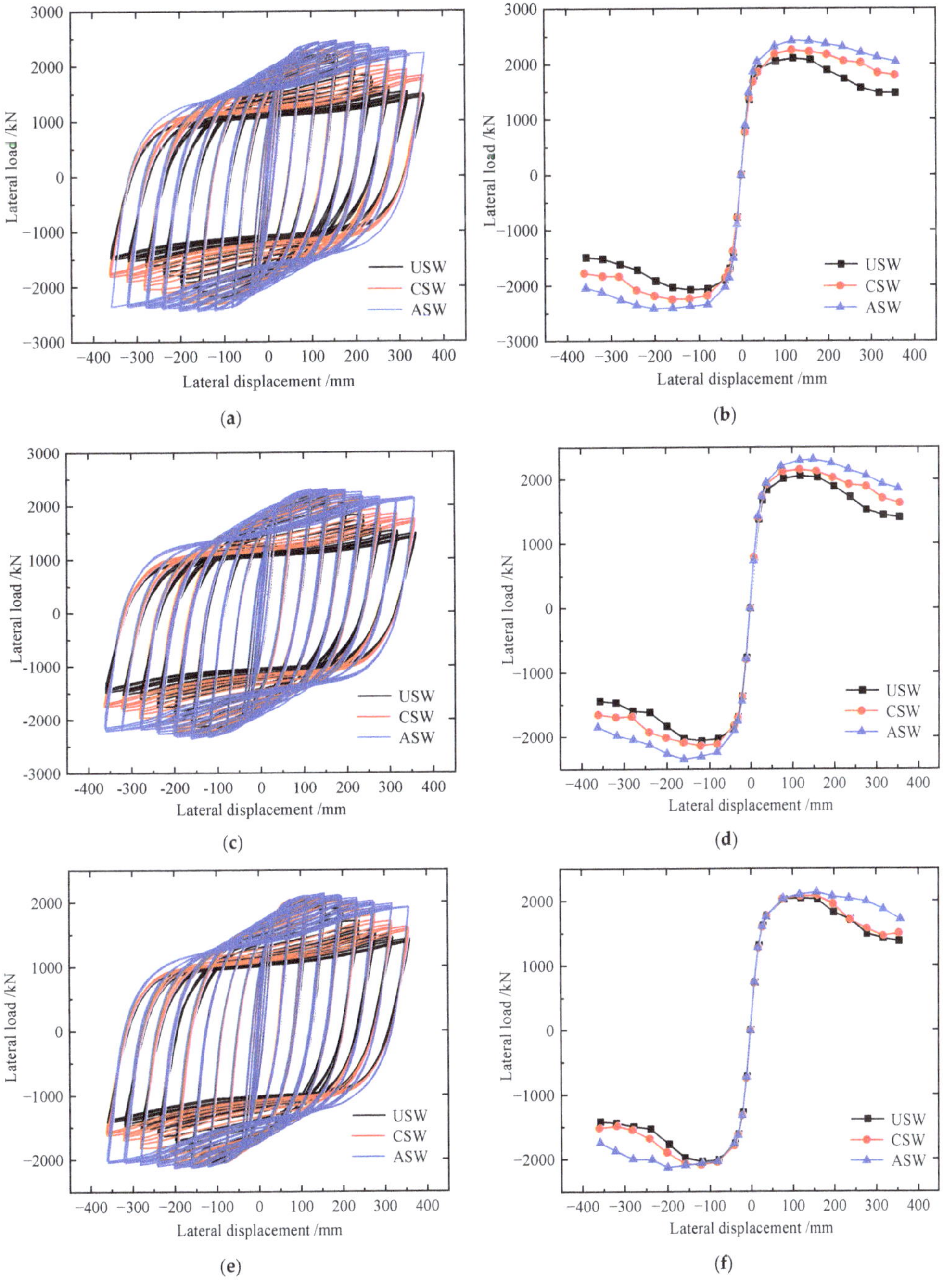

Figure 16. Load–displacement curves of corroded specimens: (**a**) hysteretic curves for corrosion 20 d; (**b**) envelope curves for corrosion 20 d; (**c**) hysteretic curves for corrosion 60 d; (**d**) envelope curves for corrosion 60 d; (**e**) hysteretic curves for corrosion 120 d; (**f**) envelope curves for corrosion 120 d.

Table 7. Characteristic results of USW, CSW and ASW for different corrosion times.

Corrosion Time/d	Specimen	K_0/(kN/mm)	P_y/kN	Δ_y/mm	P_m/kN	Δ_m/mm	μ
20	USW	78.87	1595.78	35.72	2086.18	119.25	5.81
	CSW	80.57	1868.97	40.00	2248.94	138.59	7.28
	ASW	87.93	2001.27	41.27	2413.51	159.59	8.49
60	USW	76.98	1554.87	35.39	2056.09	119.01	5.64
	CSW	78.77	1801.25	39.43	2144.82	136.30	6.15
	ASW	81.45	1947.92	39.78	2332.26	154.77	7.71
120	USW	75.06	1542.57	35.06	2039.42	118.98	5.47
	CSW	76.46	1753.13	36.40	2082.38	135.17	6.04
	ASW	79.76	1831.91	38.14	2131.54	178.43	7.63

When the corrosion time is 20 days, compared to USW, the initial stiffness, yield load, peak load and ductility factor of the CSW increase by 2.16%, 17.12%, 7.8% and 25.3%, respectively; the initial stiffness, yield load, peak load and ductility factor of ASW increase by 11.49%, 25.41%, 15.69% and 46.13%, respectively. When the corrosion time is 60 days, compared to USW, the initial stiffness, yield load, peak load and ductility factor of the CSW increase by 2.33%, 15.85%, 4.32% and 9.04%, respectively; the initial stiffness, yield load, peak load and ductility factor of ASW increase by 5.81%, 25.28%, 13.43% and 36.7%, respectively. When the corrosion time is 120 days, compared to USW, the initial stiffness, yield load, peak load and ductility factor of the CSW increase by 1.87%, 13.65%, 2.11% and 10.42%, respectively; the initial stiffness, yield load, peak load and ductility factor of the ASW increase by 6.26%, 18.76%, 4.52% and 39.49%, respectively.

Therefore, under different corrosion levels, stiffeners can improve the initial stiffness, yield load and ductility of CSPWs, increase the energy dissipation capacity and delay the stiffness degradation of CSPWs, so stiffeners can improve the hysteretic performance of CSPWs. However, with increasing corrosion time, the hysteretic performance of CSWs is affected more obviously by corrosion than that of USWs and ASWs, so asymmetric diagonal stiffeners can improve the hysteretic performance of corroded CSPWs more effectively than cross stiffeners.

5.3. Comparison of the Hysteretic Performance of CSPWs before and after Corrosion

Load–displacement curves of specimens before and after corrosion are shown in Figure 17. With increasing corrosion time, notable pinching was observed for the hysteretic loops of corroded CSWs, and the trend of stiffness degradation for corroded USW was basically consistent. Compared to the uncorroded CSW, the initial stiffness of USW after corrosion for 20, 60, and 120 days decreases by 1.78%, 4.13% and 6.53%, respectively; the peak load decreases by 2.73%, 4.14% and 4.91%, respectively; and the ductility factor de-creases by 2.68%, 5.53% and 8.38%, respectively.

With increasing corrosion time, the hysteresis curve of the corroded CSWs was pinched, and the CSW stiffness degradation trend of the uncorroded CSWs corroded for 20 and 60 days was basically consistent, but the corroded CSWs corroded for 120 days had a significant decrease in stiffness and shear resistance. Compared to the uncorroded CSW, the initial stiffness of USW after corrosion for 20, 60, and 120 days decreases by 3.6%, 5.77% and 8.53%, respectively; the peak load decreases by 6.85%, 11.35% and 13.93%, respectively; and the ductility factor decreases by 2.86%, 11.78% and 12.7%, respectively.

With increasing corrosion time, the hysteresis curve of corroded ASW was pinched more gently than that of USW and CSW, and the trend of stiffness degradation for corroded ASW was basically consistent. Compared to uncorroded ASW, the initial stiffness of ASW after corrosion for 20, 60, and 120 days decreases by 2.37%, 9.56% and 11.44%, respectively; the peak load decreases by 1.63%, 4.94% and 13.13%, respectively; and the ductility factor decreases by 2.86%, 11.78% and 12.7%, respectively.

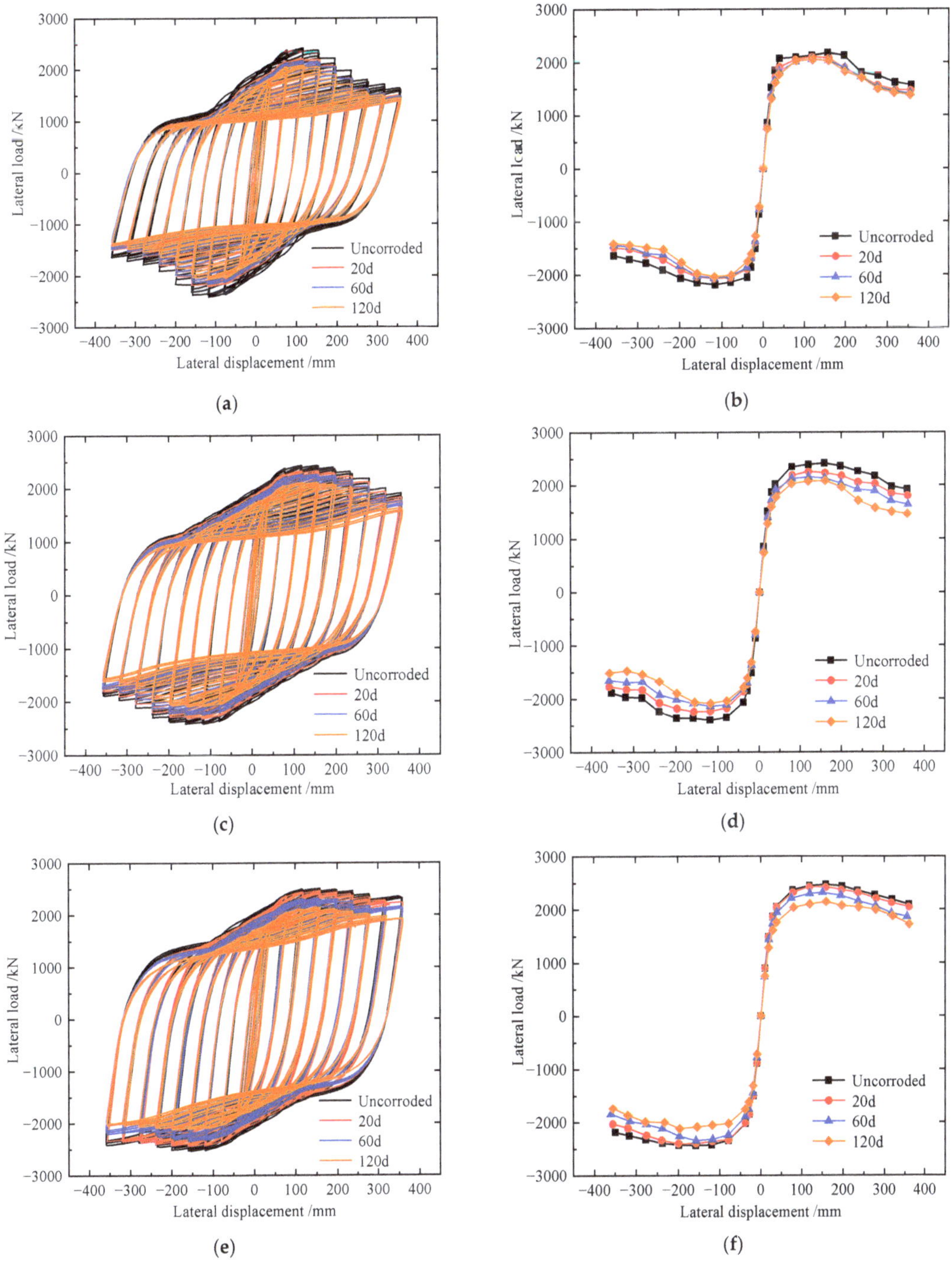

Figure 17. Load–displacement curves of specimens before and after corrosion: (**a**) hysteretic curves of USW; (**b**) envelope curves of USW; (**c**) hysteretic curves of CSW; (**d**) envelope curves of CSW; (**e**) hysteretic curves of ASW; (**f**) envelope curves of ASW.

Therefore, corrosion reduces the initial stiffness, ultimate shear resistance, ductility and energy dissipation capacity of USW, CSW and ASW. Due to the corrosion of stiffeners, the effect of stiffeners on increasing initial stiffness, yield load and delaying stiffness degradation has been reduced significantly, so the indexes of corroded CSPWs with stiffeners including CSW and ASW decreased more significantly than those of USW with increasing corrosion time, and the indexes of USW after corrosion decreased slowly with increasing corrosion time, revealing that corrosion has little effect on the hysteretic performance of USW. Meanwhile, the ultimate shear resistance and stiffness of ASW after corrosion decreased gently, and the hysteretic curve was still full, so it still has good seismic performance.

5.4. The Fitting Formulae of Ultimate Shear Resistance

The time of the accelerated corrosion test is converted to the time of actual outdoor corrosion. From Section 5.1, the times of the corrosion test are 20, 60 and 120 days, and the corresponding times of actual corrosion are 5.657, 16.972 and 33.945 years, respectively. The variation in the ultimate shear resistance of the specimens with the actual corrosion time is shown in Figure 18. The finite element results are selected to fit the formulae of ultimate shear resistance of specimens, as shown in Equations (3)–(5).

$$y = 102.809e^{-x/7.345} + 2041.223, \quad (3)$$

$$y = 296.377e^{-x/26.38} + 1997.899, \quad (4)$$

$$y = -186.533e^{x/34.021} + 2637.706, \quad (5)$$

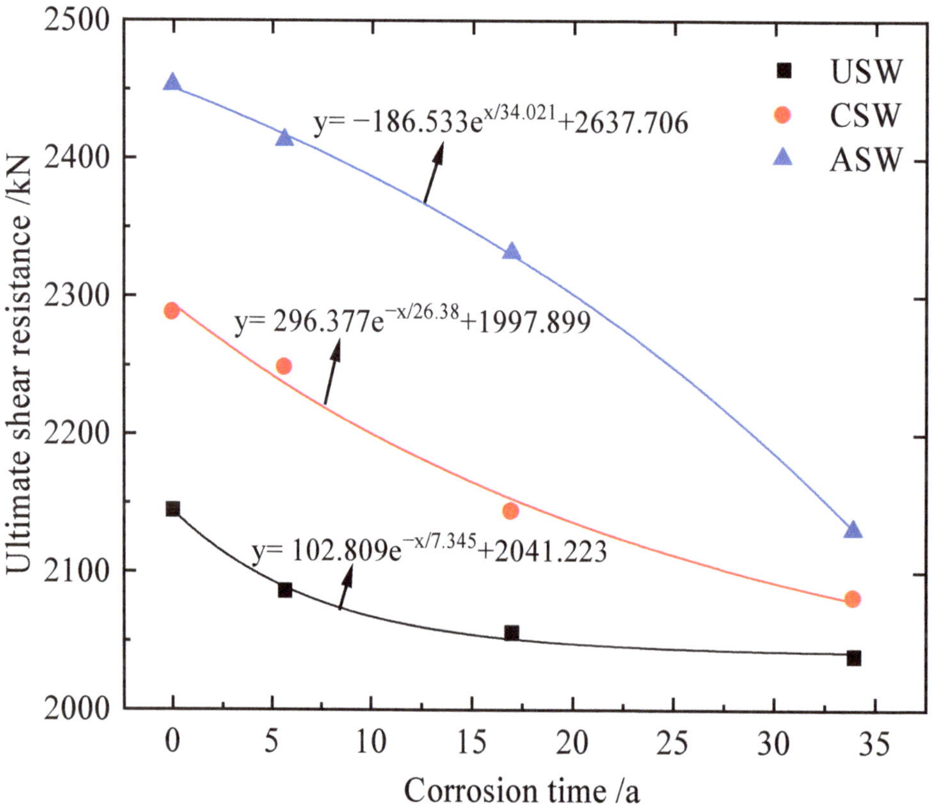

Figure 18. Fitting formulae of ultimate shear resistance of specimens.

Then FEM results of corroded specimens for 40 and 80 days are used to validate the proposed Formulas (3)–(5). According to Table 5, the actual corrosion time of specimens for outdoor accelerated corrosion time 40 and 80 days are 11.315 and 22.63 years, respectively. From the Ref. [35], the elastic modulus of corroded specimens for 40 and 80 days are 199,348 MPa and 190,493 MPa, respectively. The comparative results of ultimate shear resistance between FE analysis and proposed formulae are shown in Table 8. The ultimate shear resistance results based on the proposed theoretical model are very closed to the FEM results, and the maximum error between theoretical results and FEM results is 1.9%. The comparisons indicate that the proposed fitting formulae could effectively predict the ultimate shear resistance of corroded CSPWs.

Table 8. Comparison on ultimate shear resistance of FEM results and theoretical results.

Test Corrosion Time/d	Actual Corrosion Time/a	Specimen	Theoretical Results/kN	FEM Results/kN	Error
40	11.315	USW	2063.25	2071.51	0.4%
		CSW	2190.9	2232.45	1.9%
		ASW	2377.57	2395.23	0.7%
80	22.63	USW	2045.94	2052.79	0.3%
		CSW	2123.58	2110.09	0.6%
		ASW	2274.93	2298.25	1.01%

6. Conclusions

In this research, CSPWs with three different stiffening methods are proposed. The in-fluence of different stiffening methods on the lateral and seismic performance of CSPWs before and after corrosion is examined. The conclusions are as follows:

(1) The FEM model of specimen S-1 from the Ref. [47] is established, and test results from S-1 are selected to validate the FEM model. The results show that the initial stiffness and ultimate shear resistance analyzed by FEM are basically consistent with the test results, and the maximum error is less than 10%.

(2) Stiffeners can effectively improve the initial stiffness, ultimate shear resistance and ductility of CSPWs under monotonic loading, and the effect of asymmetric diagonal stiffeners is more significant than that of cross stiffeners. Due to asymmetric diagonal stiffeners, the maximum out-of-plane deformation of ASW is 62.02% and 39.56% of CSW and USW, respectively.

(3) Under cyclic loading, compared to CSW and USW, the ultimate shear resistance, energy dissipation capacity and ductility of ASW are larger, the amplitude of stiffness degradation is smaller, and the maximum out-of-plane deformation of ASW is 82.7% and 67.2% of CSW and USW, respectively.

(4) Under atmospheric corrosion, USW is least affected by corrosion, but compared to corroded USW and CSW, the ultimate shear resistance and energy dissipation capacity of corroded ASW are larger. It is revealed that asymmetric cross stiffeners are still more effective than cross stiffeners in inhibiting buckling and improving the seismic performance of corroded CSPWs. Meanwhile, the fitting formulae of the ultimate shear resistance of corroded CSPWs are a suitable engineering design reference.

Author Contributions: X.M.: Software, validation, investigation, data curation, writing—original draft preparation, writing—review and editing; Y.H.: Conceptualization, methodology, formal analysis, project administration, writing—review and editing; L.J. (Liqiang Jiang): Conceptualization, methodology, project administration, visualization, funding acquisition; L.J. (Lizhong Jiang): Formal analysis, supervision; G.N.: Project administration, funding acquisition; H.Z.: Methodology, supervision. All authors have read and agreed to the published version of the manuscript.

Funding: This work was supported by Scientific Research Fund of Institute of Engineering Mechanics, China Earthquake Administration (Grant No. 2020D12).

Institutional Review Board Statement: The study did not require ethical approval.

Informed Consent Statement: Not applicable.

Data Availability Statement: Data sharing not applicable.

Conflicts of Interest: The authors declare no conflict of interest.

References

1. Zhao, Q.; Li, N.; Sun, J. Analysis on lateral performance of sinusoidally corrugated steel plate shear walls. *J. Tianjin Univ.* **2016**, *49*, 152–160.
2. Jiang, R.; Jiang, L.; Hu, Y.; Jiang, L.; Ye, J. A simplified method for fundamental period prediction of steel frames with steel plate shear walls. *Struct. Des. Tall Spec. Build.* **2016**, *29*, e1718. [CrossRef]
3. Jiang, L.; Zheng, H.; Hu, Y. Experimental seismic performance of steel-and composite steel-panel wall strengthened steel frames. *Arch. Civil Mech. Eng.* **2017**, *17*, 520–534. [CrossRef]
4. Hu, Y.; Zhao, J.; Jiang, L. Seismic risk assessment of steel frames equipped with steel panel wall. *Struct. Des. Tall Spec. Build.* **2017**, *26*, e1368. [CrossRef]
5. Tong, J.; Guo, Y. Shear resistance of stiffened steel corrugated shear walls. *Thin-Walled Struct.* **2018**, *127*, 76–89. [CrossRef]
6. Jiang, L.; Zheng, H.; Hu, Y. Effects of configuration parameters on seismic performance of steel frames equipped with composite steel panel wall. *Struct. Des. Tall Spec. Build.* **2018**, *27*, e1542. [CrossRef]
7. Alinia, M.M.; Dastfan, M. Behavior of thin steel plate shear walls regarding frame members. *J. Constr. Steel Res.* **2006**, *62*, 730–738. [CrossRef]
8. Berman, J.W.; Celik, O.C.; Bruneau, M. Comparing hysteretic behavior of light-gauge steel plate shear walls and braced frames. *Eng. Struct.* **2005**, *27*, 475–485. [CrossRef]
9. Clayton, P.; Berman, J.; Lowes, L. Seismic performance of self-centering steel plate shear walls with beam-only-connected web plates. *J. Constr. Steel Res.* **2015**, *106*, 198–208. [CrossRef]
10. Berman, J.; Bruneau, M. Plastic analysis and design of steel plate shear walls. *J. Struct. Eng.* **2003**, *129*, 1148–1156. [CrossRef]
11. Berman, J.; Bruneau, M. Experimental investigation of light-gauge steel plate shear walls. *J. Struct. Eng.* **2005**, *131*, 259–267. [CrossRef]
12. Sigariyazd, M.A.; Joghataie, A.; Attari, N.K.A. Analysis and design recommendations for diagonally stiffened steel plate shear walls. *Thin-Walled Struct.* **2016**, *103*, 72–80. [CrossRef]
13. Nie, J.; Li, Y. Finite element research about the elastic buckling of six types of stiffening steel plate shear wall. *Steel Constr.* **2013**, *28*, 1–3.
14. Cao, Z.; Wang, Z.; Du, P.; Liu, H.; Fan, F. Research on steel plate shear walls stiffened with X-shaped restrainers: Hysteretic behavior and effect of height-to-thickness ratio of steel plate. *Thin-Walled Struct.* **2019**, *144*, 106316. [CrossRef]
15. Du, P.; Cao, Z.; Fan, F. Developing of steel plate shear walls braced with slidable multiple X-shaped restrainers: Hysteretic analyses and design recommendations. *Int. J. Steel Struct.* **2016**, *16*, 1227–1238. [CrossRef]
16. Khaloo, A.; Foroutani, M.; Ghamari, A. Influence of crack on the steel plate shear walls strengthened by diagonal stiffeners. *Struct. Design Tall Spec. Build.* **2020**, *29*, 102–113. [CrossRef]
17. Farzampour, A.; Laman, J.A.; Mofid, M. Behavior prediction of corrugated steel plate shear walls with openings. *J. Constr. Steel Res.* **2015**, *114*, 258–268. [CrossRef]
18. Yang, D.; Deng, E. Cyclic tests on corrugated steel plate shear walls with openings in modularized-constructions. *J. Constr. Steel Res.* **2017**, *138*, 675–691.
19. Emami, F.; Mofid, M.; Vafai, A. Experimental study on cyclic behavior of trapezoidally corrugated steel shear walls. *Eng. Struct.* **2013**, *48*, 750–762. [CrossRef]
20. Emami, F.; Mofid, M. On the hysteretic behavior of trapezoidally corrugated steel shear walls. *Struct. Design Tall Spec. Build.* **2014**, *23*, 94–104. [CrossRef]
21. Jiang, L.; Jiang, L.; Hu, Y.; Ye, J.; Zheng, H. Seismic life-cycle cost assessment of steel frames equipped with steel panel walls. *Eng. Struct.* **2020**, *211*, 110399. [CrossRef]
22. Paslar, N.; Farzampour, A.; Hatami, F. Infill plate interconnection effects on the structural behavior of steel plate shear walls. *Thin-Walled Struct.* **2020**, *149*, 106621. [CrossRef]
23. Cao, Q.; Huang, J. Experimental study and numerical simulation of corrugated steel plate shear walls subjected to cyclic loads. *Thin-Walled Struct.* **2018**, *127*, 306–317. [CrossRef]
24. Tong, J.; Guo, Y.; Zuo, J.; Gao, J. Experimental and numerical study on shear resistant behavior of double-corrugated-plate shear walls. *Thin-Walled Struct.* **2020**, *147*, 106485. [CrossRef]
25. Tong, J.; Guo, Y.; Pan, W. Ultimate shear resistance and post-ultimate behavior of double-corrugated-plate shear walls. *J. Constr. Steel Res.* **2020**, *165*, 105895. [CrossRef]
26. Wang, M.; Guo, Y.; Zhu, J.; Yang, X.; Tong, J. Sectional strength design of concrete-infilled double steel corrugated-plate walls with T-section. *J. Constr. Steel Res.* **2019**, *160*, 23–44. [CrossRef]
27. Tong, J.; Guo, Y.; Zuo, J. Elastic buckling and load-resistant behaviors of double-corrugated-plate shear walls under pure in-plane shear loads. *Thin-Walled Struct.* **2018**, *130*, 593–612. [CrossRef]

28. Dou, P.; Pi, Y.; Gao, W. Shear resistance and post-buckling behavior of corrugated panels in steel plate shear walls. *Thin-Walled Struct.* **2018**, *131*, 816–826. [CrossRef]
29. Tong, J. Theoretical and Experimental Study on Design Methods for Steel Corrugated Plate Shear Walls. Master's Thesis, Tsinghua University, Beijing, China, 2018.
30. Wang, J. Research on Behavior of Cross Stiffening Corrugated Steel Plate Shear Wall. Master's Thesis, Chang'an University, Xi'an, China, 2018.
31. Zheng, H.; Wang, W.; Yang, R.; Wang, J. Research on hysteretic properties shear walls with new cross stiffened corrugated steel plates. *Prog. Steel Build. Struct.* **2021**, *23*, 28–35.
32. Zheng, H.; Yang, R.; Wang, W.; Wang, J. Hysteresis performance analysis of shear wall with cross-stiffening corrugated steel plate under cyclic loading. *Struct. Eng.* **2021**, *37*, 15–23.
33. Khedmati, M.R.; Roshanali, M.M.; Nouri, Z.H.M.E. Strength of steel plates with both-sides randomly distributed with corrosion wastage under uniaxial compression. *Thin-Walled Struct.* **2011**, *49*, 325–342. [CrossRef]
34. Sultana, S.; Wang, Y.; Sobey, A.J.; Wharton, J.A.; Shenoi, R.A. Influence of corrosion on the ultimate compressive strength of steel plates and stiffened panels. *Thin-Walled Struct.* **2015**, *96*, 95–104. [CrossRef]
35. Zhang, X.; Zheng, S.; Zhao, X. Experimental and numerical study on seismic performance of corroded steel frames in chloride environment. *J. Constr. Steel Res.* **2020**, *171*, 106164. [CrossRef]
36. Xu, Q.; Shang, X.; Zheng, S. Cyclic behavior of steel frame joints in the offshore atmospheric environment. *Int. J. Steel Struct.* **2020**, *20*, 870–884. [CrossRef]
37. Zheng, S.; Zhang, X.; Zhao, X. Experimental investigation on seismic performance of corroded steel columns in offshore atmospheric environment. *Struct. Des. Tall Spec. Build.* **2019**, *28*, e1580. [CrossRef]
38. Wang, Y.; Shi, T.; Nie, B.; Wang, H.; Xu, S. Seismic performance of steel columns corroded in general atmosphere. *Steel Compos. Struct.* **2021**, *40*, 217–241.
39. Xu, S.; Zhang, Z.; Qin, G. Study on the seismic performance of corroded H-shaped steel columns. *Eng. Struct.* **2019**, *191*, 39–61. [CrossRef]
40. Beben, D. Backfill corrosivity around corrugated steel plate culverts. *J. Perform. Constr. Facil.* **2015**, *29*, 04014159. [CrossRef]
41. Wei, X.; Li, G.; Xiao, L.; Zhou, L.; He, K.; Han, B. Shear strength reduction of trapezoidal corrugated steel plates with artificial corrosion pits. *J. Constr. Steel Res.* **2021**, *180*, 106583.
42. Kudus, S.A. Modal analysis of corrugated plate by finite element analysis. *Int. J. Integr. Eng.* **2020**, *12*, 252–258.
43. *GB50017-2017*; Standard for Design of Steel Structures. Chinese Architecture and Industry Press: Beijing, China, 2017.
44. *JGJ/T380-2015*; Technical Specification for Steel Plate Shear Wall. Chinese Architecture and Industry Press: Beijing, China, 2016.
45. *GB50011-2010*; Code for Seismic Design of Buildings. Chinese Architecture and Industry Press: Beijing, China, 2016.
46. Hao, J. *Principle and Performance of Steel Plate Shear Wall*; Beijing Science and Technology Publishing Co., Ltd.: Beijing, China, 2020.
47. Hao, B. Cyclic Test Vertically Corrugated Steel Plate Shear Walls with Various Openings. Master's Thesis, Tianjin University, Tianjin, China, 2018.
48. Shi, W.; Tong, L.; Chen, Y.; Li, Z.; Shen, K. Experimental study on influence of corrosion on behavior of steel material and steel beams. *J. Build. Struct.* **2012**, *33*, 56–60.
49. Liu, X.; Shi, H. *Anti-Corrosive and Fireproof Paint of Steel Structures*; Chemical Industry Press: Beijing, China, 2005; pp. 20–79.

materials

MDPI

Article

Experimental Study of Aluminium-Timber Composite Bolted Connections Strengthened with Toothed Plates

Marcin Chybiński * and Łukasz Polus *

Institute of Building Engineering, Faculty of Civil and Transport Engineering, Poznan University of Technology, Piotrowo 5 Street, 60-965 Poznan, Poland

* Correspondence: marcin.chybinski@put.poznan.pl (M.C.); lukasz.polus@put.poznan.pl (Ł.P.); Tel.: +48-61-665-2477 (M.C.); +48-61-665-2098 (Ł.P.)

Abstract: This paper presents the first experimental study of the load-slip behaviour of aluminium-timber composite bolted connections reinforced with toothed plates. The effectiveness of the strengthening was evaluated in laboratory push-out tests. The push-out test samples consisted of laminated veneer lumber panels, aluminium alloy I-beams, and bolts (grade 8.8 10 mm × 125 mm and 12 mm × 135 mm bolts, grade 5.8 10 mm × 125 mm and 12 mm × 135 mm bolts). A group of 16 specimens had toothed plates as additional reinforcement, while 16 specimens had no reinforcement. The impact of the bolt diameter (10 and 12 mm) and bolt grade (5.8 and 8.8) on the behaviour of the connections was also analysed. The values of the ultimate load and the slip modulus for the bolted connections with grade 8.8 10 mm and 12 mm bolts and with grade 5.8 12 mm bolts reinforced by toothed-plate connectors were comparable to the values for the non-reinforced connections. This was because, in the case of grade 8.8 10 mm × 125 mm and 12 mm × 135 mm bolts and grade 5.8 12 mm × 135 mm bolts, the laminated veneer lumber (LVL) slabs split both in the reinforced and non-reinforced connections. The toothed-plate connectors reduced timber destruction in the bearing zones in the LVL slabs. However, they did not protect the LVL slabs against splitting. Therefore, the impact of the toothed plate connectors on the stiffness and strength of the bolted connections with grade 8.8 10 mm and 12 mm bolts and with grade 5.8 12 mm bolts analysed in this paper was found to be negligible. In the case of grade 5.8 10 mm bolts, the LVL slabs did not split. The mean slip modulus $k_{0.6}$ of the connections with grade 5.8 10 mm bolts reinforced with toothed plate connectors was 2.9 times higher than that of the non-reinforced connections. However, the strength of the connections with grade 5.8 10 mm bolts was 1.2 times lower after reinforcing. This was because the shanks of the bolts were sheared faster in the reinforced connections than in the non-reinforced connections as a result of the bolt shanks being under the bearing pressure of the aluminium flange, the LVL slab, and the toothed-plate flange. This situation did not occur for the remaining connections because they had a higher strength (grade 8.8 bolts) or a larger diameter (12 mm), and their bolts were less prone to cutting off. The investigated load–slip curves of the reinforced bolted connections can be used for designing and numerical modelling of aluminium-timber composite beams with this type of connection.

Keywords: aluminium-timber composite structures; aluminium alloy; engineering wood products; laminated veneer lumber (LVL); toothed plate; bolted connection; shear connection; push-out test

check for updates

Citation: Chybiński, M.; Polus, Ł. Experimental Study of Aluminium-Timber Composite Bolted Connections Strengthened with Toothed Plates. *Materials* **2022**, *15*, 5271. https://doi.org/10.3390/ma15155271

Academic Editors: Tomasz Garbowski and Aleksander Marek

Received: 24 June 2022
Accepted: 27 July 2022
Published: 30 July 2022

1. Introduction

1.1. Literature Review

Modern designing requires the use of sustainable solutions and is open especially to those which can help reduce the carbon footprint, such as composite structural elements and composite materials. A composite structural element consists of at least two components made of different material and permanently joined with shear connectors (e.g., steel-concrete composite beams), whereas a composite material is a combination of materials with different properties (e.g., reinforced concrete or plywood). The use of composite structural elements is continuously on the rise due to the fact that it provides for increased load-bearing capacity and helps to overcome serviceability limitations [1]. For example, the load-bearing capacity of unrestrained aluminium beams increased 7.0 times after they were joined with timber slabs [2]. The stiffness of the aluminium beam analysed in [3] increased 4.3 times after it was used together with the timber slab in a composite beam. Greater stiffness leads to lower deflections. Due to this fact, the serviceability limit state (deflection $\leq$ limit deflection) for structural beams is easier to meet. The idea behind composite elements is to take the maximum advantage of the materials they are made of [4]. Similarly, composite materials are produced to form a material with properties different from the ones of the individual components. Increasingly more often, new composite materials, such as fibre-reinforced polymer composites or carbon-epoxy composites, have been used in modern design due to their high strength-to-weight ratio and durability [5–7]. Composite materials are often applied where traditional materials tend to fail [8]. Glass, aramid, or carbon fibre-reinforced polymer sheets are used to strengthen engineered wood products [9–11]. This is also an example of two different composite materials (laminated veneer lumber and fibre-reinforced polymer) used in one solution.

Recently, new composite structural elements with timber, such as timber-concrete composite elements [12–14], steel-timber composite beams [15–17] or aluminium-timber composite beams [18] have been investigated. Timber is one of the oldest known construction materials. Ancient Romans used it to build houses, temples, and bridges [19]. Timber has been used for many years to construct bridges and churches [20–23]. Structures made of timber can be easily repaired or reinforced [24–26]. Unfortunately, solid sawn timber has some limitations, e.g., in maximum lengths or cross-sectional dimensions. These limitations can be overcome by quality controlled engineered wood products, e.g., laminated veneer lumber (LVL) [27], cross banded laminated veneer lumber (LVL-C) [28], laminated strand lumber (LSL) [29], parallel strand lumber (PSL) [30], glued-laminated timber (GLT) [31] or cross-laminated timber (CLT) [32,33]. LVL or plywood were used for making slabs in aluminium-timber composite beams [34,35]. In addition to plywood and LVL, a slab may also be made of cross-laminated timber (CLT), which was demonstrated for steel-timber composite beams [36,37] (which are similar to aluminium-timber composite beams). An aluminium-timber composite beam is a sustainable example of a composite structure with a timber structural element. The use of aluminium alloy girders increases the durability of composite beams and reduces maintenance costs due to the light weight and high corrosion resistance of aluminium alloys [38]. The use of bolts as shear connectors makes it possible to demount composite beams at the end of their structural life in a way ensuring a sustainable use of natural sources. Aluminium-timber composite beams are lighter than steel-concrete composite beams and have replaceable parts. The lightness of aluminium-timber composite beams and the fact that there is no need to wait for the hardening of concrete may speed up the construction process. The aluminium alloy girders of aluminium-timber composite beams may have different cross-sections. For example, Saleh and Jasin [39] used rectangular hollow sections in their tests. Chybiński and Polus [2] investigated composite systems with extruded aluminium alloy I-girders. Girders may also be cold-formed, which was demonstrated for steel-timber composite beams [40], which are similar to aluminium-timber composite beams. For example, cold-rolled aluminium alloy members are fabricated in Australia [41]. The aluminium-timber composite beams presented in the literature have T-shaped cross-sections. Recently, Wang et al. [42,43] demonstrated that steel-timber composite beams may have I-shaped cross-sections. The steel-timber composite beams analysed by Wang et al. consisted of timber panels, U-shaped thin-walled steel beams, and bolts and screws as connectors. Aluminium-timber composite beams may have similar cross-sections. In each composite beam, connections play a crucial role and determine its behaviour. They should be ductile and have a characteristic slip capacity exceeding 6 mm [44]. They should also show high shear resistance and stiffness. It is advisable to reinforce connections to improve their mechanical parameters. When the shear resistance of a connection is higher, the lower number of connectors may be used to obtain a full composite action in a beam. When the stiffness of a connection is higher, the longitudinal slip between the slabs and the girders is lower.

1.2. Problem Statement and the Aim of the Current Research

In this paper, the authors analysed connections to be used in aluminium-timber composite beams. The load capacity of the connections with mechanical fasteners depends on many parameters, such as timber density, loading direction, fastener spacing, and end and edge distances [45]. The connections for aluminium-timber composite beams investigated in the previous studies were screwed or bolted [2,35,46–49]. Their stiffness and strength were relatively low. For this reason, Chybiński and Polus [49] proposed to use toothed plates in screwed connections as reinforcement. In case of screwed connections, the use of toothed plate connectors was found to be effective in increasing the strength of aluminium-timber composite connections. Enhancements of 35.0% (for Geka toothed-plate connectors and 12 mm screws), 28.7% (for Bulldog toothed-plate connectors and 10 mm screws), and 23.8% (for Bulldog toothed-plate connectors and 12 mm screws) were achieved. The use of toothed plate connectors did not have an impact on the stiffness of the screwed connections. Toothed-plate connectors were only used to reinforce screwed aluminium-timber connections in the aforementioned study [49]. In this paper, toothed plates were used to reinforce bolted aluminium-timber connections for the first time. The main aim of this paper is to determine the effectiveness of reinforcing bolted connection using toothed plates. The push-out laboratory tests on 16 reinforced and 16 non-reinforced samples were conducted to determine the shear resistance and stiffness of the bolted aluminium-timber connections. Moreover, the influence of the bolt diameter (10 and 12 mm) and bolt grade (5.8 and 8.8) on the mechanical parameters of the reinforced and non-reinforced connections was analysed. Finally, the behaviour of the bolted connections analysed in this paper was compared with the behaviour of the screwed connections analysed in the literature.

2. Materials and Methods

2.1. Aluminium Alloy

The extruded aluminium alloy I-beams were made of the AW-6060 T6 aluminium alloy. The yield strength, tensile strength, and Young's modulus of this alloy were 181.5 MPa, 209.8 MPa, and 66,400 MPa, respectively [50].

2.2. LVL

Laminated veneer lumber (LVL) is made by laminating thin (3–4 mm) wood veneers using adhesives [51]. Veneers are peeled off of high-quality logs. They are oriented in a grain direction. LVL is now fabricated in the United States, Australia, Japan, New Zealand, Finland, and Poland. For the purpose of this study, STEICO LVL manufactured in Poland from Scots pine (*Pinus sylvestris* L.) and Norway spruce (*Picea abies* L. H. Karst) was used [52]. The compression strength (parallel to grain), tension strength (parallel to grain), bending strength (flatwise, parallel to grain), and Young's modulus of this engineering wood product declared by the manufacturer were 40.0 MPa, 36.0 MPa, 50.0 MPa, and 14,000 MPa, respectively [53].

2.3. Bolts

Grade 8.8 10 mm × 125 mm and 12 mm × 135 mm bolts as well as grade 5.8 10 mm × 125 mm and 12 mm × 135 mm bolts were used as shear connectors. The length of the unthreaded shanks was the same in the 10 mm bolts and in the 12 mm bolts (90 mm). The characteristic yield strength f_{yb} and the ultimate strength f_{ub} of the bolts can be determined based on the bolt grade, e.g., in case of grade 8.8 bolt the ultimate strength is 800 MPa and the yield strength is 640 MPa. To confirm the bolt grade, the yield strength and the ultimate strength of the bolts were also evaluated experimentally in accordance with [54] in tensile tests. The tests were carried out using an Instron 4483 testing machine

(Instron, Grove City, PA, USA). The mechanical parameters of two bolts per each bolt type were investigated.

2.4. Toothed Plates

Bulldog toothed-plate connectors (C2-50/M10G and C2-50/M12G) were used for reinforcing aluminium-timber bolted connections (see Figure 1). The names of the toothed plates contained information on 4 parameters: C2 represented the toothed plate type, 50 was the plate diameter in mm, M10 or M12 were the types of bolts suggested for use with the plate, and G meant that the plates were galvanised [55].

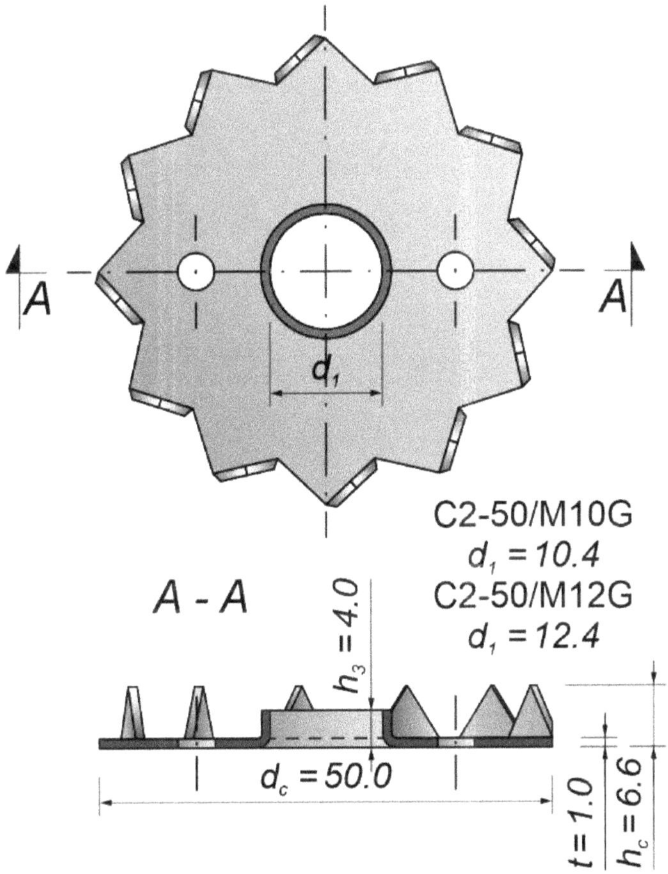

Figure 1. C2 (Bulldog) toothed-plate connector.

2.5. Push-Out Tests

An Instron 8505 Plus machine (Instron, HighWycombe, Buckinghamshire, UK) was used to investigate the load–slip behaviour of 32 specimens. Each specimen consisted of two LVL panels and an extruded aluminium alloy I-beam (see Figures 2 and 3).

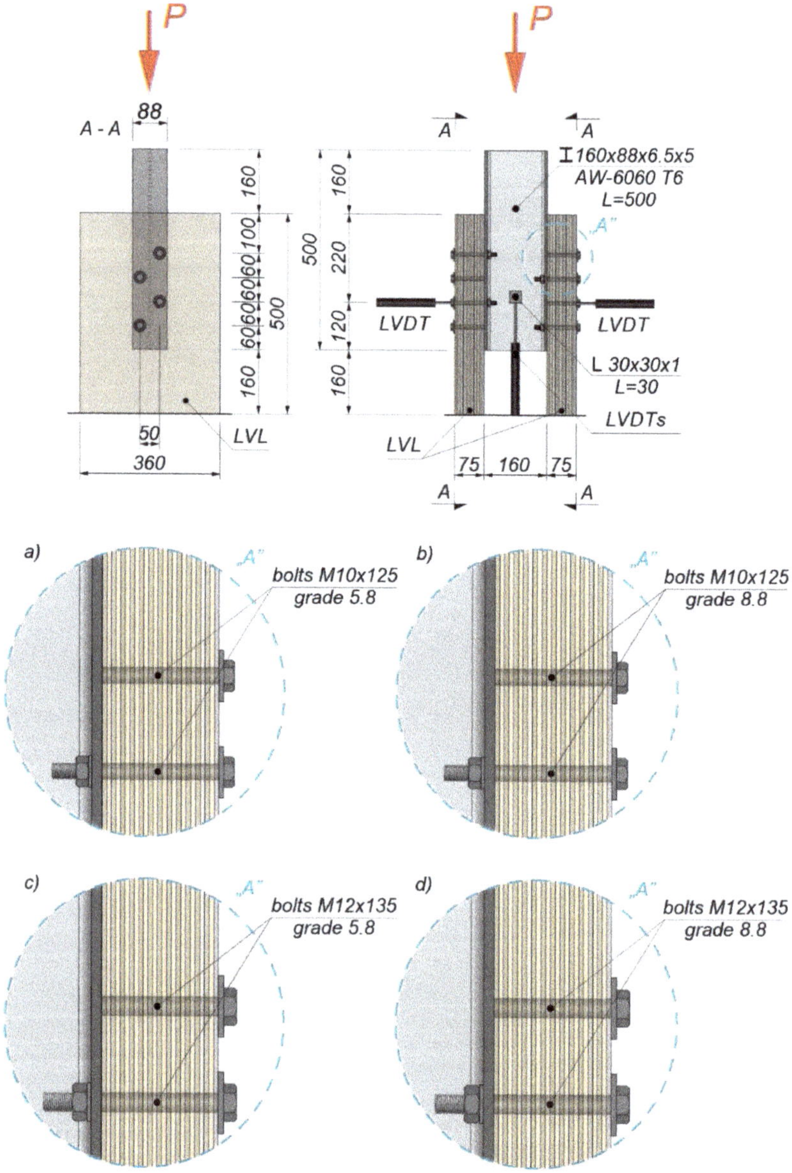

Figure 2. The tested specimens: (**a**) with 10 mm 5.8 grade bolts and without reinforcing toothed plates; (**b**) with 10 mm 8.8 grade bolts and without reinforcing toothed plates; (**c**) with 12 mm 5.8 grade bolts and without reinforcing toothed plates; and (**d**) with 12 mm 8.8 grade bolts and without reinforcing toothed plates.

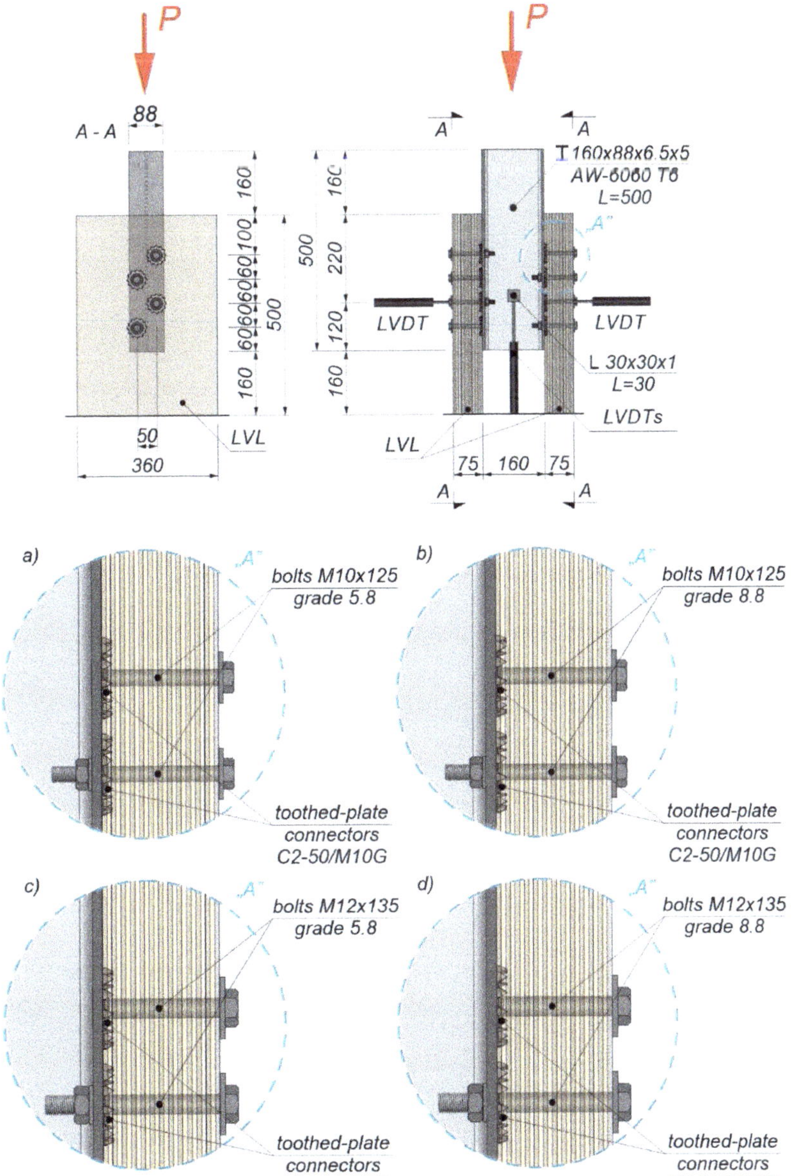

Figure 3. The tested specimens: (**a**) with 10 mm 5.8 grade bolts and reinforcing C2-50/M10G toothed plates; (**b**) with 10 mm 8.8 grade bolts and reinforcing C2-50/M10G toothed plates; (**c**) with 12 mm 5.8 grade bolts and reinforcing C2-50/M12G toothed plates; and (**d**) with 12 mm 8.8 grade bolts and reinforcing C2-50/M12G toothed plates.

The LVL panels were connected with the extruded aluminium alloy I-beams using eight variants of connections presented in Table 1.

Table 1. The variants of connections analysed in this study.

Variant	Designation of Specimens [mm]	Bolt Diameter and Length [mm]	Bolt Grade	Designation of a Reinforcing Toothed Plate
1	R8.8.10.1–R8.8.10.4	10 × 125	8.8	–
2	8.8.10.1–8.8.10.4	10 × 125	8.8	C2-50/M10G
3	R8.8.12.1–R8.8.12.4	12 × 135	8.8	–
4	8.8.12.1–8.8.12.4	12 × 135	8.8	C2-50/M12G
5	R5.8.10.1–R5.8.10.4	10 × 125	5.8	–
6	5.8.10.1–5.8.10.4	10 × 125	5.8	C2-50/M10G
7	R5.8.12.1–R5.8.12.4	12 × 135	5.8	–
8	5.8.12.1–5.8.12.4	12 × 135	5.8	C2-50/M12G

The holes in the extruded aluminium alloy I-beams and LVL panels had the same diameters as the bolts to reduce the slip between the upper aluminium girder flange and the LVL panel. In each specimen, the bolts were installed using a torque wrench (Sandvik Belzer, IZO-I-100, 10–100 Nm) (Sandvik, Portlaoise, Ireland). The loading direction was parallel to the LVL grain, and the tread–grain angle was 90°. The torque level was measured during the installation of the bolts using a torque wrench and recorded at the end of the installation process (35 Nm for the 10 mm bolt, 60 Nm for the 12 mm bolt). Before the tests, the toothed plates were pressed into the LVL panels using a compressive force of 35 kN generated by a hydraulic press. The bolts were evenly spaced out (the space between the bolts was 50 mm in the transverse direction and 60 mm in the longitudinal direction). Staggered spacing was applied to avoid the overlapping of toothed plates. The slip between the LVL panels and the extruded aluminium alloy I-beam and the horizontal move of the sample were measured using linear variable differential transformers (LVDTs) (Figures 2 and 3).

The push-out tests were conducted in accordance with [56]. A load control regime was used during the first stage of the tests to obtain a regular shape of the shear force–slip curve and to read the slip modulus of the connection for a load equal to 40% and 60% of the maximum value of the force. The constant rate of displacement was applied during the second stage of the tests. Due to this fact, the behaviour of the connections after the maximum load had been achieved could be observed. First, the load was increased from 0 to 40% of the estimated force over 2 min. Next, it remained at this level for 30 s. Subsequently, the load was decreased from 40% to 10% of the estimated force and maintained at this level for 30 s. Subsequently, the value of the load was increased from 10% to 70% of the estimated force. Up to that point, the test was conducted using a load control regime. From then on, it was conducted using a displacement control regime (5.0 mm/min). The estimated force of 195.2 kN was calculated taking into account eight bolts and the ultimate load per one M10 bolt (24.4 kN) obtained in the previous test presented in [2]. The value of the estimated force as well as the loading procedure were modified during the tests based on the previous results.

3. Results

3.1. Tensile Tests Results

The yield strength and the tensile strength of the bolts used in this study are presented in Table 2.

Table 2. The results of the tensile tests of the bolts (f_{yb}—the mean value of the yield strength of the bolts from two tests, f_{ub}—the mean value of the tensile strength of the bolts from two tests) [57].

Parameter	Bolt			
	Grade 5.8 10 mm	Grade 5.8 12 mm	Grade 8.8 10 mm	Grade 8.8 12 mm
f_{yb} [MPa]	399.0	485.5	842.0	850
f_{ub} [MPa]	483.0	564.0	935.0	908

3.2. Shear Connection Tests Results

The load–slip curves from the push-out tests are presented in Figures 4, 6, 8 and 10. The mean load–slip curves for each connection variant are shown in Figures 5, 7, 9 and 11 to describe the behavior of each variant in a simplified manner. The ultimate load per one connector (P_{ult}), the value of the slip corresponding to the ultimate load (s_{ult}), and the slip moduli per one connector ($k_{0.4}$ and $k_{0.6}$) are shown in Tables 3–10. The measurement errors for values presented in Tables 3–10 were calculated using a Student's t-distribution with three degrees of freedom and a confidence level of 0.95. The slip modulus $k_{0.4}$ per one connector was calculated as the ratio of 40% of the ultimate load per one connector to the slip corresponding to this value of the load. The slip modulus $k_{0.4}$ may be used for serviceability limit state calculations [58]. The slip modulus $k_{0.6}$ per one connector was calculated as the ratio of 60% of the ultimate load per one connector to the slip corresponding to this value of the load. The slip modulus $k_{0.6}$ may be used for the ultimate limit state calculations [59].

Table 3. The results of the push-out tests of the shear connections with 10-mm grade 8.8 bolts and without toothed-plate connectors (per one connector).

Parameter	Specimen				Mean
	R8.8.10.1	R8.8.10.2	R8.8.10.3	R8.8.10.4	(R8.8.10.1–R8.8.10.4)
P_{ult} [kN]	37.4	35.5	35.3	33.7	35.5 ± 2.4 (6.8%)
s_{ult} [mm]	47.6	48.6	49.0	46.9	48.0 ± 1.5 (3.2%)
$k_{0.4}$ [kN/mm]	5.6	5.3	5.6	5.3	5.5 ± 0.3 (5.1%)
$k_{0.6}$ [kN/mm]	3.8	4.0	3.9	4.6	4.1 ± 0.6 (14.0%)

Table 4. The results of the push-out tests of the shear connections with 10-mm grade 8.8 bolts and with toothed-plate connectors (type C2-50/M10G, Bulldog) (per one connector).

Parameter	Specimen				Mean
	8.8.10.1	8.8.10.2	8.8.10.3	8.8.10.4	(8.8.10.1–8.8.10.4)
P_{ult} [kN]	37.8	38.1	36.2	37.4	37.4 ± 1.3 (3.6%)
s_{ult} [mm]	47.4	42.7	47.4	36.7	43.6 ± 8.1 (18.5%)
$k_{0.4}$ [kN/mm]	6.0	4.8	7.5	4.7	5.8 ± 2.1 (36.2%)
$k_{0.6}$ [kN/mm]	6.7	4.9	7.3	5.1	6.0 ± 1.9 (31.4%)

Table 5. The results of the push-out tests of the shear connections with 12-mm grade 8.8 bolts and without toothed-plate connectors (per one connector).

Parameter	Specimen				Mean
	R8.8.12.1	R8.8.12.2	R8.8.12.3	R8.8.12.4	(R8.8.12.1–R8.8.12.4)
P_{ult} [kN]	38.2	37.3	36.8	38.3	37.7 ± 1.2 (3.1%)
s_{ult} [mm]	45.5	46.9	46.8	44.4	45.9 ± 1.9 (4.1%)
$k_{0.4}$ [kN/mm]	9.2	5.8	10.0	7.0	8.0 ± 3.1 (38.6%)
$k_{0.6}$ [kN/mm]	7.8	5.5	7.9	5.9	6.8 ± 2.0 (29.4%)

Table 6. The results of the push-out tests of the shear connections with 12-mm grade 8.8 bolts and with toothed-plate connectors (type C2-50/M12G, Bulldog) (per one connector).

Parameter	Specimen				Mean
	8.8.12.1	8.8.12.2	8.8.12.3	8.8.12.4	(8.8.12.1–8.8.12.4)
P_{ult} [kN]	39.1	38.2	38.0	40.4	38.9 ± 1.7 (4.5%)
s_{ult} [mm]	40.9	44.1	44.4	38.8	42.1 ± 4.3 (10.2%)
$k_{0.4}$ [kN/mm]	8.3	6.0	9.2	10.6	8.5 ± 3.1 (36.0%)
$k_{0.6}$ [kN/mm]	8.0	6.3	8.7	9.8	8.2 ± 2.3 (28.5%)

Table 7. The results of the push-out tests of the shear connections with 10-mm grade 5.8 bolts and without toothed-plate connectors (per one connector).

Parameter	Specimen				Mean
	R5.8.10.1	**R5.8.10.2**	**R5.8.10.3**	**R5.8.10.4**	**(R5.8.10.1–R5.8.10.4)**
P_{ult} [kN]	30.7	30.6	29.8	29.8	30.2 ± 0.8 (2.6%)
s_{ult} [mm]	33.1	29.4	32.6	32.3	31.9 ± 8.3 (2.7%)
$k_{0.4}$ [kN/mm]	4.8	5.7	5.2	3.9	4.9 ± 1.2 (24.7%)
$k_{0.6}$ [kN/mm]	2.1	3.1	2.7	1.7	2.4 ± 1.0 (41.2%)

Table 8. The results of the push-out tests of the shear connections with 10-mm grade 5.8 bolts and with toothed-plate connectors (type C2-50/M10G, Bulldog) (per one connector).

Parameter	Specimen				Mean
	5.8.10.1	**5.8.10.2**	**5.8.10.3**	**5.8.10.4**	**(5.8.10.1–5.8.10.4)**
P_{ult} [kN]	24.7	25.3	23.8	25.9	24.9 ± 1.4 (5.7%)
s_{ult} [mm]	19.7	20.9	20.1	20.3	20.3 ± 0.8 (3.9%)
$k_{0.4}$ [kN/mm]	6.0	6.3	5.4	6.3	6.0 ± 0.7 (11.3%)
$k_{0.6}$ [kN/mm]	6.9	7.0	6.5	7.0	6.9 ± 0.4 (5.5%)

Table 9. The results of the push-out tests of the shear connections with 12-mm grade 5.8 bolts and without toothed-plate connectors (per one connector).

Parameter	Specimen				Mean
	R5.8.12.1	**R5.8.12.2**	**R5.8.12.3**	**R5.8.12.4**	**(R5.8.12.1–R5.8.12.4)**
P_{ult} [kN]	43.6	39.1	41.3	40.5	41.1 ± 3.0 (7.3%)
s_{ult} [mm]	47.4	48.4	47.6	41.1	46.1 ± 5.4 (11.7%)
$k_{0.4}$ [kN/mm]	5.5	5.1	10.0	4.5	6.3 ± 4.0 (63.8%)
$k_{0.6}$ [kN/mm]	4.5	4.8	6.7	4.6	5.2 ± 1.7 (32.2%)

Table 10. The results of the push-out tests of the shear connections with 12-mm grade 5.8 bolts and with toothed-plate connectors (type C2-50/M12G, Bulldog) (per one connector).

Parameter	Specimen				Mean
	5.8.12.1	**5.8.12.2**	**5.8.12.3**	**5.8.12.4**	**(5.8.12.1–5.8.12.4)**
P_{ult} [kN]	42.5	39.6	40.2	38.0	40.1 ± 3.0 (7.4%)
s_{ult} [mm]	51.4	38.5	43.1	35.0	42.0 ± 11.3 (26.9%)
$k_{0.4}$ [kN/mm]	5.6	10.0	8.8	4.7	7.3 ± 4.0 (55.3%)
$k_{0.6}$ [kN/mm]	5.2	8.3	7.8	5.4	6.7 ± 2.6 (38.2%)

In the case of grade 8.8 10 mm bolts, the values of the ultimate load per one connector and of the slip modulus per one connector $k_{0.4}$ for the specimens with toothed-plate connectors were comparable to the values for the specimens without toothed-plate connectors (compare Tables 3 and 4).

The mean value of the slip modulus per one connector $k_{0.6}$ for the specimens with toothed-plate connectors was insignificantly higher than for the specimens without toothed-plate connectors (compare Tables 3 and 4). The energy accumulated in specimens 8.8.10.1–8.8.10.4 with toothed-plate connectors was higher than in specimens R8.8.10.1–R8.8.10.4 without toothed-plate connectors, due to the fact that all the load–slip curves of the reinforced connections had a higher load than the non-reinforced connections for the same displacement (see Figures 4 and 5).

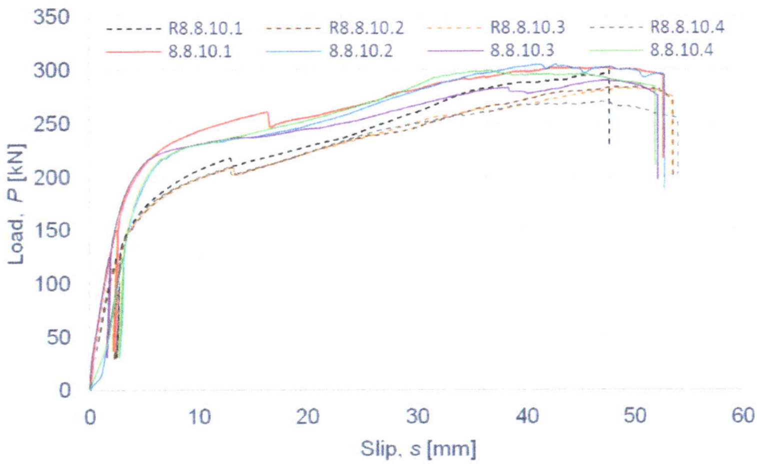

Figure 4. The load–slip curves from the push-out tests of the shear connections with 10-mm grade 8.8 bolts and with toothed-plate connectors (type C2-50/M10G, Bulldog) in specimens 8.8.10.1–8.8.10.4 or without toothed-plate connectors in specimens R8.8.10.1–R8.8.10.4.

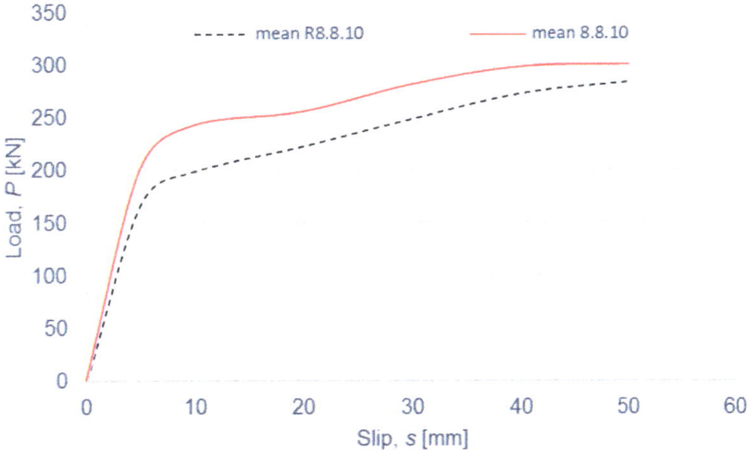

Figure 5. The mean load–slip curves for specimens 8.8.10.1–8.8.10.4 and R8.8.10.1–R8.8.10.4.

In the case of grade 8.8 12-mm bolts, the same conclusions can be drawn as for grade 8.8 10-mm bolts (see Tables 5 and 6 and Figures 6 and 7). Upon comparing the load-carrying capacities and the slip moduli of the tested connections with grade 8.8 bolts, it was observed that the use of toothed plate connectors was ineffective in improving both the load-carrying capacity and the stiffness of aluminium-timber composite connections.

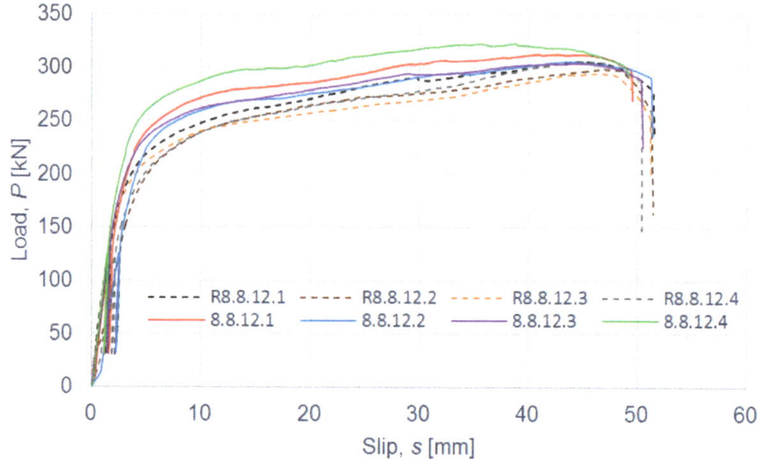

Figure 6. The load–slip curves from the push-out tests of the shear connections with 12-mm grade 8.8 bolts and with toothed-plate connectors (type C2-50/M12G, Bulldog) in specimens 8.8.12.1–8.8.12.4 or without toothed-plate connectors in specimens R8.8.12.1–R8.8.12.4.

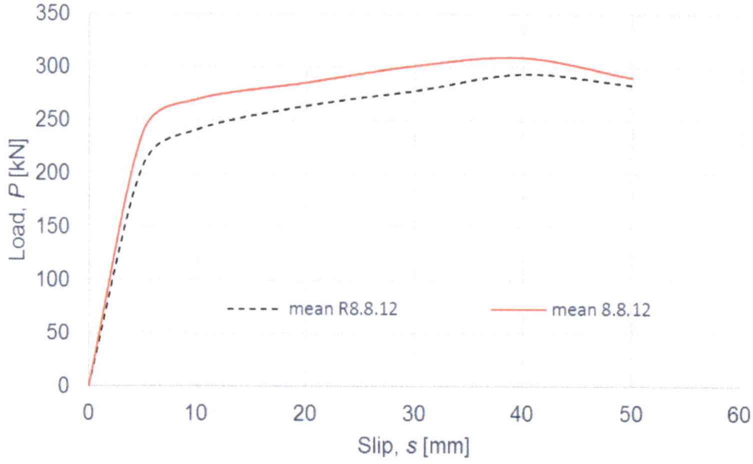

Figure 7. The mean load–slip curves for specimens 8.8.12.1–8.8.12.4 and R8.8.12.1–R8.8.12.4.

The connections with grade 5.8 10-mm bolts reinforced with toothed plate connectors showed higher stiffness ($k_{0.6}$ = 6.9 kN/mm per one connector) than the non-reinforced connections ($k_{0.6}$ = 2.4 kN/mm per one connector) (compare Tables 7 and 8). However, their strength was 1.2 times lower after reinforcing (see Figures 8 and 9). After comparing the behaviour of the non-reinforced connections and the reinforced connections with grade 5.8 10-mm bolts, the following conclusions may be drawn. The toothed-plate connectors reduced timber destruction in the bearing zones in the LVL slabs, because some part of the load was transferred by the teeth. However, the shanks of the bolts were sheared faster in the reinforced connections due to the fact that the bolt shanks were under the bearing pressure from the aluminum flange and the LVL slab as well as the toothed-plate flange. In the case of non-reinforced connections, the bolts were more tensioned than sheared, whereas in the case of reinforced connections, it was the opposite. For these reasons, the reinforced connections were both stiffer and weaker than the non-reinforced connections with grade 5.8 10-mm bolts. The above did not occur for the remaining connections because they had a higher strength (grade 8.8 bolts) or a larger diameter (12 mm). Additionally, the LVL slabs were split in the reinforced and non-reinforced connections with grade 8.8 10 mm × 125 mm and 12 mm × 135 mm bolts, and grade 5.8 12 mm × 135 mm bolts.

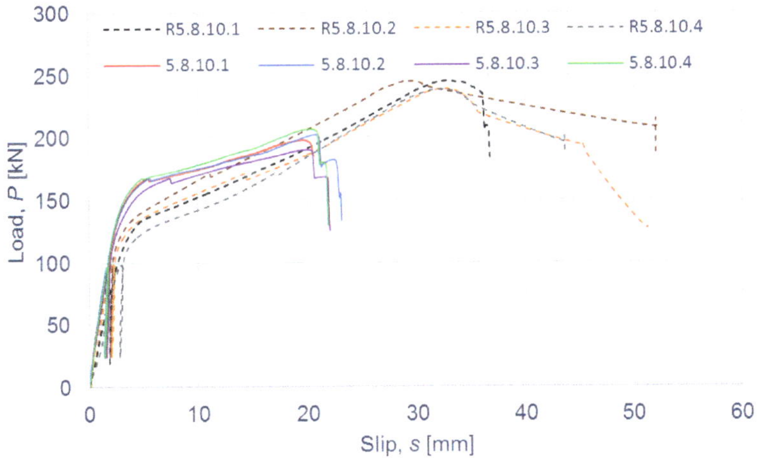

Figure 8. The load–slip curves from the push-out tests of the shear connections with 10-mm grade 5.8 bolts and with toothed-plate connectors (type C2-50/M10G, Bulldog) in specimens 5.8.10.1–5.8.10.4 or without toothed-plate connectors in specimens R5.8.10.1–R5.8.10.4.

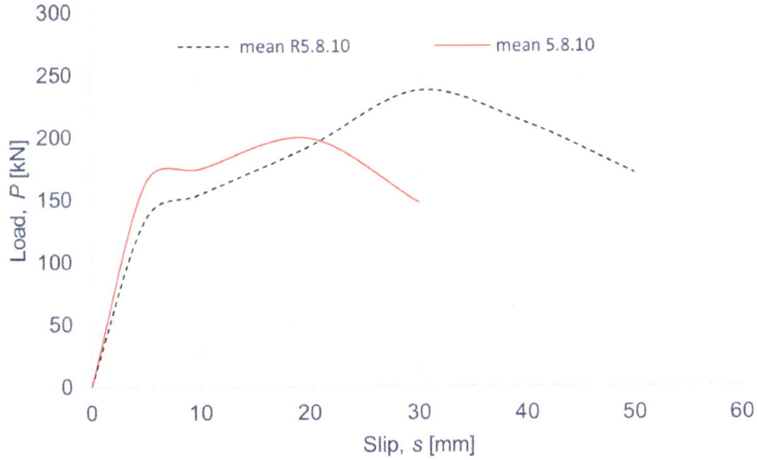

Figure 9. The mean load–slip curves for specimens 5.8.10.1–5.8.10.4 and R5.8.10.1–R5.8.10.4.

In the case of grade 5.8 12-mm bolts, it was observed that the use of toothed plate connectors was ineffective in improving both the load-carrying capacity and the stiffness of aluminium-timber composite connections (see Tables 9 and 10 and Figures 10 and 11).

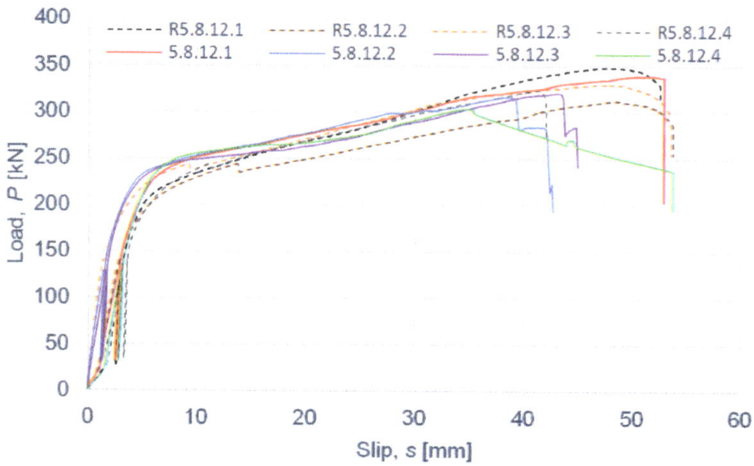

Figure 10. The load–slip curves from the push-out tests of the shear connections with 12-mm grade 5.8 bolts and with toothed-plate connectors (type C2-50/M12G, Bulldog) in specimens 5.8.12.1–5.8.12.4 or without toothed-plate connectors in specimens R5.8.12.1–R5.8.12.4.

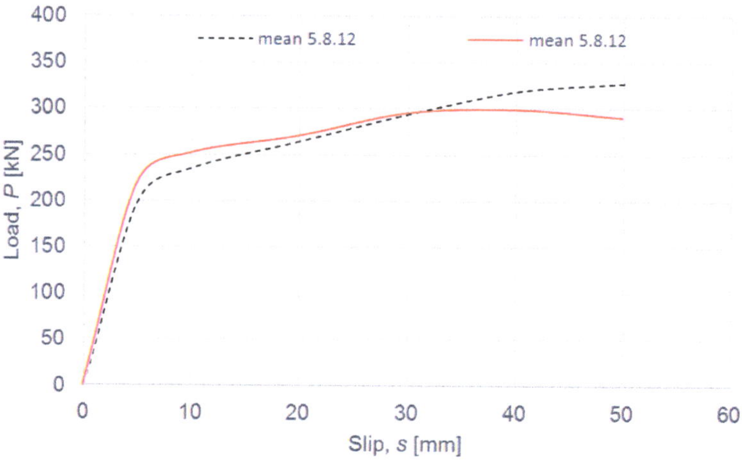

Figure 11. The mean load–slip curves for specimens 5.8.12.1–5.8.12.4 and R5.8.12.1–R5.8.12.4.

Connections are ductile if their characteristic slip capacity is at least 6 mm [60]. The value of the characteristic slip capacity exceeded 6 mm in all tested connections, and therefore they were all ductile.

The modes of failure of the tested bolted connections with or without Bulldog toothed-plate connectors (C2-50/M10G, C2-50/M12G) are presented in Figures 12–19. The authors observed the formation of two plastic hinges within the bolt, the crushing of LVL near the bolts, hole ovalisation in the flange of the aluminium alloy beam, bent teeth of toothed plates, and yielded washers due to washer pressure. Some of the bolts were additionally sheared. Furthermore, the LVL slabs were split both in the reinforced and non-reinforced connections with grade 8.8 10 mm × 125 mm and 12 mm × 135 mm bolts, and grade 5.8 12 mm × 135 mm bolts. This also explains why reinforcing was ineffective for the connections with these bolts. The LVL slabs in the connections with 5.8 10 mm × 125 mm bolts were not split.

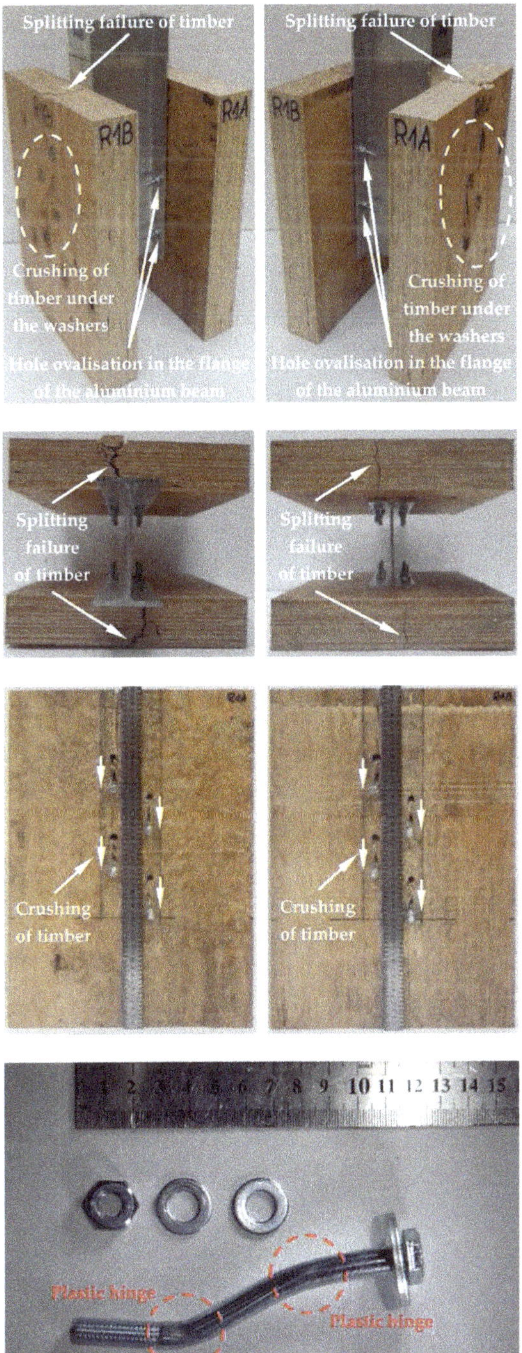

Figure 12. The failure mode of the aluminium-timber connections with 10-mm 8.8 grade bolts and without reinforcing toothed plates.

Figure 13. The failure mode of the aluminium-timber connections with 10-mm 5.8 grade bolts and without reinforcing toothed plates.

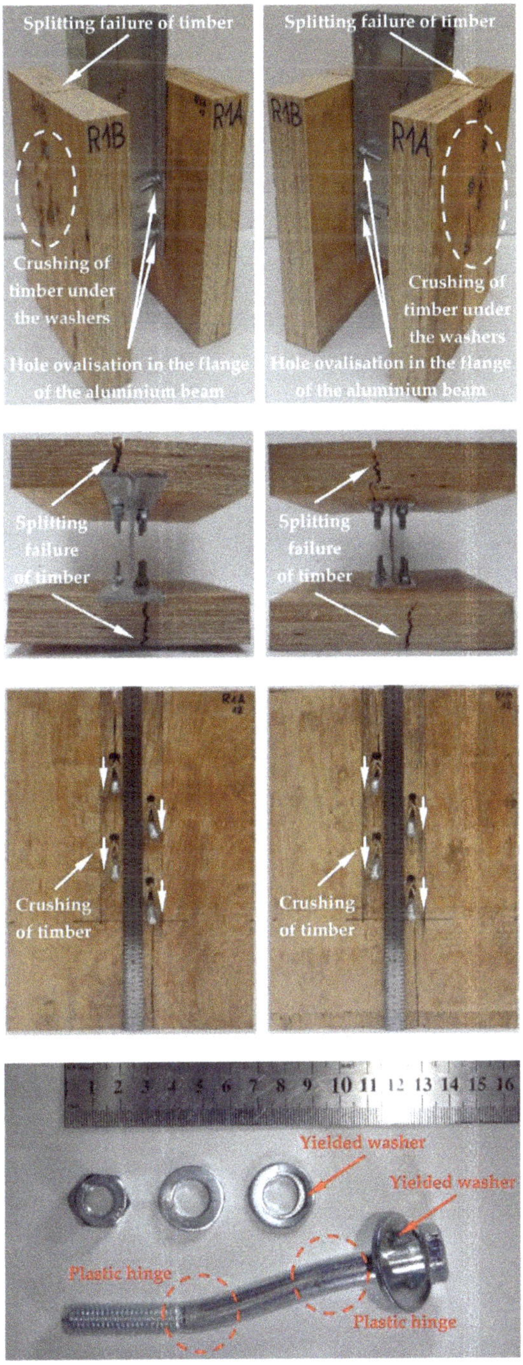

Figure 14. The failure mode of the aluminium-timber connections with 12-mm 8.8 grade bolts and without reinforcing toothed plates.

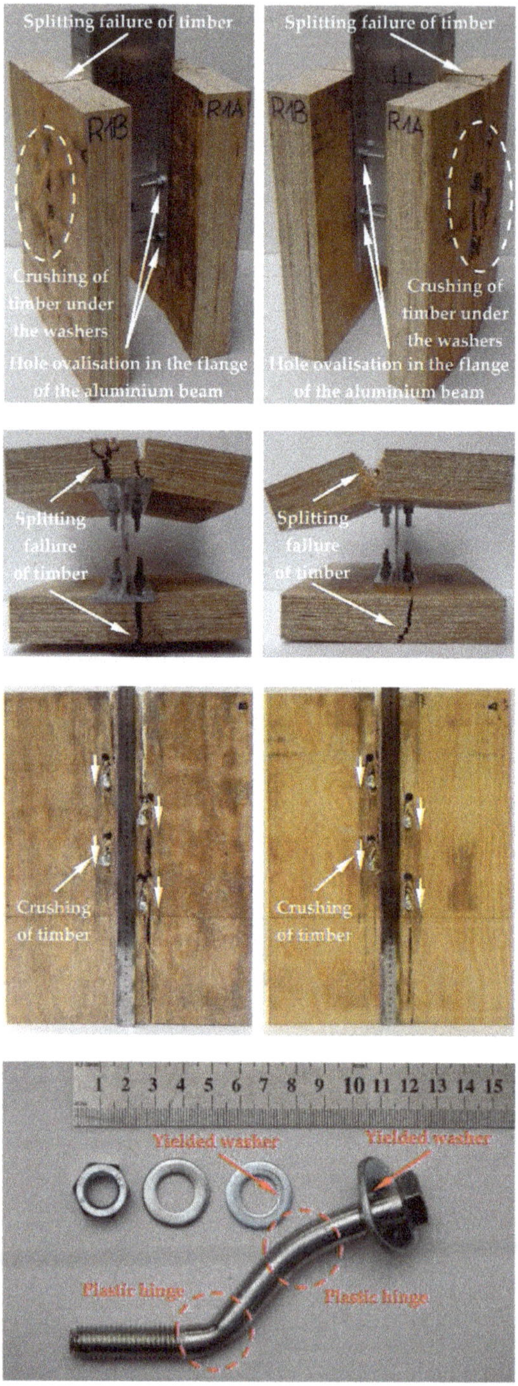

Figure 15. The failure mode of the aluminium-timber connections with 12-mm 5.8 grade bolts and without reinforcing toothed plates.

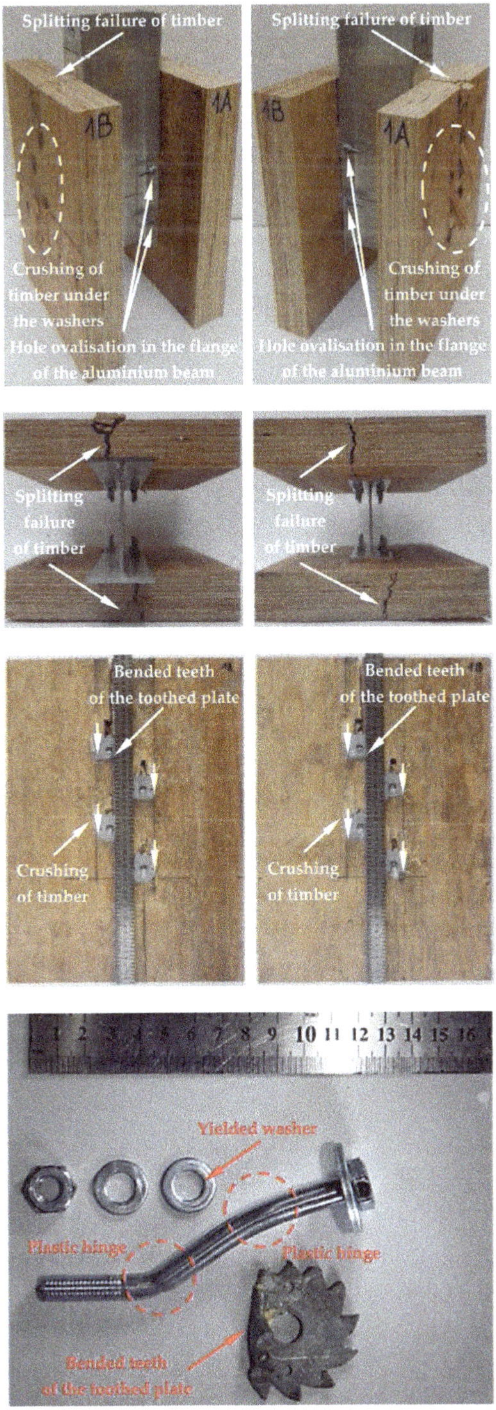

Figure 16. The failure mode of the aluminium-timber connections with 10-mm 8.8 grade bolts and with reinforcing toothed plates (type C2-50/M10G, Bulldog).

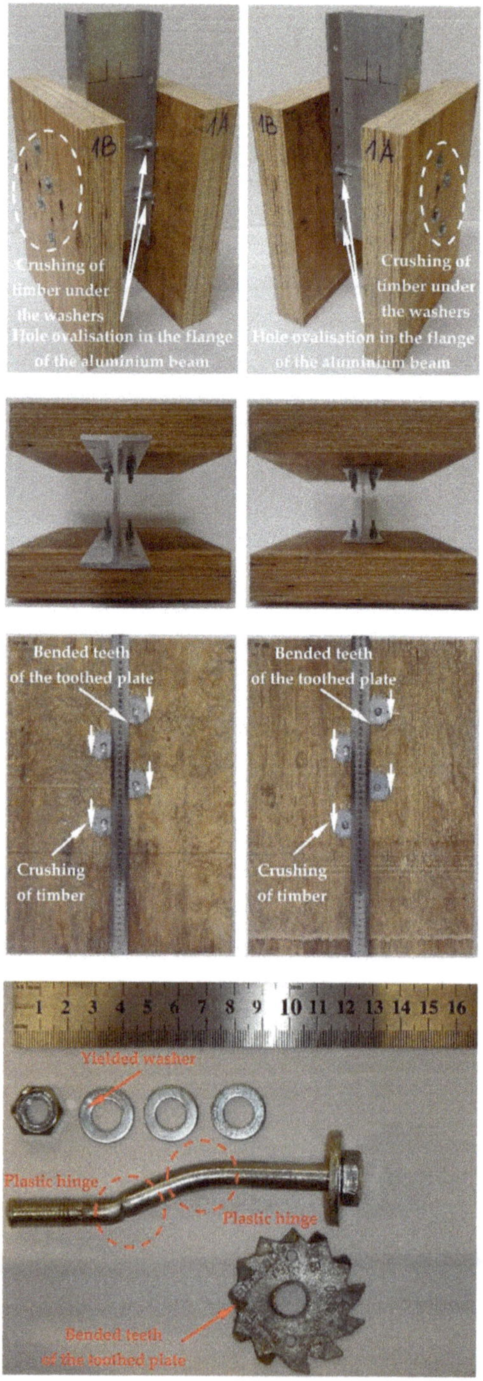

Figure 17. The failure mode of the aluminium-timber connections with 10-mm 5.8 grade bolts and with reinforcing toothed plates (type C2-50/M10G, Bulldog).

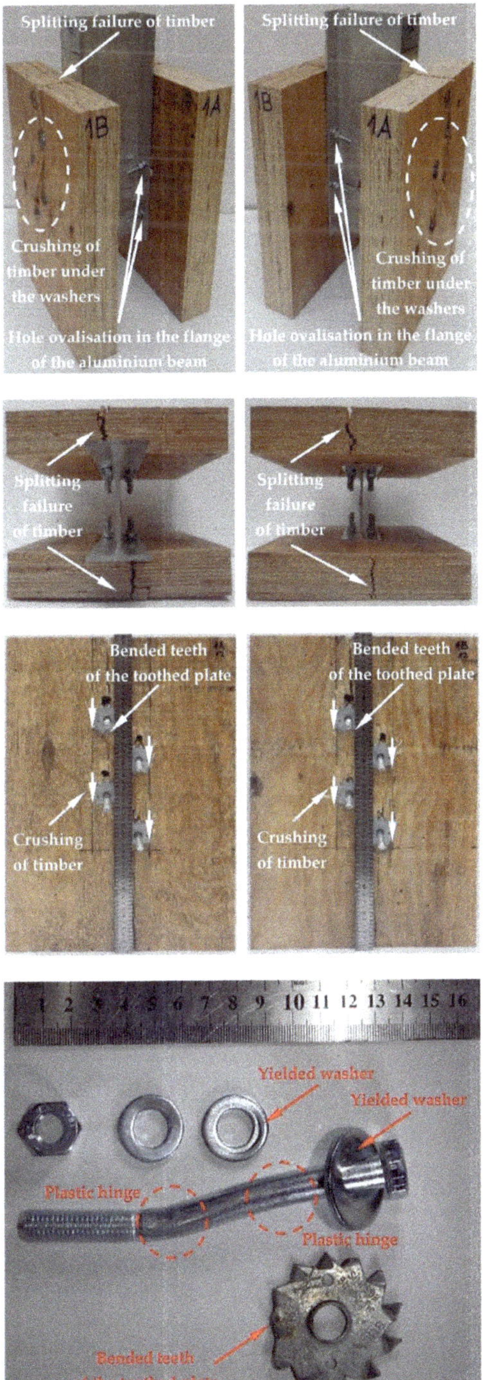

Figure 18. The failure mode of the aluminium-timber connections with 12-mm 8.8 grade bolts and with reinforcing toothed plates (type C2-50/M12G, Bulldog).

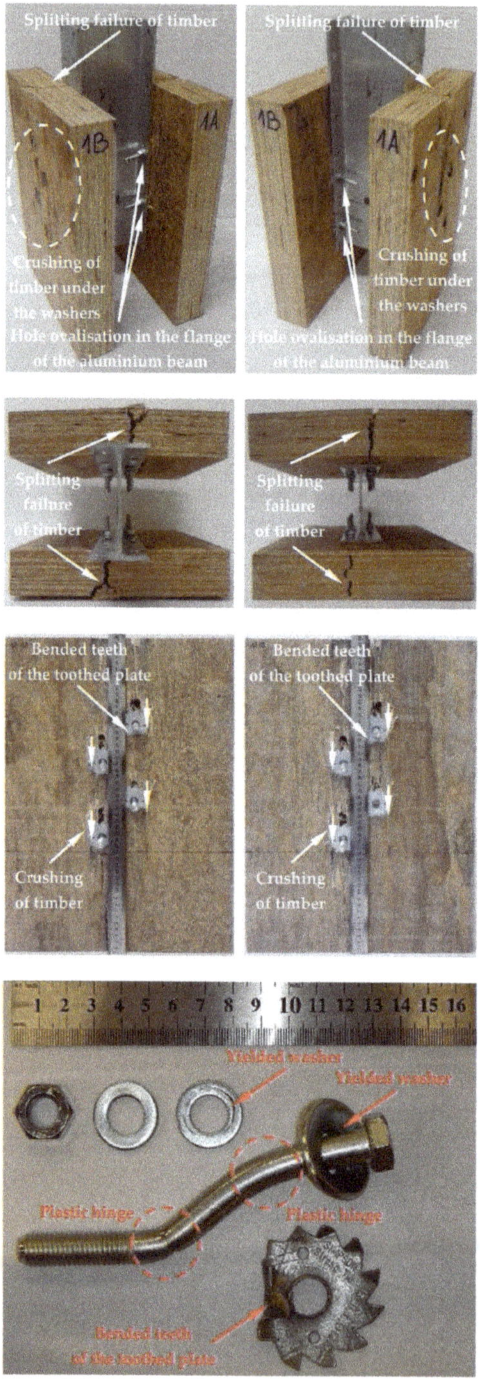

Figure 19. The failure mode of the aluminium-timber connections with 12-mm 5.8 grade bolts and with reinforcing toothed plates (type C2-50/M12G, Bulldog).

4. Discussion

4.1. A Comparison of the Obtained Results with the Literature

In this paper, bolts were used to connect LVL slabs with aluminium beams. Another option is to use screws as shear connectors [49]. The push-out samples with bolts analysed in this article had the same geometry as the push-out samples with screws investigated in [49]. The screws had the same diameter and grade as the bolts. Moreover, the screwed connections were reinforced using the same toothed plates as the bolted connections. This made it possible to compare the structural behaviour of such connections (Table 11).

Table 11. A comparison of the obtained results with the literature [49].

Connection	Connector Dimensions [mm]	Connector Grade	Reinforcing Toothed Plate	P_{ult} [kN]	s_{ult} [mm]	$k_{0.4}$ [kN/mm]	$k_{0.6}$ [kN/mm]
Screwed	10 × 80	5.8	–	16.7	16.7	4.3	4.7
Screwed	10 × 80	5.8	C2-50/M10G	21.5	12.8	6.4	5.9
Screwed	12 × 80	5.8	–	22.3	24.0	8.5	7.1
Screwed	12 × 80	5.8	C2-50/M12G	27.6	12.8	7.5	7.3
Bolted	10 × 125	5.8	–	30.2	31.9	4.9	2.4
Bolted	10 × 125	5.8	C2-50/M10G	24.9	20.3	6.0	6.9
Bolted	12 × 135	5.8	–	41.1	46.1	6.3	5.2
Bolted	12 × 135	5.8	C2-50/M12G	40.1	42.0	7.3	6.7

In the case of the screwed connections, the strength increased 1.3 times (10 mm) or 1.2 times (12 mm) after reinforcing. The splitting of timber was not observed. The use of toothed-plate connectors reduced timber destruction in the bearing zones and provided for a strength increase. In the case of the bolted connections with grade 5.8 12-mm bolts, the strength hardly changed after reinforcing. The connection strength was limited by the splitting strength of timber, and the toothed-plate connectors did not protect the timber slabs against splitting. In the bolted connections, the washers located on the LVL panels prevented the withdrawal of the bolts, and the LVL panels were split by the bolt shanks ("knife effect"). For this reason, the LVL panels in the bolted connections were more prone to splitting than in the screwed connections in which the withdrawal was prevented by the screw threads. In the case of the bolted connections with grade 5.8 10-mm bolts, the strength was 1.2 times lower after reinforcing. In these connections, the bolt shanks were under the bearing pressure of the aluminium flange and the LVL slab as well as the toothed-plate flange, and the bolts were sheared faster in the reinforced connections. This situation did not occur for the bolted connections with grade 8.8 bolts and with bolts with a larger diameter (12 mm), as they had higher shear resistance. The strength of the connections with grade 5.8 10-mm bolts was limited by the shear resistance of the bolts. Moreover, the grade 5.8 10-mm bolts also demonstrated a lower ultimate strength (483 MPa) than the 5.8 10-mm screws (553.9 MPa) [49], and they were more prone to cutting off during the push-out tests. For these reasons, in the connections with grade 5.8 10-mm bolts, the LVL slabs did not split. Both for the screwed and the bolted connections, the increase of the connector diameter provided for the increase of the connection shear strength. The non-reinforced bolted connections had a 1.8 times higher shear strength than the non-reinforced screwed connections.

4.2. Future Research and Possible Applications of the Results of This Study

The behaviour of composite beams depends on the mechanical parameters of their connections. The results of the double-shear tests presented in this study, such as shear resistance, can be used to evaluate the number of connectors necessary to achieve the full composite action in beams. Moreover, the obtained load–slip curves from the laboratory push-out tests can be used in finite element models of aluminium-timber composite beams to model the behaviour of connections which are discrete and represented by spring elements. Finite element analyses may be used to complement laboratory tests. Models with spring elements can reflect the behaviour of a real structure [61]. Moreover, spring elements provide for high computational speed. Discrete modelling of connections was used for example in the numerical models presented in [62–64].

Further experimental shear connection tests should be performed. The impact of the shear connector length on its strength may be examined. The influence of the toothed-plate connectors on the behaviour of the connections, the strength of which does not depend on the splitting strength of LVL (e.g., connections with bolts of a smaller diameter or of wider spacing than those used in this study), is still worth analysing.

5. Conclusions

In this paper, eight groups of push-out tests were conducted to investigate the load-carrying capacity, stiffness, load–slip response, failure modes, and ductility of aluminium-timber bolted connections strengthened or unstrengthened with toothed plates.

The results of the double-shear tests conducted in the study lead to some important conclusions. The toothed plate connectors were found ineffective in improving the stiffness and the strength of the bolted connections with grade 8.8 10 mm × 125 mm and 12 mm × 135 mm bolts, and grade 5.8 10 mm × 125 mm and 12 mm × 135 mm bolts. In the case of grade 8.8 10 mm × 125 mm and 12 mm × 135 mm bolts and grade 5.8 12 mm × 135 mm bolts, the LVL slabs split both in the reinforced and non-reinforced connections. The toothed-plate connectors reduced timber destruction in the bearing zones in the LVL slabs, but they did not protect the LVL slabs against splitting. The ultimate load of the connections with grade 8.8 10 mm × 125 mm and 12 mm × 135 mm bolts and grade 5.8 12 mm × 135 mm bolts mainly depended on the splitting strength of timber. In the case of grade 5.8 10-mm bolts, the LVL slabs did not split. The reinforced connections were both stiffer and weaker than the non-reinforced connections with grade 5.8 10-mm bolts. This was because the shanks of the bolts were sheared faster in the reinforced connections than in the non-reinforced connections. The reason for this was that the bolt shanks were under the bearing pressure of the aluminium flange, the LVL slab, and the toothed-plate flange. This situation did not occur for the remaining connections because they had a higher strength (grade 8.8 bolts) or a larger diameter (12 mm).

The results of the push-out tests of the bolted connections were compared with the results of the shear connection tests of the screwed connections presented in [49]. The non-reinforced bolted connections had 1.8 times higher shear strength than the non-reinforced screwed connections. In the case of the screwed connections, the reinforcing with toothed-plate connectors provided for a shear strength increase, whereas in the case of the bolted connections, it did not, as their strength was limited by the splitting strength of timber.

Author Contributions: Conceptualization, M.C. and Ł.P.; methodology, M.C. and Ł.P.; investigation, M.C. and Ł.P.; specimen preparation, M.C. and Ł.P.; writing—original draft preparation, M.C. and Ł.P.; writing—review and editing, M.C. and Ł.P.; visualization, M.C. and Ł.P. All authors have read and agreed to the published version of the manuscript.

Funding: This research was funded by the Polish Ministry of Science and Higher Education under grants 0412/SBAD/0050 and 0412/SBAD/0060.

Institutional Review Board Statement: Not applicable.

Informed Consent Statement: Not applicable.

Data Availability Statement: All data are contained within the article.

Acknowledgments: The authors wish to thank STEICO company for the LVL panels.

Conflicts of Interest: The authors declare no conflict of interest.

References

1. Szeptyński, P. Closed-form analytical solution to the problem of bending of a multilayer composite beam—Derivation and verification. *Compos. Struct.* **2022**, *291*, 115611. [CrossRef]
2. Chybiński, M.; Polus, Ł. Experimental and numerical investigations of aluminium-timber composite beams with bolted connections. *Structures* **2021**, *34*, 1942–1960. [CrossRef]
3. Szumigala, M.; Chybinski, M.; Polus, L. Stiffness of Composite Beams with Full Shear Connection. *IOP Conf. Ser. Mater. Sci. Eng.* **2019**, *471*, 052083. [CrossRef]
4. Szewczyk, P.; Szumigała, M. Optimal Design of Steel–Concrete Composite Beams Strengthened under Load. *Materials* **2021**, *14*, 4715. [CrossRef]
5. Alajarmeh, O.; Zeng, X.; Aravinthan, T.; Shelley, T.; Alhawamdeh, M.; Mohammed, A.; Nicol, L.; Vedernikov, A.; Safonov, A.; Schubel, P. Compressive behaviour of hollow box pultruded FRP columns with continuous-wound fibres. *Thin-Walled Struct.* **2021**, *168*, 108300. [CrossRef]
6. Teter, A.; Debski, H.; Samborski, S. On buckling collapse and failure analysis of thin-walled composite lipped-channel columns subjected to uniaxial compression. *Thin-Walled Struct.* **2014**, *85*, 324–331. [CrossRef]
7. Tucci, F.; Vedernikov, A. Design Criteria for Pultruded Structural Elements. *Encycl. Mater. Compos.* **2021**, *3*, 51–68. [CrossRef]
8. Garbowski, T. Mechanics of Corrugated and Composite Materials. *Materials* **2022**, *15*, 1837. [CrossRef]
9. Bakalarz, M.M.; Kossakowski, P.G.; Tworzewski, P. Strengthening of Bent LVL Beams with Near-Surface Mounted (NSM) FRP Reinforcement. *Materials* **2020**, *13*, 2350. [CrossRef]
10. Bakalarz, M.M.; Kossakowski, P.G. Ductility and Stiffness of Laminated Veneer Lumber Beams Strengthened with Fibrous Composites. *Fibers* **2022**, *10*, 21. [CrossRef]
11. Bakalarz, M.; Kossakowski, P. Mechanical Properties of Laminated Veneer Lumber Beams Strengthened with CFRP Sheets. *Arch. Civ. Eng.* **2019**, *65*, 57–66. [CrossRef]
12. Szumigała, M.; Szumigała, E.; Polus, Ł. Laboratory tests of new connectors for timber-concrete composite structures. *Eng. Trans.* **2018**, *66*, 161–173.
13. Bao, Y.; Lu, W.; Yue, K.; Zhou, H.; Lu, B.; Chen, Z. Structural performance of cross-laminated timber-concrete composite floors with inclined self-tapping screws bearing unidirectional tension-shear loads. *J. Build. Eng.* **2022**, *55*, 104653. [CrossRef]
14. Otero-Chans, D.; Estévez-Cimadevila, J.; Suárez-Riestra, F.; Martín-Gutiérrez, E. Perforated board shear connector for timber-concrete composites. *Wood Mater. Sci. Eng.* **2022**, 1–14. [CrossRef]
15. Chybiński, M.; Polus, Ł.; Szwabiński, W.; Niewiem, P. Fe analysis of steel-timber composite beams. *AIP Conf. Proc.* **2019**, *2078*, 020061. [CrossRef]
16. Yang, R.; Wan, J.; Zhang, X.; Sun, Y. Modelling of steel-timber composite beams: Validation of finite element model and parametric study. *Wood Res.* **2021**, *66*, 806–820. [CrossRef]
17. Aspila, A.; Heinisuo, M.; Mela, K.; Malaska, M.; Pajunen, S. Elastic design of steel-timber composite beams. *Wood Mater. Sci. Eng.* **2022**. [CrossRef]
18. Szumigała, M.; Chybiński, M.; Polus, Ł. Preliminary Analysis of the Aluminium-Timber Composite Beams. *Civ. Environ. Eng. Rep.* **2017**, *27*, 131–141. [CrossRef]
19. Pyrak, S.; Włodarczyk, W. *Posadowienie Budowli, Konstrukcje Murowe i Drewniane*; WSiP: Warsaw, Poland, 2000. (In Polish)
20. Zobel, H.; Al-Khafaji, T. *Mosty Drewniane*; Wydawnictwa Komunikacji i Łączności: Warsaw, Poland, 2006. (In Polish)
21. Karolak, A.; Jasieńko, J.; Raszczuk, K. Historical scarf and splice carpentry joints: State of the art. *Herit. Sci.* **2020**, *8*, 105. [CrossRef]
22. Witomski, P.; Krajewski, A.; Kozakiewicz, P. Selected mechanical properties of Scots pine wood from antique churches of Central Poland. *Eur. J. Wood Prod.* **2014**, *72*, 293–296. [CrossRef]
23. Kotwica, E.; Krzosek, S. Timber bridges—Revive of old and new bridges built in Switzerland. Annals of Warsaw University of Life Sciences—SGGW. *For. Wood Technol.* **2015**, *92*, 207–210.
24. Rapp, P. Application of adhesive joints in reinforce-ment and reconstruction of weakened wooden elements loaded axially. *Drewno* **2016**, *59*, 59–73. [CrossRef]
25. Rapp, P. The numerical modeling of adhesive joints in reinforcement of wooden elements, subjected to bending and shearing. *Drewno* **2015**, *60*, 21–36. [CrossRef]
26. Burawska-Kupniewska, I.; Zbiec, M.; Tomusiak, A.; Beer, P. Local Reinforcement of Timber with Composite and Lignocellulosic Materials. *BioResources* **2015**, *10*, 457–468. [CrossRef]
27. Purba, C.Y.C.; Pot, G.; Viguier, J.; Ruelle, J.; Denaud, L. The influence of veneer thickness and knot proportion on the mechanical properties of laminated veneer lumber (LVL) made from secondary quality hardwood. *Eur. J. Wood Prod.* **2019**, *77*, 393–404. [CrossRef]
28. Nguyen, H.H.; Gilbert, B.P.; McGavin, R.L.; Bailleres, H.; Karampour, H. Embedment strength of mixed-species laminated veneer lumbers and cross-banded laminated veneer lumbers. *Eur. J. Wood Prod.* **2020**, *78*, 365–386. [CrossRef]
29. Aro, M.D.; Wang, X.; McDonald, D.E.; Begel, M. Tensile strength of thermally modified laminated strand lumber and laminated veneer lumber. *Wood Mater. Sci. Eng.* **2017**, *12*, 228–235. [CrossRef]
30. Ahmad, M.; Kamke, F.A. Properties of parallel strand lumber from Calcutta bamboo (*Dendrocalamus strictus*). *Wood Sci. Technol.* **2011**, *45*, 63–72. [CrossRef]
31. Dietsch, P.; Tannert, T. Assessing the integrity of glued-laminated timber elements. *Constr. Build. Mater.* **2015**, *101*, 1259–1270. [CrossRef]

32. Jeleč, M.; Varevac, D.; Rajčić, V. Cross-laminated timber (CLT)—A state of the art report. *J. Croat. Assoc. Civ. Eng.* **2018**, *70*, 75–95. [CrossRef]
33. Ringhofer, A.; Brandner, R.; Blaß, H.J. Cross laminated timber (CLT): Design approaches for dowel-type fasteners and connections. *Eng. Struct.* **2018**, *171*, 849–861. [CrossRef]
34. Saleh, S.M.; Jasim, N.A. Structural behavior of timber aluminum composite beams under static loads. *Int. J. Eng. Res. Technol.* **2014**, *3*, 1166–1173.
35. Chybiński, M.; Polus, Ł. Theoretical, experimental and numerical study of aluminium-timber composite beams with screwed connections. *Constr. Build. Mater.* **2019**, *226*, 317–330. [CrossRef]
36. Hassanieh, A.; Valipour, H.; Bradford, M. Load-slip behaviour of steel-cross laminated timber (CLT) composite connections. *J. Constr. Steel Res.* **2016**, *122*, 110–121. [CrossRef]
37. Heinisuo, M.; Mela, K.; Pajunen, S.; Malaska, M. New Steel-Timber Composite Beam, Nordic System. *CE/Papers* **2019**, *3*, 193–202. [CrossRef]
38. Zhang, Z.-J.; Han, F.-X.; Liu, Q. Experimental Study on Axial Compression Performance of 7A04 High Strength Aluminum Alloy Circular Tube Concrete Short Column. *Geofluids* **2022**, *2022*, 8270471. [CrossRef]
39. Saleh, S.M.; Jasim, N.A. Structural behavior of timber aluminum composite beams under impact loads. *Int. J. Sci. Eng. Res.* **2014**, *5*, 865–873.
40. Kyvelou, P.; Gardner, L.; Nethercot, D.A. Design of Composite Cold-Formed Steel Flooring Systems. *Structures* **2017**, *12*, 242–252. [CrossRef]
41. Pham, N.H.; Pham, C.H.; Rasmussen, K.J. Global buckling capacity of cold-rolled aluminium alloy channel section beams. *J. Constr. Steel Res.* **2021**, *179*, 106521. [CrossRef]
42. Wang, X.; Su, P.; Liu, J.; Chen, Z.; Khan, K. Seismic performance of light steel-natural timber composite beam-column joint in low-rise buildings. *Eng. Struct.* **2022**, *256*, 113969. [CrossRef]
43. Wang, X.; Zhang, J.; Wu, P.; Li, Y. Experimental investigation on the flexural and shear behaviour of LVL I-beam strengthened with steel channels. *Constr. Build. Mater.* **2022**, *341*, 127719. [CrossRef]
44. Johnson, R. *Designers Guide to Eurocode 4: Design of Composite Steel and Concrete Structures*, 2nd ed.; EN 1994-1-1; ICE Publishing: London, UK, 2012. [CrossRef]
45. Fonseca, E.M.M.; Leite, P.A.S.; Silva, L.D.S.; Silva, V.S.B.; Lopes, H.M. Parametric Study of Three Types of Timber Connections with Metal Fasteners Using Eurocode 5. *Appl. Sci.* **2022**, *12*, 1701. [CrossRef]
46. Zhou, Y.; Zhao, Y.; Wang, C.-L.; Zhou, Y.; Zheng, J. Experimental study of the shear performance of H-shaped aluminum-timber composite connections. *Constr. Build. Mater.* **2022**, *334*, 127421. [CrossRef]
47. Chybiński, M.; Polus, L. Withdrawal strength of hexagon head wood screws in laminated veneer lumber. *Eur. J. Wood Prod.* **2022**, *80*, 541–553. [CrossRef]
48. Szumigała, M.; Chybiński, M.; Polus, Ł. Composite beams with aluminium girders—A review. In *Modern Trends in Research on Steel, Aluminium and Composite Structures: Proceedings of the XIV International Conference on Metal Structures (ICMS2021), Poznan, Poland, 16–18 June 2021*; Giżejowski, M.A., Kozłowski, A., Chybiński, M., Rzeszut, K., Studziński, R., Szumigała, M., Eds.; Routledge: Leiden, The Netherlands, 2021; pp. 249–255.
49. Chybiński, M.; Polus, Ł. Mechanical Behaviour of Aluminium-Timber Composite Connections with Screws and Toothed Plates. *Materials* **2022**, *15*, 68. [CrossRef] [PubMed]
50. Chybiński, M.; Polus, Ł.; Ratajczak, M.; Sielicki, P.W. The Evaluation of the Fracture Surface in the AW-6060 T6 Aluminium Alloy under a Wide Range of Loads. *Metals* **2019**, *9*, 324. [CrossRef]
51. Porteous, J.; Kermani, A. *Structural Timber Design to Eurocode 5*; Wiley-Blackwell: Chichester, UK, 2013.
52. Chybiński, M.; Polus, Ł. Experimental and numerical investigations of laminated veneer lumber panels. *Arch. Civ. Eng.* **2021**, *67*, 351–372. [CrossRef]
53. Komorowski, M. *Podręcznik Projektowania i Budowania w Systemie STEICO. Podstawy. Fizyka Budowli. Zalecenia Wykonawcze*; Forestor Communication: Warsaw, Poland, 2017. (In Polish)
54. *EN ISO 6892-1*; Metallic materials—Tensile testing—Part 1: Method of test at room temperature. European Committee for Standardization: Brussels, Belgium, 2020.
55. *EN 912*; Timber Fasteners, Specifications for Connectors for Timbers. European Committee for Standardization: Brussels, Belgium, 2011.
56. *EN 26891*; Timber Structures—Joints Made with Mechanical Fasteners—General Principles for the Determination of Strength and Deformation Characteristics. European Committee for Standardization: Brussels, Belgium, 1991.
57. The Łukasiewicz Research Network, Metal Forming Institute. *Research Report: 34/22/BB.903.0287.06*; The Łukasiewicz Research Network, Metal Forming Institute: Poznań, Poland, 2022.
58. Lukaszewska, E.; Johnsson, H.; Fragiacomo, M. Performance of connections for prefabricated timber–concrete composite floors. *Mater. Struct.* **2008**, *41*, 1533–1550. [CrossRef]
59. Ceccotti, A.; Fragiacomo, M.; Giordano, S. Long-term and collapse tests on a timber-concrete composite beam with glued-in connection. *Mater. Struct.* **2007**, *40*, 15–25. [CrossRef]
60. *EN 1994-1-1*; Eurocode 4, Design of Composite Steel and Concrete Structures—Part 1-1: General Rules and Rules for Buildings. European Committee for Standardization: Brussels, Belgium, 2004.

61. Wróblewski, T.; Pełka-Sawenko, A.; Abramowicz, M.; Berczyński, S. Parameter identification of steel-concrete composite beams by finite element method. *Diagnostyka* **2013**, *14*, 43–46.
62. Kyvelou, P.; Gardner, L.; Nethercot, D.A. Finite element modelling of composite cold-formed steel flooring systems. *Eng. Struct.* **2018**, *158*, 28–42. [CrossRef]
63. Studziński, R.; Ciesielczyk, K. Connection stiffness between thin-walled beam and sandwich panel. *J. Sandw. Struct. Mater.* **2019**, *21*, 2042–2056. [CrossRef]
64. Polus, L.; Szumigala, M. Finite Element Modelling of the Connection for Timber-concrete Composite Beams. *IOP Conf. Ser. Mater. Sci. Eng.* **2019**, *471*, 052081. [CrossRef]

materials

MDPI

Article

Experimental Investigation of the Failure Scenario of Various Connection Types between Thin-Walled Beam and Sandwich Panel

Katarzyna Ciesielczyk * and **Robert Studziński ***

Institute of Building Engineering, Poznan University of Technology, Piotrowo 5 Street, 60-965 Poznan, Poland
* Correspondence: katarzyna.ciesielczyk@put.poznan.pl (K.C.); robert.studzinski@put.poznan.pl (R.S.); Tel.: +48-61-665-3325 (K.C.); +48-61-665-2091 (R.S.)

Abstract: The paper presents failure scenarios for various types of connections between a thin-walled beam and a sandwich panel. In addition to standard connections used in civil engineering applications, that is, self-drilling fasteners for sandwich panels, the study examined the use of bolts, blind rivets, and double-sided acrylic tape applied linearly and pointwise. The connections were subjected to the horizontal load applied with constant eccentricity with respect to the plane of the connection surface. This load arrangement simulates the behaviour of a free flange of the thin-walled beam in bending while lateral-torsional buckling occurs. In this way, the research covers the determination of the lateral stiffness of the thin-walled beam-free flange, while the other flange is connected to the sandwich panel using various connection systems.

Keywords: sandwich panels; thin-walled beam; failure scenario; connections; laboratory experiment

check for updates

Citation: Ciesielczyk, K.; Studziński, R. Experimental Investigation of the Failure Scenario of Various Connection Types between Thin-Walled Beam and Sandwich Panel. *Materials* **2022**, *15*, 6277. https://doi.org/10.3390/ma15186277

Academic Editor: Michele Bacciocchi

Received: 13 August 2022
Accepted: 7 September 2022
Published: 9 September 2022

1. Introduction

Claddings are inseparable elements of industrial halls. The purpose of the claddings is three-fold. Firstly, the claddings protect the structure of the building from environmental actions such as wind, rain, and snow [1]. Secondly, the claddings ensure the desired temperature and moisture content in the interior of the building [2]. Thirdly, the claddings transmit the mechanically (dead load, climatic action, and useful loads) and thermally (gradient temperature between the facings) induced loads to the main or secondary structural elements [3]. Nowadays, the cladding systems in steel industrial halls are increasingly made of sandwich panels [4]. The sandwich panel is a composite element consisting of the two external facings and the internal core [5]. In civil engineering applications, the facings are made of high strength and of a high stiffness material [6–8] such as steel, aluminium, or less frequently composite laminate. Their thickness is approximately 0.5 mm in the case of metallic facings and 0.7 mm in the case of laminate facings. The core, from a mechanical point of view, can be defined as a soft and flexible material of low weight and excellent thermal insulation properties. In building construction, the core, which is usually made of mineral wool, polyurethane foam, or styrofoam layer, is significantly thicker than the facings. Its thickness ranges from 40.0 mm to 200.0 mm (depending on the desirable level of thermal insulation). Usually, the core is assumed to be isotropic; nevertheless, recent research [9,10] also investigates its anisotropy.

Sandwich panels are usually attached to the main structural elements and secondary elements (hot-rolled or thin-walled cross-sections) via self-drilling fasteners. The size and type of load transferred by the fasteners (from surface load to the linear load) depend on the assumed 'structural class' according to Eurocode 3 [11] associated with the failure consequences defined in Eurocode 0 [12]. Structural classes provide information on the reliability level of cold-formed elements (thin-walled beams) and claddings (corrugated metal plates). The authors of this paper, according to the guidebook titled European Recommendations on the Stabilization of Steel Structures by Sandwich Panels [13], published by the European Convention for Constructional Steelwork, assumed that structural classes can be extended to structures with sandwich panels used as cladding. The mentioned document [13] is the basis for the new standard in development prEN 14509-2 [14]. According to Eurocode 3 [11], three construction classes are distinguished. Construction class I refers to construction where cold-formed members and cladding are designed to contribute to the overall strength and stability of a structure. Construction class II refers to construction where cold-formed members and cladding are designed to contribute to the strength and stability of individual structural members. Construction class III refers to the construction in which the cladding is used as an element that only transfers loads to the main structure. From a mechanical point of view, the fasteners in each construction class participate in transferring the surface load from plate elements (claddings) to their supports, i.e., beam elements. Depending on the direction of the support load in relation to the support plane and the angle of its inclination in relation to the ground plane, the fastener is subjected to axial and/or shear force. Note that compression is mainly transferred from the cladding element directly to the beam flange due to the contact surface. However, in the case of construction classes I and II, it is taken into account that the fasteners transfer the additional membrane forces (forces perpendicular to the fastener axis) that arise from the lateral-torsional stabilization provided to the thin-walled beam elements by the claddings [15–18]. Additionally, in [19], the hot-rolled sections were considered as beam elements stabilized by sandwich panels. Nowadays, this type of interaction is often used by structural engineering designers in their analyses. Please note that the connection mechanical properties (stiffness, ultimate capacity, and deformation capacity) are also important in other structural elements used in building applications. For example, in [20,21] the authors investigated the various connections in the aluminium–timber composite beam (ATC) that consist of the aluminium beam and laminated veneer lumber (LVL).

The experimental research presented in this article simulates the behaviour of the free flange of the thin-walled Z beam in bending. The other flange is laterally restrained by the sandwich panel. This laboratory model refers to the calculation method described in Section 10.1 of Eurocode 3, Parts 1–3 [11], where the free flange of the thin-walled beam is considered as a beam on an elastic foundation. The research covers the determination of the lateral stiffness of the thin-walled beam-free flange, while the other flange is connected to the sandwich panel using various connection systems: self-drilling fasteners for sandwich panels, blind rivets, bolts, and double-sided acrylic tape. In addition, the equilibrium path (force-displacement), initial and secant stiffness, ultimate capacity, and deformation capacity were determined. Furthermore, the research allows determining failure scenarios of various connection types between a thin-walled beam and a sandwich panel.

Currently, as part of the work of the ECCS working group TWG 7.9 named Sandwich Panels and related Structures, the new document is being prepared. This document covers the experimental aspects of the determination of the rotational restraint provided by sandwich panels. The ECCS document, among others, was prepared on the basis of the following articles [22,23]. The laboratory tests presented in the manuscript are in line with the scope of this document, in which the sandwich panel can provide rotational restraint for thin-walled elements by means of various types of connector.

Note that in the research, the thin-walled beams used in the experiments were without openings. The results can be easily extended, for example, on the thin-walled beams with perforations discussed in References [24,25]. In these papers, the authors proposed a method for determining the equivalent stiffness of beam with and without openings or perforations along its length. To be specific in the presented homogenization method, there is no need to provide formal analysis such as solving the system of equations. The method uses the 3D representation of a beam modelled with shell finite elements and global stiffness matrix of the representative volumetric element (RVE).

2. Materials and Methods

2.1. Problem Formulation

In Figure 1, the scheme of the test bed is depicted. The test bed consisted of one sandwich panel with the following dimensions: width 1100 mm, length 1100 mm, and thickness 80 mm. The sandwich panel was made of two steel facings with a thickness equal to: 0.545 mm (external–upper) and 0.491 mm (internal–bottom) and a soft core made of polyisocyanurate foam (PIR). The mechanical properties of the sandwich panel layers were determined by previously conducted research. Young's modulus of facings was equal to 190.0 MPa [26], and the value of the shear modulus of the polyurethane foam core was equal to 3.43 MPa [27].

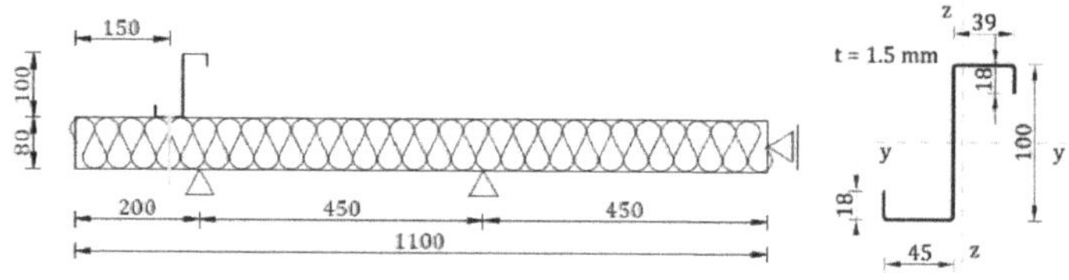

Figure 1. Test bed scheme—geometry of the sandwich panel and the thin-walled Z-beam.

The thin-walled steel beam with a Z-section was fastened to the sandwich panel at an axial distance of 150 mm from its free edge. The dimensions of the thin-walled beam cross section were equal to 18 × 39 × 100 × 45 × 18 mm with constant walls thicknesses of 1.5 mm. The properties values of the thin-walled beam material were determined on the basis of laboratory tests [18,28]. Young's modulus was equal to 198 MPa, the value of the upper yield strength was equal to 348.87 MPa, and the value of the ultimate tensile strength was equal to 402.72 MPa. The following five types of connectors to the authors' knowledge can be used to connect the thin-walled section to the sandwich panel:

- Bolts (B) fully threaded with a diameter equal to 6 mm and length equal to 100 mm; see Figure 2a;
- Self-drilling fasteners for fastening sandwich panels to steel construction (F) with a diameter equal to 6.3 mm and length equal to 110 mm; see Figure 2b;
- Pulled blind rivets (BT) (threefold aluminium blind rivets of a diameter equal to 4.75 mm and clamping arm's length measured after pulling equal to 6 mm); see Figure 2c;
- Tightened blind rivets (FB) (fourfold steel/aluminium blind rivets of a diameter equal to 7.80 mm and clamping arm's length measured after pulling equal to 10 mm); see Figure 2d;
- Double-sided acrylic foam tape (TL—applied continuously, TP—applied pointwise) with thickness equal to 1.5 mm and width equal to 38 mm; see Figure 2e.

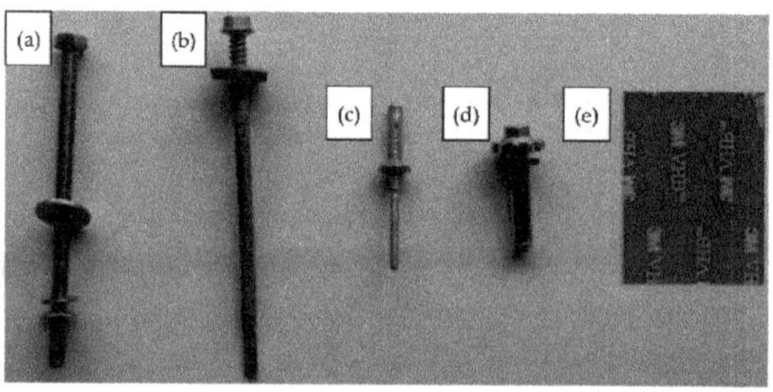

Figure 2. Connectors used to join the thin-walled beam and the sandwich panel (description in the text): (**a**) bolt (B), (**b**) self-drilling fastener dedicated for sandwich panels (F), (**c**) pulled blind rivet (BT), (**d**) tightened blind rivet (FB), (**e**) acrylic tape (TL, TP).

Sandwich panels used in structural engineering applications are usually attached to the supporting structure by three fasteners per width. Therefore, in the experiment, three mechanical connectors per width (B, F, BT, and FB) were used, that is, in the middle of the beam and 500 mm on the left and on the right from the middle point. In the case of acrylic tape, two methods of joining were used. In the first method of joining (TP), three short sections of tape measuring 38 × 50 mm were used. In the second method (TL), the tape was glued linearly along the entire length of the thin-walled beam. Double-sided acrylic foam tape, used in the experiments, may be used to bond a variety of materials, such as aluminium, steel, glass, plastics, painted, or powder-coated surfaces.

In the experiment, the Z beam-free flange was subjected to a horizontal load perpendicular to the central beam axis (to the beam web). The load was applied by a loading cell with a speed of 5 mm/min. The exact position of the loading cell is described in Figure 3 through a bold dot. Force was measured using a force transducer (U93) of 10 kN capacity of 0.5 class (it gives the accuracy of force measurement equal to 0.05 kN). During the test, vertical and horizontal beam displacements were measured at three points: in the middle of the beam span and at the outer edges of the beam. The precise position of the inductive displacement transducer with nominal displacements of 50 mm and 100 mm (WA 50 and WA 100) is depicted in Figure 3.

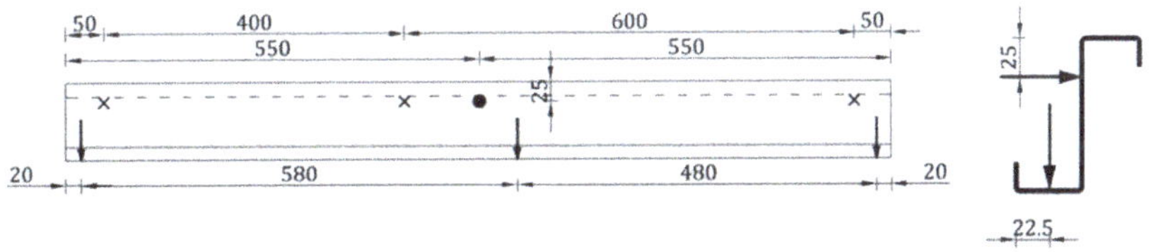

Figure 3. Position of the force and inductive displacement transducers along the thin-walled beam (description in the text).

In Figure 3, '×' represents the position of horizontal displacement transducers, vertical arrows represent the position of vertical displacement transducers, and 'bold dot' represents the position of the horizontal force transducer. Figure 4 presents the view of the test bed where **1** represents a sandwich panel with polyisocyanurate foam core and steel lightly profiled facings, **2** represents a thin-walled Z beam, **3** represents force transducer U93, **4** represents the electromechanical actuator (assembled by Archimedes Ltd., Toruń, Poland), **5** represents the inductive displacement transducers WA 100 (range 100 mm), **6** represents the inductive displacement transducers WA 50 (range 50 mm), **F** represents the typical self-drilling fastener, **B** represents the bolt, **FB** represents the tightened rivet, and **BT** represents the pulled blind rivet. It is worth noting that the innovative use of pulled blind rivets was presented in References [29,30] where laboratory and numerical analyses of the beam-to-column connection in cold formed steel frames were carried out.

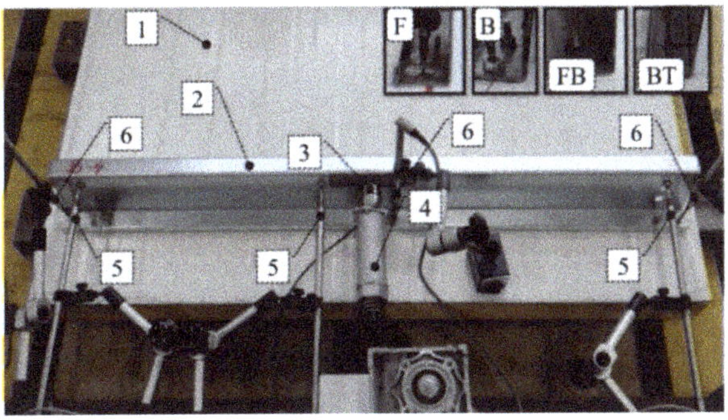

Figure 4. View of a test bed (description in text).

2.2. *Experimental Investigation*

The connection methods considered in the paper can be divided into three groups. The first group is represented by mechanical connectors that penetrate all layers of the sandwich panel: self-drilling fasteners (F) and bolts (B). The second group is represented by mechanical connectors: tightened blind rivets (FB) and pulled blind rivets (BT) that are attached only to one sandwich panel facing. Finally, the third group represents the non-penetrating connection method using continuous (TL) and pointwise (TP) tape arrangements. Each of the connection methods was investigated experimentally. The testbed setup was described in Section 2 and is depicted in Figure 4. The equilibrium load-displacement paths, obtained for each type of connection, basically represent the two-stage response: linear and nonlinear. The linear response was determined using the iterative linear regression formula. The ultimate point is defined as a point in which the derivative of the load-displacement curve is less than zero. The equilibrium load-displacement paths obtained provide information on the linear and ultimate resistance, the linear and ultimate deformation capacity, and the linear and secant stiffness, which are graphically depicted in Figure 5.

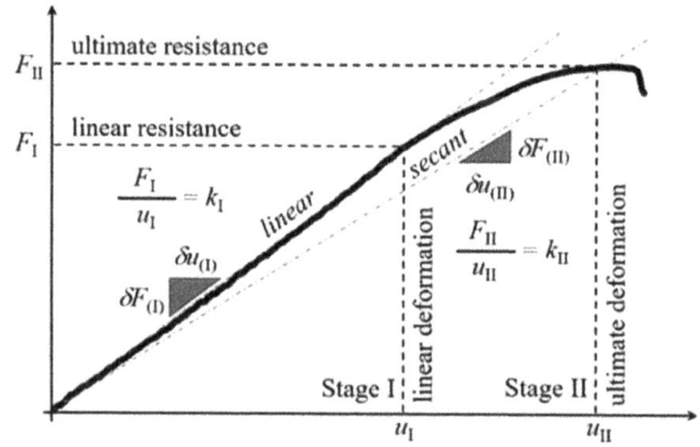

Figure 5. Connections' parameters.

3. Results

Please note that the mechanical response of penetrating connectors also consists of the initial response, which refers to the initial loading phase (up to 0.1 kN). It arises from the initial stiffness of the section when the connection is tightly connected with the sandwich panel from the beginning of the loading process (F, B). In all figures that present the equilibrium load-displacement curves from the experiments, the vertical line represents the boundary between the linear (I) and non-linear (II) stages. Additionally, in the figures mentioned, the thick black continuous lines represent the mean value of the individual results of the connector type considered. The thin grey lines represent the load-displacement paths recorded during laboratory experiments. Note that in the experiment, the horizontal force (*F*) applied in the middle length of the element (Figures 3 and 4) is transferred by the support conditions of the thin-walled element as a pair of normal forces (*N*) to the surface of the sandwich panel and horizontal forces (*S*). The shear resistance of the connectors transfers the top facing of the horizontal force to the sandwich panel. The normal force, located along the rotation line, is directed into (N_c) the sandwich panel facing while the other one, localized along with the connectors, is directed outward (N_t); see Figure 6. As presented in Figure 6, the force distribution is general for all investigated connection types.

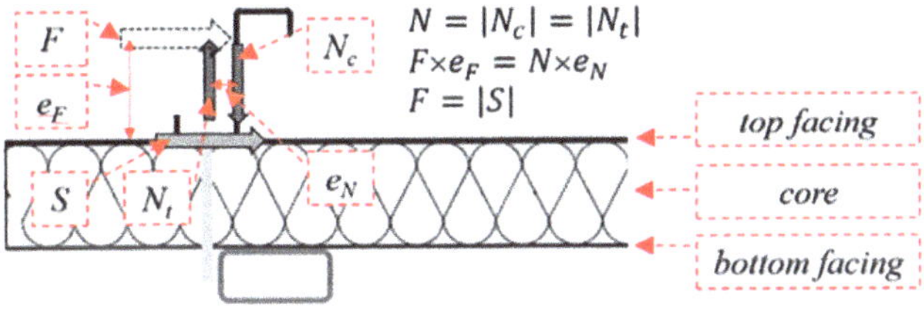

Figure 6. General scheme of the forces in the investigated connections.

3.1. Failure Scenarios and Equilibrium Load Displacement Paths

3.1.1. Connectors Penetrating Whole Sandwich Panel Depth

Self-drilling (F) fasteners are commonly used in civil engineering applications to connect thin-walled elements to sandwich panels. The use of self-drilling fasteners was also recently investigated in [26] where the test bed was used for the double lap shear test. This type of connection does not require predrilling of the hole and therefore is easy and fast to install. However, during installation, the substructure is below the sandwich panel and thus is not visible to the cladding fitter. This leads to the problem of achieving collinearity of the fasteners during sandwich panel installation; see Figure 7.

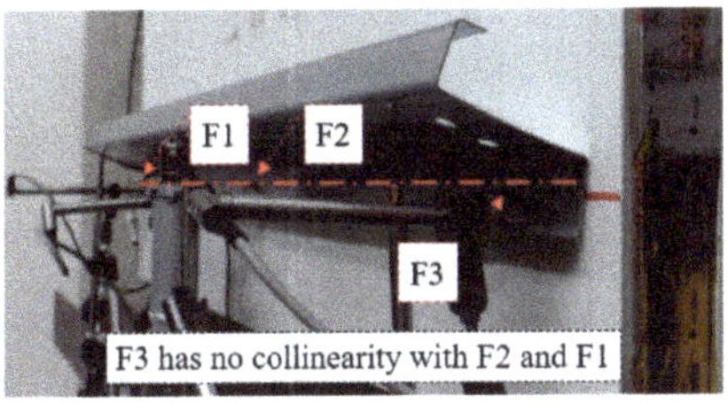

Figure 7. Lack of the collinearity of the self-drilling fasteners (F).

In Figure 8a, the equilibrium load-displacement paths of self-drilling fasteners (F) are presented. It was observed during the experiments, that up to $F \approx 0.05$ kN the load is compensated by the stiffness of the cross-section. It is manifested as an elastic bow deformation of the web along with its depth without cross-section rotation; see Figure 8b. With increasing load, the section starts to rotate. The rotation line is located on the fold between the web and the bottom flange, see Figure 8c. The observed web bow of the cantilever-buckling shape was kept to the end of the linear part of the equilibrium path (stage I). Thus, the initial angle between the bottom flange and the web remains constant during the linear stage (I).

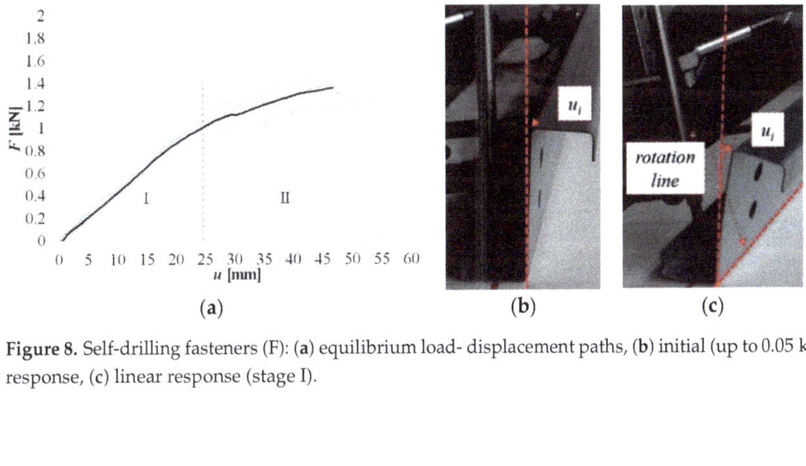

Figure 8. Self-drilling fasteners (F): (**a**) equilibrium load- displacement paths, (**b**) initial (up to 0.05 kN) response, (**c**) linear response (stage I).

An additional increase in the load (stage II—non-linear behaviour) increases the deformation of the web bow and the section rotation. Rotation of the cross section results in an indentation of the bottom flange in the vicinity of the fasteners (see Figure 9) and its slip over the fastener thread. Additionally, the cross-section rotation is compensated by the bottom flange bow deformation depicted in Figure 9. It should be noted that the position of the connectors with respect to the cross-sectional web is important in preventing rotation. The closer the fasteners are to the web, the higher the cross-sectional rotation is observed.

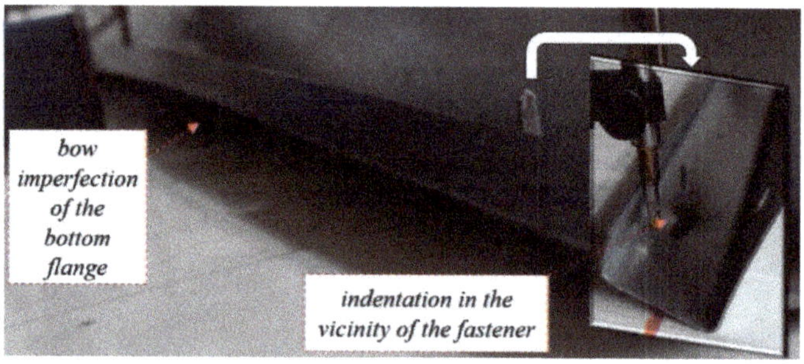

Figure 9. Failures in the nonlinear stage of the kinematical response of the connection (stage II) using self-drilling fasteners (F).

Figure 10a shows the equilibrium load-displacement paths of the connection between the thin-walled beam and sandwich panel made with the **bolts** with a nut and two steel washers. This connection requires pre-drilling of all layers of the sandwich panel (both facings and core) and the bottom flange of the thin-walled section. Thus, the potential problem of collinearity is easier to eliminate. It is worth noticing that this connection is characterized by a similar kinematical response to the self-drilling fasteners, i.e., the same stages can be distinguished. That is, it was observed that initially (up to $F \approx 0.1$ kN) the section compensates the load by an elastic bow deformation of the web along with its depth without cross-section rotation; see Figure 10b.

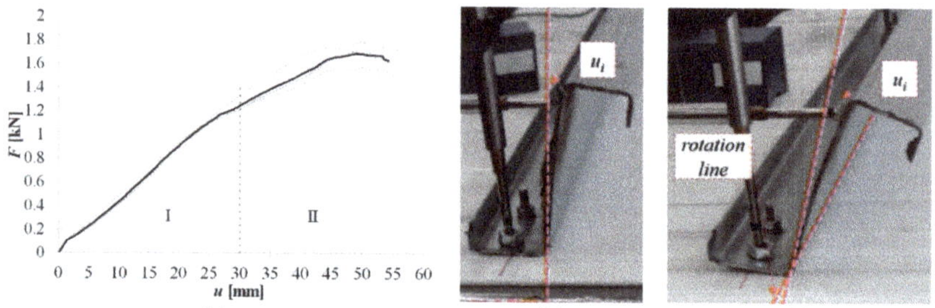

(a) (b) (c)

Figure 10. Self-drilling fasteners (F): (**a**) equilibrium load-displacement paths, (**b**) initial (up to 0.05 kN) response, (**c**) linear response (stage I).

An increase in the load (above stage I) leads to an increase in the rotation of the cross section. The use of bolts with nuts prevents the bottom flange of the beam from being separated from the top facing of the sandwich panel. However, at stage II, the bow deformation of the bottom flange is also observed; see Figure 11. Additionally, the washer between the nut and the bottom flange prevents local indentation in the bottom cross section flange, which was observed in the case of self-drilling fasteners. However, the uplift force N_t (see Figure 6) causes the indentation of the bottom facing in the vicinity of the head and the unscrewing of the nut, see Figure 11.

Figure 11. Failure mechanisms in stage II in connection using bolts: indentation at the bottom facing and bow deformation of the bottom flange.

3.1.2. Connectors Attached to One Sandwich Panel Facing—Blind Rivets

The installation of both types of blind rivets considered requires predrilling of the holes. Thus, similarly to the bolts, the problem with their collinearity is easier to eliminate. After predrilling, the rivets are inserted into the holes, and the bottom flange of the cross section and one sandwich panel facing are tied. Note that in this case, the core layer is not penetrated; thus, no point thermal bridges are created. The diameter of the tightened and pulled blind rivets, as well as their length, varies, and thus, their results are presented separately. However, general conclusions can be drawn, namely, that the blind rivet clamping arm significantly influences connection strength, deformation capacity, and failure mechanisms. Figure 12a shows the equilibrium load-displacement paths of the connection realized by the **tightened blind rivets**. The initial stiffness of the cross section, which occurred with fasteners and bolts, was not observed in this connection. It means that from the beginning, the rotation of the cross section and the web bow deformation occurred; see Figure 12b.

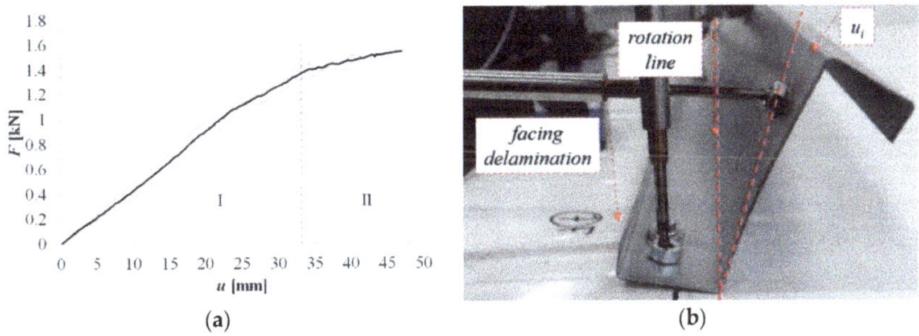

Figure 12. Tightened blind rivet: (**a**) equilibrium load-displacement paths, (**b**) connection behaviour in stage I.

Furthermore, facing delamination was observed on the extension of the blind rivet line (perpendicular to the length of the thin-walled beam). The further load increase increases cross-section rotation and facing delamination that reaches the edge of the sandwich panel, see Figure 13.

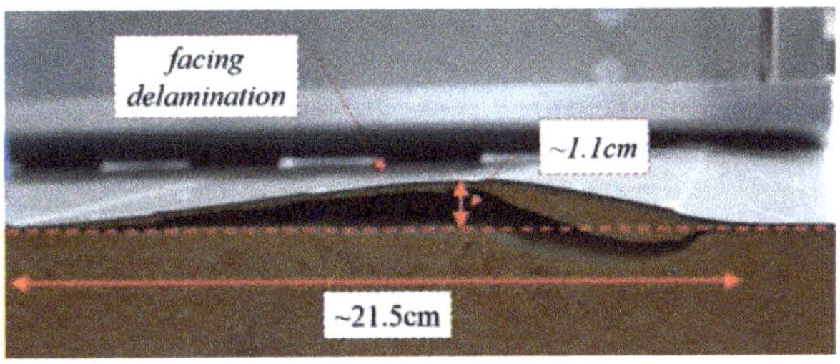

Figure 13. Tightened blind rivet–facing delamination in stage II.

Figure 14a shows the equilibrium load-displacement paths of the connection realized by **pulled blind rivets**. In the case of these connectors, it was observed that the bow deformation of the cross-sectional web was negligible, see Figure 14b. The deformation of the connection is manifested by the cross-sectional rotation which, in stage II, leads to the punching shear failure. The punching shear failure mechanism is connected to the plastic deformation of the clamping arms of the blind rivet.

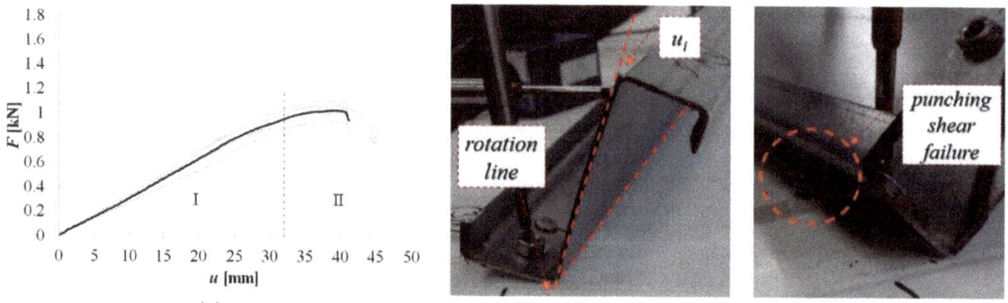

(a) (b) (c)

Figure 14. Pulled blind rivet: (**a**) equilibrium load-displacement paths, (**b**) cross-section rotation at stage I, (**c**) punching shear failure at stage II.

3.1.3. Non-Penetrating Connectors

The last method of connection is performed with double-sided acrylic tape. In Figure 15a,b the equilibrium load-displacement paths of the pointwise (TP) and continuous tape (TL) arrangement, respectively. Please note that the ultimate strength, stiffness, and deformation capacity of the connection achieved with the use of tape applied pointwise is significantly lower than with the use of tape applied tape.

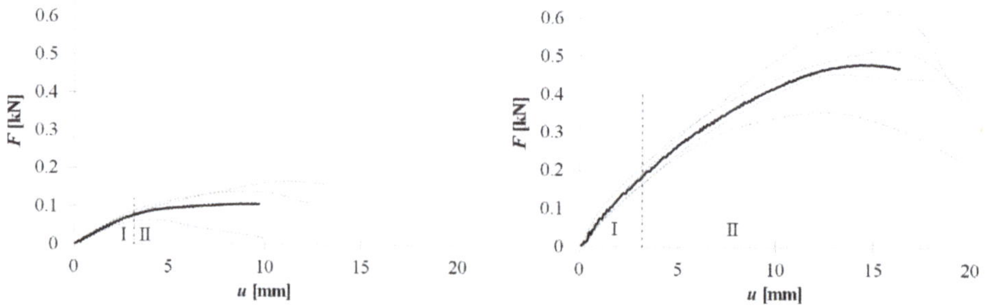

(a) (b)

Figure 15. Equilibrium load-displacement paths of non-penetrating connectors: (**a**) pointwise tape (TP), (**b**) continuous tape (TL).

In the case of pointwise tape and continuous tape, the cross section remains straight during stage I and stage II, that is, no bow deformation of the cross section web was observed; see Figure 16a,b, respectively. It was due to the very small capacity of the connection. In this case, only the tape detachment governed the load-displacement behaviour of the connection.

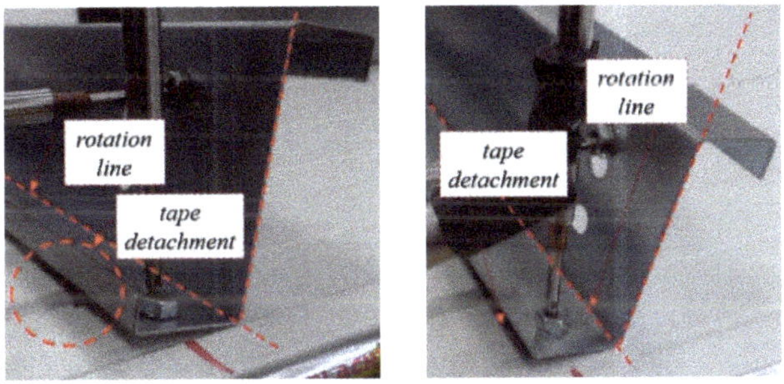

(a) (b)

Figure 16. Cross-section rotation without bow web deformation: (**a**) pointwise tape (TP), (**b**) continuous tape (TL).

4. Discussion

In Table 1, data collected from all tests are presented. These data represent the load-horizontal displacement relation of the middle point located in the beam web. The parameters are illustrated graphically in Figure 5. Nevertheless, for the record in Table 1, the following parameters are listed:

- F_I and F_{II} represent linear and ultimate resistance, respectively;
- u_I and u_{II} represent linear and ultimate deformation capacities, respectively;
- k_I and k_{II} represent linear and secant stiffness, respectively;
- A_I and A_{II} represent the area below the curve for linear and nonlinear part, respectively.

Table 1. Data from laboratory test.

Parameter	Unit	F	B	BT	FB	TL	TP
F_I	[kN]	1.02 ± 0.06	1.25 ± 0.06	0.95 ± 0.07	1.37 ± 0.05	0.18 ± 0.02	0.08 ± 0.01
u_I	[mm]	24.63 ± 3.73	30.03 ± 1.44	32.08 ± 2.20	33.13 ± 1.83	3.20 ± 0.17	3.20 ± 0.07
k_I	[kN/m]	41.3	41.6	29.9	41.3	57.5	23.9
A_I	[kNm]	0.0135	0.0201	0.0161	0.0240	0.0003	0.0001
F_{II}	[kN]	1.35 ± 0.09	1.68 ± 0.09	1.01 ± 0.07	1.52 ± 0.05	0.47 ± 0.12	0.11 ± 0.05
u_{II}	[mm]	46.18 ± 3.64	48.16 ± 1.86	40.75 ± 0.80	43.47 ± 2.37	16.32 ± 2.00	9.72 ± 4.11
k_{II}	[kN/m]	29.3	34.9	24.9	34.9	28.9	10.9
A_{II}	[kNm]	0.0396	0.0471	0.0248	0.0390	0.0054	0.0008
n	[-]	5	4	4	4	4	3

Resistances and horizontal deformations represent the mean values for each group of connectors (n represents the statistical sample size). Please note that, in the case of nonmechanical connectors (tapes), the scatter of the results is significant; therefore, they should not be considered in this type of usage.

In Table 1, AI and AII represent the area below the linear and nonlinear parts of the load-displacement curves. These areas can be interpreted as the strain energy capacity of the connection for its linear and nonlinear mechanical response. In Figure 17, the mean representations of each connection type are plotted. One can observe that the bolt connection (B) and tightened blind rivets connection (FB) are characterized by exact behaviour in stage one (linear response) and comparable in stage 2 (nonlinear response). The typically used self-drilling fasteners (F) are characterized by the same linear stiffness as bolts (B) and tightened blind rivets (FB) but are also characterized by 20% lower ultimate resistance and secant stiffness. Nevertheless, the deformation capacity of bolts (B), tightened blind rivets (FB), and self-drilling fasteners (F) is comparable. The mechanical response of the pulled blind rivet (BT) connection is significantly smaller than the response of other mechanical connectors. This is because its clamping arm's length and diameter are smaller than in the case of the tightened blind rivet. This leads to a failure by pulling the rivets from the facing. Non-mechanical connectors—tapes—are characterized by their lowest resistance and deformation capacity.

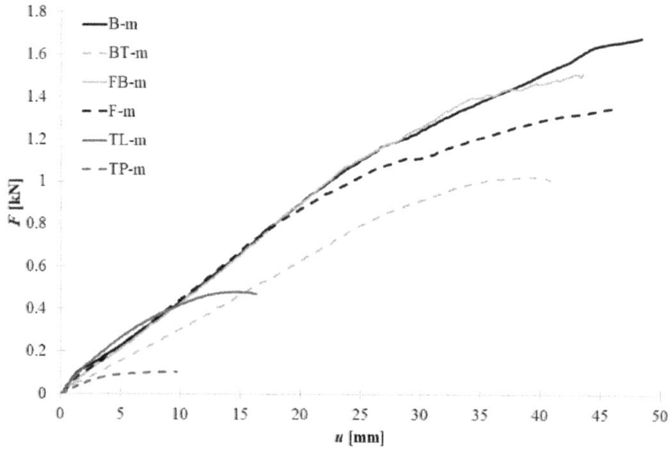

Figure 17. Mean equilibrium load-displacement paths of the various method of connection between thin-walled beam and sandwich panel.

Due to the fact that self-drilling fasteners are used in the case of civil engineering applications, the results for this connector were considered as reference results for the other methods of connecting sandwich panels with thin-walled beams. In Figure 18a,b, the changes of three selected parameters for the linear and non-linear stages are depicted, respectively. This comparison sustains the conclusions defined above, i.e., that the bolts (B) and tightened blind rivets (FB) are characterized by higher resistance and deformation capacity than self-drilling fasteners.

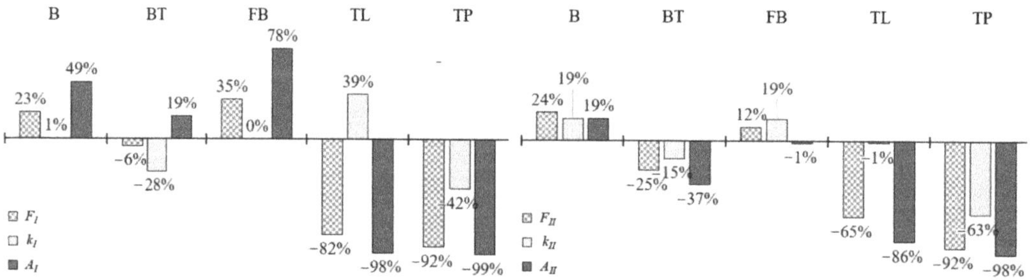

(a) (b)

Figure 18. Change of the resistance (F_i), stiffness (k_i), and strain energy capacity (A_i) concerning the self-drilling fasteners (F): (**a**) linear stage, (**b**) non-linear stage.

5. Conclusions

The laboratory results presented in the paper, give quantitative information on the possible usage of the different connectors to connect the thin-walled members to sandwich panels. These results can be considered as a starting point when full-scale laboratory tests are being prepared in the area of stabilization of the thin-walled members by sandwich panels. The research was limited to a core layer material, i.e., polyisocyanurate foam core (PIR foam). However, in the case of the through-drilling connectors (fasteners, bolts) and the tapes, the failure mechanisms were not connected with the mechanical properties of the core layer, but rather with its thickness. Therefore, the results of these types of connectors could be extended to other core layer materials, for example, styrofoam or mineral wool.

It cannot be assumed only for blind rivets because the observed failure mechanism was related to the core layer delamination. In that case, further considerations could be directed to the use of analytical solutions described in [31,32].

Author Contributions: Conceptualization, R.S. and K.C.; methodology, R.S. and K.C.; software, K.C.; validation, R.S. and K.C.; formal analysis, K.C.; investigation, R.S.; resources, R.S. and K.C.; data curation, K.C.; writing—original draft preparation, R.S.; writing—review and editing, K.C.; visualization, R.S. and K.C.; supervision, R.S.; project administration, R.S. and K.C.; funding acquisition, R.S. All authors have read and agreed to the published version of the manuscript.

Funding: The research was financially supported by Poznan University of Technology Grant no. 0412/SBAD/0050 and 0412/SBAD/0060.

Institutional Review Board Statement: Not applicable.

Informed Consent Statement: Not applicable.

Data Availability Statement: Not applicable.

Acknowledgments: We would like to kindly acknowledge support from the Inter Institute Laboratory of Building Structures especially to Justyna Grzymisławska, Marek Kałużny and Roman Dąbrowski.

Conflicts of Interest: The authors declare no conflict of interest.

References

1. Hassinen, P.; Misiek, T.; Naujoks, B. Cladding systems for sandwich panels–Refurbishment of walls and roof. In Proceedings of the Eurosteel 2011, Budapest, Hungary, 31 August–2 September 2011; pp. 2199–2204.
2. *EN 14509:2007*; Self-Supporting Double Skin Metal Faced Insulating Panels–Factory Made Products–Specifications. iTeh, Inc.: Newark, DE, USA, 2007.
3. *European Recommendations for the Determination of Loads and Actions on Sandwich Panels, International Council for Research and Innovation in Building and Construction*; CIB: Kanata, ON, Canada, 2015.
4. Vălean, C.; Şoşdean, C.; Marşavina, L.; Linula, E. Mechanical characterization of lightweight foam-based sandwich panels. *Mater. Today Proc.* **2021**, *45*, 4166–4170. [CrossRef]
5. Faidzi, M.K.; Abdullah, S.; Abdullah, M.F.; Azman, A.H.; Hui, D.; Singh, S.S.K. Review of current trends for metal-based sandwich panel: Failure mechanisms and their contribution factors. *Eng. Fail. Anal.* **2021**, *123*, 105302. [CrossRef]
6. Allen, H.G. *Analysis and Design of Structural Sandwich Panels*; Pergamon Press: Oxford, UK, 1969.
7. Zenkert, D. *An Introduction to Sandwich Construction*; Engineering Materials Advisory Services Ltd.: Worcester, UK, 1997.
8. Birman, V.; Kardomateas, G.A. Review of current trends in research and applications of sandwich structures. *Compos. Part B* **2018**, *142*, 221–240. [CrossRef]
9. Chuda-Kowalska, M.; Garstecki, A. Experimental study of anisotropic behavior of PU foam used in sandwich panels. *Steel Compos. Struct.* **2016**, *20*, 43–56. [CrossRef]
10. Pozorski, Z.; Pozorska, J.; Kreja, I.; Smakosz, Ł. On Wrinkling in Sandwich Panels with an Orthotropic Core. *Materials* **2021**, *14*, 5043. [CrossRef]
11. *EN 1993-1-3:2006*; Eurocode 3: Design of Steel Structures—Part 1–3: General Rules—Supplementary Rules for Cold-Formed Members and Sheeting. CEN European Committee for Standardization: Brussels, Belgium, 2006.
12. *EN 1990:2002*; Eurocode—Basis of Structural Design. iTeh, Inc.: Newark, DE, USA, 2002.
13. *European Recommendations on the Stabilization of Steel Structures by Sandwich Panels, No. 135*, 1st ed.; ECCS–European Convention for Constructional Steelwork: Brussels, Belgium, 2014; ISBN 978-92-9147-118-8.
14. *prEN 14509-2:2016*; Double Skin Metal Faced Insulating Panels. Factory Made Products. Specifications. Part 2: Structural Applications–Fixings and Potential Uses of Stabilization of Individual Structural Elements. 2019; *not published*.
15. Sokół, L. Stability of Cold Formed Purlins Braced by Steel Sheeting. *Thin-Walled Struct.* **1996**, *25*, 247–268. [CrossRef]
16. Li, L.Y. Lateral-torsional buckling of cold-formed zet purlins partial laterally restrained by metal sheeting. *Thin-Walled Struct.* **2004**, *42*, 995–1011. [CrossRef]
17. Dürr, M. Die Stabilisierung Biegedrillknickgefährdeter Träger Durch Sandwichelemente und Trapezbleche. Ph.D. Dissertation, Fakultät für Bauingenieur-, Geo- und Umweltwissenschaften, Universität Fridericiana zu Karlsruhe (TH), Karlsruhe, Germany, 2008.
18. Ciesielczyk, K.; Studziński, R. Experimental and numerical investigation of stabilization of thin-walled Z-beams by sandwich panels. *J. Constr. Steel Res.* **2017**, *133*, 77–83. [CrossRef]
19. Górski, M.; Kozłowski, A. Behaviour of hot-rolled purlins connected with sandwich panels. *Arch. Civ. Eng.* **2021**, *67*, 249–267. [CrossRef]

20. Chybiński, M.; Polus, Ł. Experimental and numerical investigations of aluminium-timber composite beams with bolted connections. *Structures* **2021**, *34*, 1942–1960. [CrossRef]
21. Chybiński, M.; Polus, Ł. Mechanical Behaviour of Aluminium-Timber Composite Connections with Screws and Toothed Plates. *Materials* **2022**, *15*, 68. [CrossRef] [PubMed]
22. Balázs, I.; Melcher, J.; Belica, A. Experimental Investigation of Torsional Restraint Provided to Thin-walled Purlins by Sandwich Panels under Uplift Load. *Procedia Eng.* **2016**, *161*, 818–824. [CrossRef]
23. Balazs, I.; Melcher, J.; Karmazinova, M.; Belica, A.; Oly, R.; Misiek, T. Experimental setups for the measurement of the rotational restraint provided to cold-formed Z-purlins by sandwich panels. *Proc. Eurosteel* **2017**, *2017*, 1677–1686. [CrossRef]
24. Yang, H.; Wang, H.; Qian, H.; Jin, X.; Chen, D.; He, Y.; Li, Q.; Fan, F. Mechanical performance of pre-engineered closed section beam-column connections in CFS frames: Experimental investigation. *Structures* **2022**, *39*, 164–174. [CrossRef]
25. Wang, H.; Yang, H.; Qian, H.; Chen, D.; Jin, X.; Fan, F. Static experimental analysis and optimization of innovative pre-engineered tubular section beam-column connections in cold-form steel frames. *J. Build. Eng.* **2022**, *48*, 103989. [CrossRef]
26. Ciesielczyk, K.; Studziński, R. Influence of type and orientation of thin-walled beams on the interaction effectiveness with sandwich panel. In Proceedings of the Modern Trends in Research on Steel, Aluminium and Composite Structures: The XIV International Conference on Metal Structures (ICMS2021), Poznan, Poland, 16–18 June 2021; pp. 208–214. [CrossRef]
27. Studziński, R.; Pozorski, Z.; Garstecki, A. Structural behavior of sandwich panels with asymmetrical boundary conditions. *J. Constr. Steel Res.* **2015**, *104*, 227–234. [CrossRef]
28. Studziński, R.; Ciesielczyk, K. Connection stiffness between thin-walled beam and sandwich panel. *J. Sandw. Struct. Mater.* **2017**, *21*, 2042–2056. [CrossRef]
29. Staszak, N.; Gajewski, T.; Garbowski, T. Shell-to-Beam Numerical Homogenization of 3D Thin-Walled Perforated Beams. *Materials* **2022**, *15*, 1827. [CrossRef] [PubMed]
30. Gajewski, T.; Staszak, N.; Garbowski, T. Parametric Optimization of Thin-Walled 3D Beams with Perforation Based on Homogenization and Soft Computing. *Materials* **2022**, *15*, 2520. [CrossRef]
31. Sahmani, S.; Fattahi, A.M.; Ahmed, N.A. Size-dependent nonlinear forced oscillation of self-assembled nanotubules based on the nonlocal strain gradient beam model. *J. Braz. Soc. Mech. Sci. Eng.* **2019**, *41*, 239. [CrossRef]
32. Sahmani, S.; Fattahi, A.M.; Ahmed, N.A. Develop a refined truncated cubic lattice structure for nonlinear large-amplitude vibrations of micro/nano-beams made of nanoporous materials. *Eng. Comput.* **2020**, *36*, 359–375. [CrossRef]

materials

MDPI

Review

Review on Stress-Fractional Plasticity Models

Pengfei Qu [1,2], Yifei Sun [2] and Wojciech Sumelka [3,*]

1 School of Management Science and Engineering, Shandong Technology and Business University, Yantai 264005, China
2 Key Laboratory of Geomechanics and Embankment Engineering of Ministry of Education, Hohai University, Nanjing 210098, China
3 Institute of Structural Analysis, Poznan University of Technology, Piotrowo 5, 60-965 Poznan, Poland
* Correspondence: wojciech.sumelka@put.poznan.pl

Abstract: Fractional calculus plays an increasingly important role in mechanics research. This review investigates the progress of an interdisciplinary approach, fractional plasticity (FP), based on fractional derivative and classic plasticity since FP was proposed as an efficient alternative to modelling state-dependent nonassociativity without an additional plastic potential function. Firstly, the stress length scale (SLS) is defined to conduct fractional differential, which influences the direction and intensity of the nonassociated flow of geomaterials owing to the integral definition of the fractional operator. Based on the role of SLS, two branches of FP, respectively considering the past stress and future reference critical state can be developed. Merits and demerits of these approaches are then discussed, which leads to the definition of the third branch of FP, by considering the influences of both past and future stress states. In addition, some specific cases and potential applications of the third branch can be realised when specific SLS are adopted.

Keywords: fractional derivative; fractional plasticity; nonassociated; state dependence

Citation: Qu, P.; Sun, Y.; Sumelka, W. Review on Stress-Fractional Plasticity Models. *Materials* **2022**, *15*, 7802. https://doi.org/10.3390/ma15217802

Academic Editors: Tomasz Garbowski and Aleksander Marek

Received: 6 October 2022
Accepted: 1 November 2022
Published: 4 November 2022

1. Introduction

Geomaterials, such as clay, sand, ballast and rock, are often encountered or used in practical engineering [1,2]. Before the designing and construction of infrastructure, site investigation was usually carried out, to have a full understanding of the mechanical properties of the underlying geomaterial. It was found that the constitutive responses of geomaterials were state-dependent and non-associated, due to the frictional nature [3–6]. The associated plasticity developed for metals could not be simply employed for modelling the stress and strain behaviour of geomaterials [7,8]. Instead, the non-associated plasticity within the framework of critical state soil mechanics was often suggested [9]. However, the classic non-associated plastic models required the incorporation of an additional plastic potential function and a state parameter, to capture the state-dependent non-associated behaviour of geomaterials, which inevitably resulted in the complexity of the developed model. Recently, nonconventional mechanical approaches using fractional calculus [10–13] have attracted increasing attention. Inspired by the fractional viscoplasticity (FVP) originally proposed by Sumelka [14,15], Sun and Sumelka [16], Lu et al. [17,18] and Qu [19,20] developed a series of fractional plasticity (FP) models, to solve this limitation. Without using an additional plastic potential function, the developed approach can be used to characterise the state-dependent non-associated stress-dilatancy behaviour of geomaterials.

In this study, a comprehensive introduction to the development and application of the FP for geomaterials will be provided, in terms of the role of SLS for carrying out the fractional differentiation. Three branches of the FP will be proposed and discussed. This study is intended to provide potential guidance for those who have an interest in this research branch of stress-fractional mechanics.

2. Progress in FP

The FP was originally established by incorporating the stress-fractional operator into the associated plasticity. It was inspired by the pioneering work of Sumelka [14,15] on the FVP. Nevertheless, these two types of research have differences with regard to the initial definition of the stress-fractional operator. According to Sumelka et al. [21,22], the FVP was developed based on the 'short memory principle', where the close virtual neighbourhood of a stress state (σ'_{ij}) (at a material point of interest) influences the fractional viscoplastic strain ($d\varepsilon_{ij}^{vp}$) direction of the material, such that:

$$d\varepsilon_{ij}^{vp} = \Lambda \frac{{}^{RC}_{a}D^{\alpha}_{b} f\left(\sigma'_{ij}\right)}{\left\| {}^{RC}_{a}D^{\alpha}_{b} f\left(\sigma'_{ij}\right) \right\|} \tag{1}$$

where $i, j = 1, 2, 3$; Λ is the intensity of viscoplastic flow (provided as a material function, as in original Perzyna [23] approach); D indicates partial differential; f is yielding function; a and b denote the close virtual neighbourhood of a stress state (σ'_{ij}); α is the fractional-order, with $\alpha \in (n-1, n]$ and n the positive integer; $\| \ \|$ indicates the norm of a tensor. The superscript, RC, denotes the Riesz–Caputo fractional derivative, where in FVP it was defined by using the 'short memory principle' as:

$${}^{RC}_{a}D^{\alpha}_{b} f\left(\sigma'_{ij}\right) = \frac{1}{2}\left[{}^{C}_{a}D^{\alpha}_{\sigma'_{ij}} f\left(\sigma'_{ij}\right) + (-1)^n\, {}^{C}_{\sigma'_{ij}}D^{\alpha}_{b} f\left(\sigma'_{ij}\right)\right] \tag{2}$$

in which the superscript, C, indicates the Caputo fractional derivative; the definition of the Caputo fractional derivative can be found in the Appendix A. Note that ${}^{C}_{a}D^{\alpha}_{\sigma'_{ij}} f\left(\sigma'_{ij}\right)$ is the left-sided fractional derivative, whereas ${}^{C}_{\sigma'_{ij}}D^{\alpha}_{b} f\left(\sigma'_{ij}\right)$ is the right-sided fractional derivative.

Compared with FVP, the FP was developed based on the 'long memory principle', where the initial stress onset (σ'_0) or the targeted future stress $\left(\sigma'_{cij}\right)$ influences the fractional plastic flow of the material at the current stress state. Thus, it is defined as:

$$d\varepsilon_{ij}^{p} = d\lambda\, {}^{C,RL}_{\sigma'_0}D^{\alpha}_{\sigma'_{ij}} f\left(\sigma'_{ij}\right), \quad \sigma'_{ij} > \sigma'_0 \tag{3}$$

or

$$\begin{cases} d\varepsilon_{ij}^{p} = d\lambda\, {}^{C,RL}_{\sigma'_{ij}}D^{\alpha}_{\sigma'_{cij}} f\left(\sigma'_{ij}\right), & \sigma'_{cij} > \sigma'_{ij} \\ d\varepsilon_{ij}^{p} = d\lambda\, {}^{C,RL}_{\sigma'_{cij}}D^{\alpha}_{\sigma'_{ij}} f\left(\sigma'_{ij}\right), & \sigma'_{ij} > \sigma'_{cij} \end{cases} \tag{4}$$

where the superscript (C, RL) indicates Caputo fractional derivative or Riemann–Liouville's fractional derivatives; $d\lambda$ is the non-negative plastic multiplier; σ'_0 and σ'_{cij} are the integral limits. Equation (3) indicates the first type of the FP, here denoted as FP-n (n stands for nonassociated), which was adopted by researchers [17,19,24] for capturing the nonassociated plastic flow of granular soil, while Equation (4) indicates the second type of FP, here denoted as FP-sn (sn stands for state-dependent nonassociated), which was defined in [25] for modelling the state-dependent nonassociated behaviour of granular soil. It is noted that the FP-n based on Equation (3) assumes that the past loading history $\left(\sigma'_0 \to \sigma'_{ij}\right)$ plays a role in the nonassociated plastic flow of geomaterial; the FP-sn based on Equation (4) assumes an effect of the future reference critical state, i.e., the distance $\left(\sigma'_{ij} - \sigma'_{cij}\right)$ from the current stress state σ'_{ij} to the corresponding future reference critical state σ'_{cij}, on the current plastic flow direction of geomaterial. Note that the future reference critical state is a state which can be reached by soils subjected to sufficient shearing. This state was experimentally and numerically evidenced in many reported researches [3,4,6,7,26–28], and characterized by the critical state lines in the mean effective stress v.s. deviator stress space and the mean effective stress v.s. void ratio space. The current state is moving along the yield curve until reaching the critical state line. Although the FP was developed based on using fractional derivatives with power-law kernel, it can be also developed by other definitions using, for example, the exponential kernel, as long as it has analytical solutions of the yielding function. However, no matter which definition is used, the basic constitutive relation for FP-n and FP-sn should be the same.

Reformulating Equations (3) and (4), one can have a unified description for the FP as:

$$d\varepsilon_{ij}^{P} = d\lambda \frac{\partial^{\alpha} f\left(\sigma'_{ij}, \hbar_{ij}\right)}{\partial \sigma'^{\alpha}_{ij}} \tag{5}$$

where $\hbar_{ij}$ denotes the hardening variable of the yielding function. Then, one needs to determine $d\lambda$ for model application. Through applying the consistency condition at the yielding surface, one has:

$$df = \frac{\partial f\left(\sigma'_{kl}, \hbar_{kl}\right)}{\partial \sigma'_{kl}} d\sigma'_{kl} + \frac{\partial f\left(\sigma'_{kl}, \hbar_{kl}\right)}{\partial \hbar_{kl}} d\hbar_{kl} = 0 \tag{6}$$

where the hardening variable $d\hbar_{kl} = \frac{\partial \hbar_{kl}}{\partial \varepsilon^{P}_{ab}} d\varepsilon^{P}_{ab}$. Substituting Equation (5) into Equation (6), one has:

$$d\lambda = -\frac{\frac{\partial f\left(\sigma'_{kl}, \hbar_{kl}\right)}{\partial \sigma'_{kl}} d\sigma'_{kl}}{\frac{\partial f\left(\sigma'_{kl}, \hbar\right)}{\partial \hbar_{kl}} \frac{\partial \hbar_{kl}}{\partial \varepsilon^{P}_{ab}} \frac{\partial^{\alpha} f\left(\sigma'_{ab}, \hbar_{ab}\right)}{\partial \sigma'^{\alpha}_{ab}}} \tag{7}$$

Then, substituting Equation (7) into Equation (5), one has the following constitutive relation for the FP:

$$d\varepsilon^{P}_{ij} = \frac{1}{H} n_{ij} m_{kl} d\sigma'_{kl} \tag{8}$$

where the hardening modulus (H), plastic flow tensor (n_{ij}), and plastic loading tensor (m_{kl}) can be derived as:

$$H = \frac{-\frac{\partial f\left(\sigma'_{kl}, \hbar_{kl}\right)}{\partial \hbar_{kl}} \frac{\partial \hbar_{kl}}{\partial \varepsilon^{P}_{ab}} \frac{\partial^{\alpha} f\left(\sigma'_{ab}, \hbar_{ab}\right)}{\partial \sigma'^{\alpha}_{ab}}}{\left\| \frac{\partial^{\alpha} f\left(\sigma'_{rs}, \hbar_{rs}\right)}{\partial \sigma'^{\alpha}_{rs}} \right\| \left\| \frac{\partial f\left(\sigma'_{ct}, \hbar_{ct}\right)}{\partial \sigma'_{ct}} \right\|} \tag{9}$$

$$n_{ij} = \frac{\frac{\partial^{\alpha} f\left(\sigma'_{ij}, \hbar_{ij}\right)}{\partial \sigma'^{\alpha}_{ij}}}{\left\| \frac{\partial^{\alpha} f\left(\sigma'_{rs}, \hbar_{rs}\right)}{\partial \sigma'^{\alpha}_{rs}} \right\|} \tag{10}$$

$$m_{kl} = \frac{\frac{\partial f\left(\sigma'_{kl}, \hbar_{kl}\right)}{\partial \sigma'_{kl}}}{\left\| \frac{\partial f\left(\sigma'_{ct}, \hbar_{ct}\right)}{\partial \sigma'_{ct}} \right\|} \tag{11}$$

Figure 1 modified from [29] schematically shows the plastic flow and loading directions calculated using Equations (10) and (11), where a deviation of the plastic flow direction from the plastic loading direction can be observed, indicating a nonassociated plastic flow rule in the developed FP. Based on Equation (8), a series of FP models were developed for the constitutive descriptions of different geomaterials. Depending on the definition of the adopted fractional operator, these FP models can be categorised into two branches, i.e., the ones considering the role of 'past' SLS [18,20,24] and the others considering the role of 'future' critical-state SLS [25]. These two branches will be respectively introduced in the next two sections.

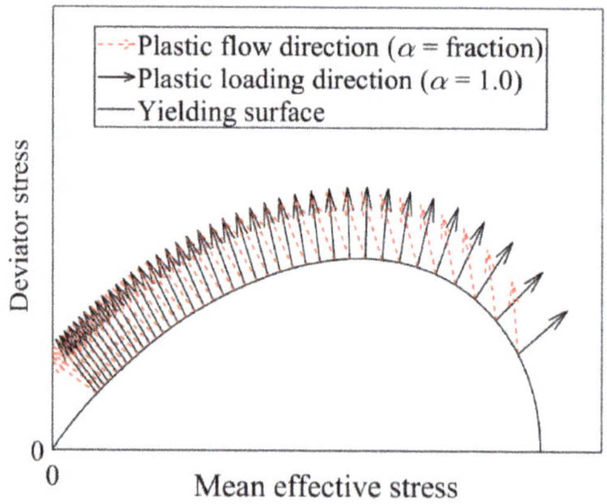

Figure 1. Schematic show of the loading and plastic flow directions.

2.1. FP-n: The Role of Past SLS

2.1.1. Modelling of Soils

As indicated by Equation (3), compared with the previous works [30,31], the SLS is characterized by the length from current stress state to the reference stress state. Sun et al. [32] pointed out that in geotechnical engineering if one took the consolidation pressure instead of the zero-stress state as the initial stress state (σ_0'), the developed model could predict a higher strain. However, this prediction difference could be reduced by tuning the values of some model parameters. For the sake of simplicity, σ_0' = 0 kPa was thus assumed for model derivation in most cases, cf. [18–20,33–35]. Through this assumption, the developed fractional plastic flow rule can be simple and yet flexible in constitutive modelling.

Specifically speaking, to capture the stress-dilatancy behaviour of granular soil, the following fractional-order dilatancy ratio (d_g) based on the Modified Cam-clay (MCC) function was proposed [33]:

$$d_g = \frac{{}_0D^{\alpha}_{p'}(p')}{{}_0D^{\alpha}_{q}f(q)} = \frac{M^2 - (1-\alpha/2)(\eta^2 + M^2)}{\eta^{2-\alpha}} \tag{12}$$

where $p' = 1/3\sigma'_{ij}\delta_{ij}$ and $q = \sqrt{3/2\left(\sigma'_{ij} - p'\delta_{ij}\right)\left(\sigma'_{ij} - p'\delta_{ij}\right)}$, are the mean effective stress and deviatoric stress, respectively; δ_{ij} is the Kronecker delta; M and η denote the critical-state and current-state stress ratios, respectively. Unlike other critical state parameters, the critical-state stress ratio (M) can be influenced by many factors, e.g., the particle shape [27], but it should not be affected by fines content [26,36] or shearing mode [28,37]. The effect of α on the stress-dilatancy behaviour of granular soil can be observed in Figure 2. It is found that with the increase of α, the dilatancy ratio at the same level of stress ratio increases, while with the increase of the stress ratio (η), the dilatancy ratio at the same α decreases.

It is worthwhile to mention that Equation (12) does not consider the dependence of stress-dilatancy on the material state in its current form unless a state-dependent parameter is introduced. However, state-dependent stress-dilatancy is a common phenomenon in granular soils, e.g., sand and rockfill, where the stress-dilatancy behaviour is determined by not only the current stress state but also the material state (i.e., void ratio, e, and pressure, p'). To consider this, an empirical correlation of the fractional-order with the state parameter $\psi(= e - e_c)$ can be suggested, such that

$$\alpha = \exp(-\langle -\Delta\psi \rangle) \tag{13}$$

where Δ is a material parameter; e_c is the void ratio at the critical state; $\langle\ \rangle$ is Macauley brackets. Based on Equation (13), the state-dependent stress-dilatancy or plastic flow direction can be captured. There are two chances for Equation (12) to be equal to zero: one is at the phase transformation state where $d_g = 0$, the other is at the critical state where $\eta = M, \psi = 0$ and $\alpha = 1$, which ensures that Equation (12) conforms to the basic restrictions of the Critical State Soil Mechanics (CSSM) [38].

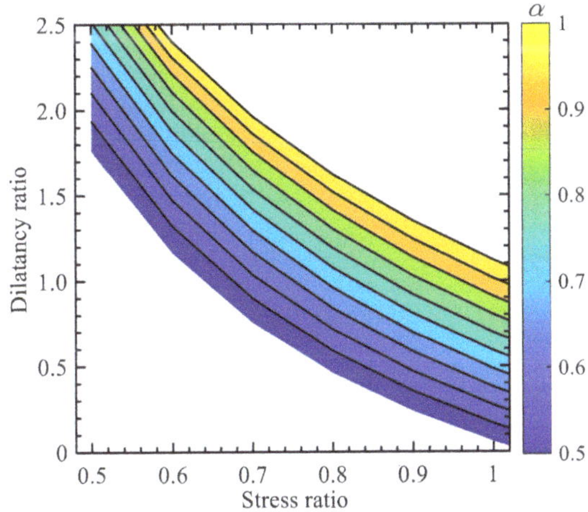

Figure 2. Effect of α on the stress-dilatancy relation.

Note that if a constant α was used, Equation (12) would either overestimate or underestimate the critical-state strength of the material, because, at the critical state, $d_g = 0$, such that Equation (12) can be solved as:

$$\eta_c = \sqrt{\frac{\alpha}{2-\alpha}}M \tag{14}$$

where η_c indicates the calculated critical-state stress ratio. In critical state soil mechanics, $\eta_c = M$ should always exist at the critical state. However, this can only be true if $\alpha = 1$ at the critical state or one uses a different constant instead of M in Equation (12). The latter option was introduced by Liang, et al. [39] and Lu, et al. [17] to consider the effect of multiaxial loading on soft soils, e.g., clay, where they developed the multiaxial stress-dilatancy relation by using a well-established characteristic stress concept, such that Equation (12) can be reformulated as [40]:

$$d_g = \frac{{}_0D^{\alpha}_{\tilde{p}'} f(\tilde{p}')}{{}_0D^{\alpha}_{\tilde{q}} f(\tilde{q})} = \frac{N^2 - (1-\alpha/2)(\tilde{\eta}^2 + N^2)}{\tilde{\eta}^{2-\alpha}} \tag{15}$$

where $\tilde{p}'$, $\tilde{q}$ and $\tilde{\eta}$ are the characteristic stress components of p', q and η; N is a material parameter, different from M in the original FP model. Then, $\eta_c = M$ at the critical state can be guaranteed via a constant fractional-order shown in Equation (14), i.e., $\eta_c = \sqrt{\alpha/(2-\alpha)}N$. To consider the dependence of dilatancy on material state, a dependence of d_g on the state parameter, for example, the relation in Equation (13) may be further introduced. However, there is one other option: that is to use the stress ratio at the phase transformation state, i.e., M_{pt}, to determine the fractional order as suggested by Liang, et al. [40]. As suggested by Nguyen and his coworkers [26–28,36,37], the phase-transformation-state parameter (M_{pt}) and strain hardening parameter (M_p) are also a function of M and the state parameter.

Despite the above successful applications, the FP-n models based on MCC function usually predicted much higher dilatancy for sand at the same stress level when compared to the corresponding test data [41]. This can be attributed to the larger elastic region of the MCC yielding surface at the 'dry' side (Figure 3 modified from [42]) of the critical state line in the $p' - q$ plane [43]. For modelling sand, the original Cam-clay (CC) function [38] with reduced elastic region seems to work better. A new fractional stress-dilatancy relation based on the original CC function can be proposed by using the RL definition:

$$d_g = \{[f_d(2) - f_d(2-\alpha)]M - \eta\}|\eta|^{\alpha-1}\frac{1}{\alpha} + \left(\frac{1}{\alpha} - 1\right)|\eta|^{\alpha} \tag{16}$$

where f_d denotes the digamma function, which can be defined as $f_d = D^1(\ln\Gamma)$, with Γ the gamma function. It is easily found that $d_g = 0$ at the critical state. However, at the phase transformation state, $d_g = 0$ will result in a much complex condition for determining the fractional order from laboratory test data. For example, iteration should be required for parameter identification. Therefore, from the perspective of practical application, one may ask if a simplified version of the CC-based fractional dialtancy equation can be suggested, which can lead to a much easier way, i.e., directly measuring from test data, to determine the fractional order. In fact, during model development, the RL derivatives of constants can be omitted due to its limited influence on the dilatancy equation [44]. Because such influence can be compensated through further calibration of model parameters, e.g., the fractional order. Thus, a modified fractional stress-dilatancy relation for granular soil and soil-structure interface can be derived as:

$$d_g = \{[f_d(2) - f_d(2-\alpha)]M - \eta\}|\eta|^{\alpha-1} \tag{17}$$

In addition, one can also derive Equation (17) by using the Caputo definition, as shown in [45]. Through such simplification, the RL definition and Caputo definition can lead to the same expression of d_g. To consider the state dependence, the fractional order can be also correlated to the state parameter via Equation (13). Equations (16) and (17) conform to the CSSM, as $d_g = 0$ at both the phase transformation state and critical state.

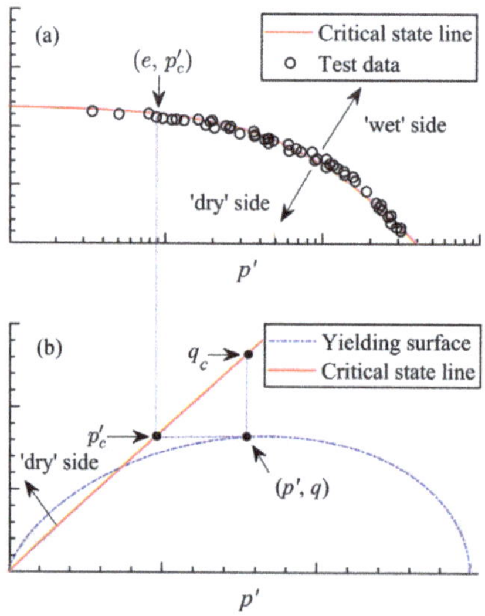

Figure 3. Relative position between current state and critical state in the (**a**) $e - p'$ plane and (**b**) $p' - q$ plane (data cited from Verdugo and Ishihara (1996)).

2.1.2. Modelling of Rocks

In addition to the application of FP-n in modelling granular or soft soils, several attempts have been also made to capture the stress-strain behaviours of rocks [19,20,46,47] and rock-like materials [18]. In these applications, different constitutive models with a fractional plastic flow rule were proposed based on the diverse problems that were focused on. For the purpose of describing the volumetric compression/dilation transition phenomenon of soft and hard rocks, Qu et al. [19] developed an elastoplastic model with fractional-order plastic flow where a unified hardening/sofening function κ_{p} was proposed as follows:

$$\kappa_{\mathrm{p}} = \kappa_{\mathrm{p}}^{0} + (1-\kappa_{\mathrm{p}}^{0})\frac{\Pi\xi}{\Pi+\xi^{\Pi}-1}, \quad \xi = \gamma^{\mathrm{p}}/\gamma_{\mathrm{c}}^{\mathrm{p}} \tag{18}$$

with

$$\gamma^{\mathrm{p}} = \int\sqrt{\frac{2}{3}\mathrm{d}e_{ij}^{\mathrm{p}}\mathrm{d}e_{ij}^{\mathrm{p}}}, \quad e_{ij}^{\mathrm{p}} = \varepsilon_{ij}^{\mathrm{p}} - \frac{1}{3}\mathrm{tr}(\varepsilon^{\mathrm{p}})\delta_{ij} \tag{19}$$

in which $\varepsilon_{ij}^{\mathrm{p}}$ denotes the plastic part of total strain ε_{ij}; γ^{p} is equivalent plastic shear strain; $\gamma_{\mathrm{c}}^{\mathrm{p}}$ indicates the generalized plastic shear strain at peak stress; $\Pi > 1$ represents the model parameter; κ_{p}^{0} means the initial value corresponding to $\gamma^{\mathrm{p}} = 0$. Moreover, the maximum value $\kappa_{\mathrm{p}} = 1$ is obtained at the critical state $\gamma^{\mathrm{p}} = \gamma_{\mathrm{c}}^{\mathrm{p}}$. To calibrate the fractional order α, Qu et al. [19] derived the formulation:

$$-\frac{\partial f}{\partial\sigma_{ij}}\mathrm{D}_{ijkl}\mathrm{d}\varepsilon_{kl}A\kappa_{p}p^{1-\alpha} = \left(\frac{\partial f}{\partial\sigma_{ij}}\mathrm{D}_{ijmn}\frac{\partial^{\alpha} f}{\partial\sigma_{mn}^{\alpha}} - \frac{\partial f}{\partial\kappa_{p}}\frac{\partial\kappa_{p}}{\partial\gamma_{p}}\frac{\partial^{\alpha} f}{\partial q^{\alpha}}\right)\delta_{rs}\mathrm{D}_{rskl}^{-1}\mathrm{d}\varepsilon_{kl}\Gamma(2-\alpha) \tag{20}$$

where A defines the friction coefficient; D_{ijkl} is the fourth-order elasticity tensor; p and q are the mean stress and deviatoric stress, respectively. In the process of determining α, compressibility/dilation boundary $\mathrm{d}\varepsilon_{v} = 0$ of claystone subjected to conventional triaxial compression tests was employed and plotted in Figure 4.

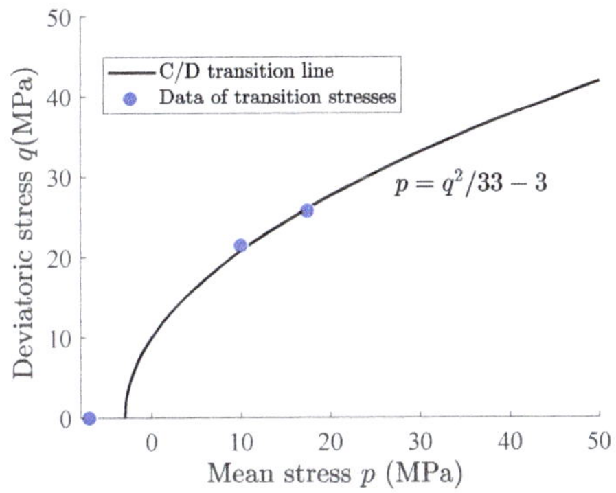

Figure 4. Compressibility/dilation (C/D) boundary of claystone subjected to conventional triaxial compression.

Note that the influence of micro-crack growth on plastic volume was not considered in [19]. Aiming to provide a new insight for investigating the complicated effect of plastic flow direction on damage evolution, Qu and Zhu [48] take the following damage evolution function $\mathcal{G}_{\mathrm{d}}$:

$$\mathcal{G}_{\mathrm{d}}(\varepsilon^{\mathrm{p}}, d) = d - d_{\mathrm{c}}\left[1-\exp\left(v\varepsilon_{\mathrm{v}}^{\mathrm{p}}\right)\right] = 0 \tag{21}$$

with d_{c} being the asymptotic damage value in the residual stage, and v indicates the material parameter controlling the velocity of the damage growth. Note that the variation of plastic volumetric strain $\varepsilon_{\mathrm{v}}^{\mathrm{p}}$ is related to the fractional order α as demonstrated in the following relation:

$$\mathrm{d}\varepsilon_{\mathrm{v}}^{\mathrm{p}} = \Lambda\frac{\partial^{\alpha} f}{\partial p^{\alpha}} \tag{22}$$

with p denoting the mean stress. As such, the numerical simulation of Beishan granite subjected to the confining pressure of 10 MPa is displayed in Figure 5. It can be observed from Figure 5 that the developed fractional plastic damage model has the potential to reproduce the damage evolution under the loading process. Moreover, it can be found from [48] that the fractional order α plays a critical role in the damage growth. To further account for the influence of the fractional plastic flow on the micromechanics for quasi-brittle rocks, a friction criterion regarding local stresses was adopted as follows [20]:

$$f(\sigma^c) = \|s^c\| - \tilde{A}p^c \leq 0 \tag{23}$$

with

$$s^c = s - \frac{1}{\gamma_2 \omega} 2\mu^m \Gamma, \; p^c = p + \frac{1}{\gamma_1 \omega} k^m \beta \tag{24}$$

where p^c and s^c denote the hydrostatic part and the deviatoric part of the local stress σ^c, respectively; $\tilde{A}$ is the generalized friction coefficient; s and p means the macroscopic deviatoric stress; k^m and μ^m represent the bulk and shear moduli of the matrix, respectively; ω indicates microscopic damage internal variable; Γ and $\beta = \varepsilon^p : \delta$ describe the relative slip degree between microcrack surfaces and the degree of microcrack's opening, respectively. γ_1 and γ_2 associated with the Poisson's ratio of the solid matrix ν^m and can be written as:

$$\gamma_1 = \frac{16}{9}\frac{1-(\nu^m)^2}{1-2\nu^m}, \; \gamma_2 = \frac{32}{45}\frac{(1-\nu^m)(5-\nu^m)}{2-\nu^m} \tag{25}$$

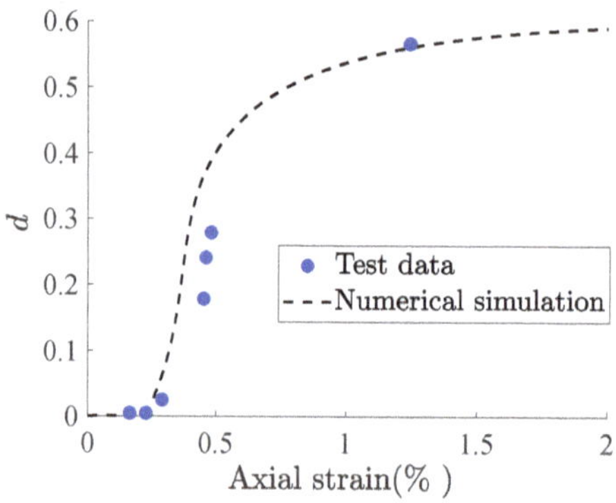

Figure 5. Numerical simulation of damage evolution versus axial strain under triaxial compression of (σ_3 = 10 MPa).

Based on Equation (23), the yield surface in the local stress space can be given in Figure 6, which is a conical surface with the diagonal of the space as the axis.

Figure 7 shows the influence of the fractional order α on plasctic flow with the case of $\eta = 1$ where dotted arrows represent the orthogonal direction and the solid arrows denote fractional plastic flow direction. It can be observed from Figure 7 that the fractional order brings a significant influence on the plasctic flow direction, especially under the high hydrostatic pressure. In Figure 7, the decrease of α results in a larger deviation from the orthogonal direction in the case of $\alpha < 1$. In Figure 7, the deviation from the loading direction is larger with an increase of α in the case of $\alpha > 1$. Note that when $\alpha = 1$, the fractional plastic flow direction degenerates to the classical associated plastic flow as shown in Figure 7. Hence, it is found that the change of the fractional order can capture plastic flow direction more flexibly without the additional plastic potentical.

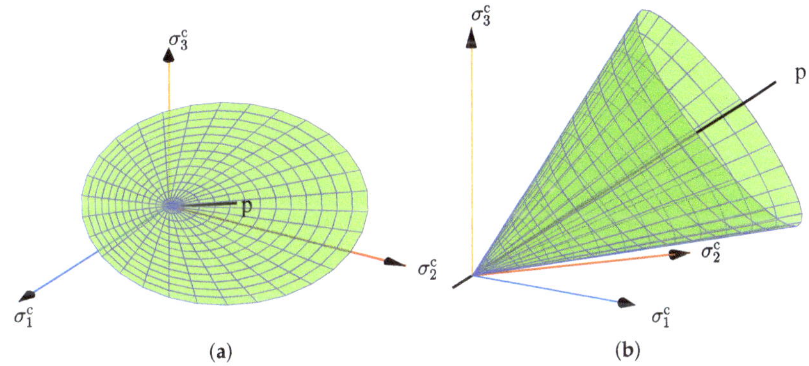

Figure 6. Yield surface in the local stress space: (**a**) front view, (**b**) lateral view.

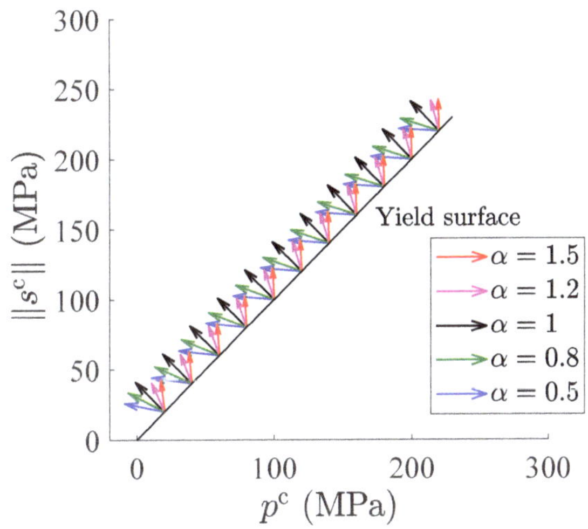

Figure 7. Influence of the fractional order on the plastic flow direction in $p^c - \|s^c\|$ plane with $\eta = 1$.

In [20], comparisons between test data and simulation results on Beishan granite under the confining pressure of $\sigma_3 = 0, 5, 10$ and 20 MPa are displayed in Figure 8. Numerical results of the fractional model are in good agreement with test data. By comparing the traditional associated model and the fractional model, it can be found that the fractional model have better performance on reproducing the main features of mechanical behaviors of Beishan granite, especially in the softening phase.

In addition, Li et al. [46] established a fractional constitutive model of soft rock considering temperature effect where model parameter m related to dilatancy characteristics was introduced. In this model, the relation between the fractional order α and similarity factor R is given by:

$$\alpha = \frac{2R^{2m}}{1 + R^{2m}} \tag{26}$$

Based on the microstructure of porous matrix-inclusion, Shen et al. [47] developed an elastoplastic damage constitutive model with a fractional plastic flow where the yield criterion can be applied as follows:

$$f = \frac{\frac{1+2\ell/3}{\tilde{A}^2} + \frac{2}{3}\rho\left(\frac{3\ell}{2\tilde{A}^2} - 1\right)}{\frac{4\tilde{A}^2-12\ell-9}{6\tilde{A}^2-13\ell-6}\rho + 1} q^2 + \left(\frac{3\ell}{2\tilde{A}^2} - 1\right)p^2 + 2(1-\ell)hp - \frac{3+2\ell+3\ell\rho}{3+2\ell}(1-\ell)^2h^2 = 0 \tag{27}$$

where ℓ and ρ represent the volume fraction of pores and the volume fraction of inclusions, respectively; h is the hydrostatic tensile strength. This study [47] shows that when considering the material microstructure information including the porosity, the inclusion and the solid phase, the introduction of the fractional plasticity is still effective.

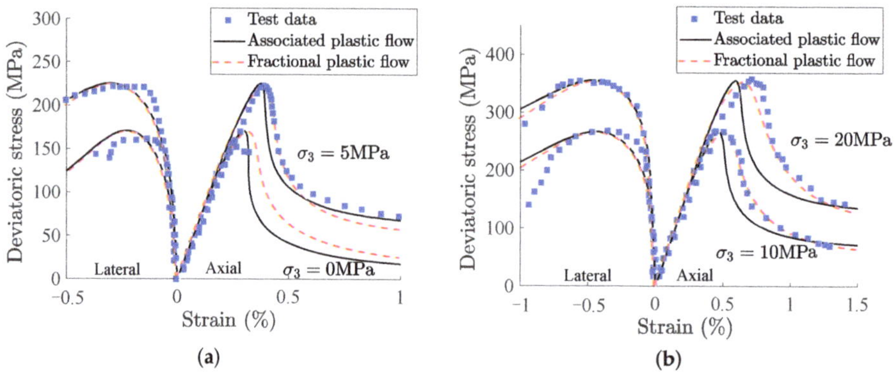

Figure 8. Comparisons between test data and the model predictions of Beishan granite under triaxial compression tests with different confining pressures: (**a**) σ_3 = 0 and 5 MPa, (**b**) σ_3 = 10 and 20 MPa.

To better simulate the direction and magnitude of $d\varepsilon_{ij}^{p}$ for rock-like material, i.e., concrete, Lu et al. [18] proposed a three-dimensional fractional elastoplastic constitutive model in which the expression of fractional plastic flow direction is as follows:

$$\mathbf{n} = \left[(\tilde{\mathbf{n}}^{\alpha})^T : \left(\frac{\partial\tilde{p}}{\partial\tilde{\sigma}}, \frac{\partial\tilde{q}}{\partial\tilde{\sigma}}, \frac{\partial\tilde{\theta}}{\partial\tilde{\sigma}}\right)^T : \frac{\partial\tilde{\sigma}}{\partial\sigma}\right]^T \tag{28}$$

where $\tilde{\sigma}$ is the transformed stress tensor; $\tilde{p}$, $\tilde{q}$ and $\tilde{\theta}$ are the hydrostatic pressure, the deviatoric stress and the Lode angle in the transformed stress space, respectively. The fractional gradient of yield function $\tilde{\mathbf{n}}^{\alpha}$ in the transformed stress space can be expressed as:

$$\tilde{\mathbf{n}}^{\alpha} = \left(\frac{\partial^{\alpha_1} f}{\partial\tilde{p}^{\alpha_1}}, \frac{\partial^{\alpha_2} f}{\partial\tilde{q}^{\alpha_2}}, \frac{\partial^{\alpha_3} f}{\partial\tilde{\theta}^{\alpha_3}}\right)^T = \left(\frac{\partial^{\alpha} f}{\partial\tilde{p}^{\alpha}}, \frac{\partial^{\alpha} f}{\partial\tilde{q}^{\alpha}}, \frac{\partial^{\alpha} f}{\partial\tilde{\theta}^{\alpha}}\right)^T \tag{29}$$

in which $\alpha_1 = \alpha_2 = \alpha_3 = \alpha$ for the simplification of the developed model. Finally, the corresponding stress-dilatancy relationship can be obtained, such that:

$$d_g = -\frac{\left(\frac{\partial^{\alpha} f}{\partial\tilde{p}^{\alpha}}, \frac{\partial^{\alpha} f}{\partial\tilde{q}^{\alpha}}, \frac{\partial^{\alpha} f}{\partial\tilde{\theta}^{\alpha}}\right)\left(\frac{\partial\tilde{p}}{\partial q}, \frac{\partial\tilde{q}}{\partial q}, \frac{\partial\tilde{\theta}}{\partial q}\right)^T}{\left(\frac{\partial^{\alpha} f}{\partial\tilde{p}^{\alpha}}, \frac{\partial^{\alpha} \tilde{f}}{\partial\tilde{q}^{\alpha}}, \frac{\partial^{\alpha} f}{\partial\tilde{\theta}^{\alpha}}\right)\left(\frac{\partial\tilde{p}}{\partial p}, \frac{\partial\tilde{q}}{\partial p}, \frac{\partial\tilde{\theta}}{\partial p}\right)^T} \tag{30}$$

Together with damage feature of concrete materical, a 3D non-orthogonal plastic damage model is developed in [49] where α can be obtained based on the following equation of phase transformation:

$$\mathrm{d}\varepsilon_{\mathrm{v}}^{\mathrm{p}} = \mathrm{d}\lambda \tilde{p}^{-\alpha}\left[\frac{a_0}{\Gamma(1-\alpha)} + \frac{a_1\Gamma(2)\tilde{p}}{\Gamma(2-\alpha)} + \frac{a_2\Gamma(3)\tilde{p}^2}{\Gamma(3-\alpha)} + \frac{a_3\Gamma(4)\tilde{p}^3}{\Gamma(4-\alpha)} + \frac{a_4\Gamma(5)\tilde{p}^4}{\Gamma(5-\alpha)}\right] = 0 \quad (31)$$

where $a_k(k = 0,1,2,3,4)$ are coefficients of power functions for $\tilde{p}$. Subsequently, Lu et al. [50] developed a cohesion-friction combined hardening plastic model of concrete based on the fractional flow rule. Moreover, this model is implemented with the help of an open-source user defined material subroutine UMAT in the framework of the implicit return mapping algorithm.

2.1.3. Numerical Schemes

Integration algorithms significantly influence computation accuracy and efficiency in the process of the implementation of constitutive equations. For the fractional model as presented in [49], the Next Increment Corrects Error [51] approach were adopted where the workflow of the NICE algorithm can be summarized in Algorithm 1. In this Algorithm, n and $n+1$ are the current step and the previous step; $\overline{\sigma}$ is the effective stress; $\overline{\sigma}^{\mathrm{trial}}$ denotes the trial stress; $\mathbf{D}_0$ indicates the undamaged elastic stiffness matrix; $\mathbf{r}$ refers to the direction of plastic strain increment.

Algorithm 1: Flowchart of the NICE algorithm for the fractional model

Input: $\varepsilon_n, \varepsilon_n^{\mathrm{p}}, \gamma_n^p, \Delta\varepsilon, \sigma_n$
Output: $\varepsilon_{n+1}, \varepsilon_{n+1}^{\mathrm{p}}, \gamma_{n+1}^p, \sigma_{n+1}$
Elastic prediction:
$d_n = d\left(\gamma_n^p\right), \overline{\sigma}_n = \frac{\sigma_n}{(1-d_n)}, \Delta\overline{\sigma}^{\mathrm{trial}} = \mathbf{D}_0 : \Delta\varepsilon, \overline{\sigma}_{n+1}^{\mathrm{trial}} = \overline{\sigma}_n + \Delta\overline{\sigma}^{\mathrm{trial}}$, and
$f_{n+1}^{\mathrm{trial}} = f\left(\overline{\sigma}_{n+1}^{\mathrm{trial}}, \gamma_n^{\mathrm{p}}\right)$
if $f_{n+1}^{\mathrm{trial}} > 0$ **then**
  $f_n = f\left(\overline{\sigma}_n, \gamma_n^{\mathrm{p}}\right), \Xi = \frac{f_n}{f_{\bar{\sigma},n}:\mathbf{D}_0:\Delta\varepsilon}$
  $\Delta\lambda_n^{\mathrm{NICE}} = \frac{(1+\Xi)f_{\bar{\sigma},n}:\mathbf{D}_0:\Delta\varepsilon}{f_{\sigma,n}:\mathbf{D}_0:\mathbf{r}_n - f_{\gamma^p}:f_{q,n}^{\alpha}}$
  $\mathbf{D}_0^{\mathrm{ep}} = \mathbf{D}_0 - \frac{(1+\Xi)(\mathbf{D}_0:\mathbf{r}_n)\otimes(f_{œ,n}:\mathbf{D}_0)}{f_{œ,n}:\mathbf{D}_0:\mathbf{r}_n - f_{\gamma^p}:f_{\tilde{q},n}^{\alpha}}$
  $\varepsilon_{n+1} = \varepsilon_n + \Delta\varepsilon, \varepsilon_{n+1}^{\mathrm{p}} = \varepsilon_n^{\mathrm{p}} + \Delta\lambda_n^{\mathrm{NICE}}\mathbf{r}_n$
  $\gamma_{n+1}^p = \gamma_n^{\mathrm{p}} + \Delta\lambda_n^{\mathrm{NICE}} f_{\tilde{q},n}^{\alpha}, \overline{\sigma}_{n+1} = \overline{\sigma}_n + \mathbf{D}^{\mathrm{ep}} : \Delta\varepsilon$
  $d_{n+1} = d\left(\gamma_{n+1}^p\right), \sigma_{n+1} = (1-d_{n+1})\overline{\sigma}_{n+1}$
else
  $\varepsilon_{n+1} = \varepsilon_n + \Delta\varepsilon, \varepsilon_{n+1}^{\mathrm{p}} = \varepsilon_n^{\mathrm{p}}, \gamma_{n+1}^{\mathrm{p}} = \gamma_n^{\mathrm{p}}$
  $\overline{\sigma}_{n+1} = \overline{\sigma}_{n+1}^{\mathrm{trial}}, \sigma_{n+1} = (1-d_n)\overline{\sigma}_{n+1}$

In addition, Qu and Zhu [48] proposed a semi-implicit return mapping (SRM) algorithm for the implementation of a novel fractional plastic damage model as illustrated in Figure 9. Aiming to more efficiently conduct a micromechanics-based fractional frictional damage model, an explicit return mapping algorithm was put forward in [20] and is given in Figure 10. Furthermore, it can be found that the numerical solutions are consistent with the analytical ones when increment step is enough large. Compared to the plasticity-damage decoupling correction (PDDC) algorithm proposed by [52], the explicit return mapping algorithm has a better performance in computational efficiency.

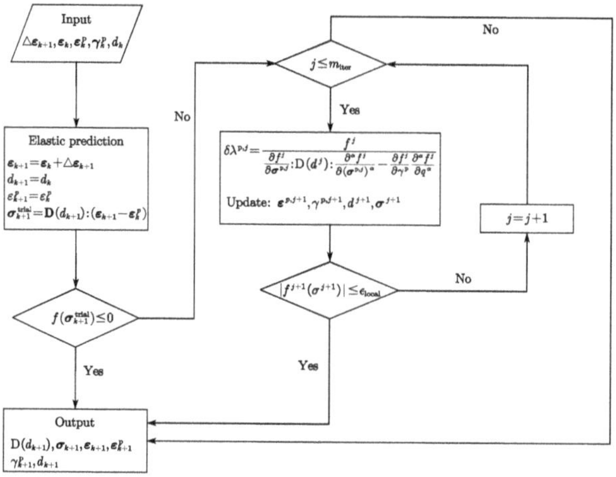

Figure 9. Flowchart of SRM algorithm.

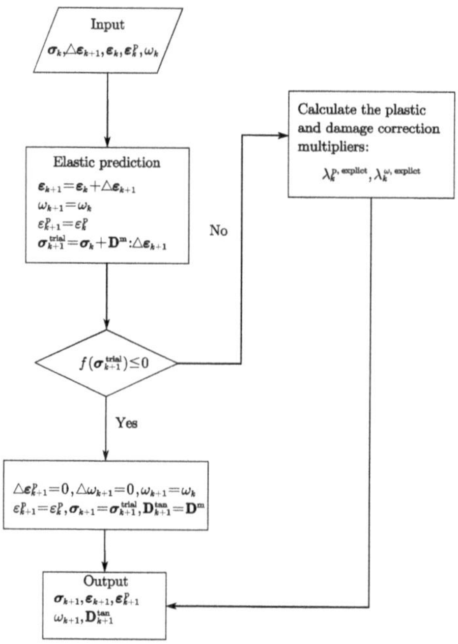

Figure 10. Flowchart of the explicit return mapping algorithm.

2.2. *FP-sn: The Role of Future Reference Critical State*

In this section, an introduction of the FP-sn models based on Equation (4) is made. It was observed in experimental tests that the volumetric dilatancy of soils, e.g., sand and over-consolidated clay, depends on not only the current state (e, p') but also the distance $(e - e_c$ or $p' - p'_c)$ from current state to future reference critical state (e_c, p'_c).

After revisiting the CSSM, one can find that soils under shearing would finally reach the critical state represented by the critical-state void ratio (e_c), mean effective stress (p'_c) and deviator stress (q_c). Here, $p'_c = 1/3\sigma'_{cij}\delta_{ij}$ and $q_c = \sqrt{3/2\left(\sigma'_{cij} - p'_c\delta_{ij}\right)\left(\sigma'_{cij} - p'_c\delta_{ij}\right)}$. Then, it can be assumed that the future critical-state stresses (p'_c, q_c) can serve as the integral limit (σ'_{cij}) in Equation (4). Substituting the MCC function into Equation (4) with RL and Caputo derivatives, one can obtain the following state-dependent stress-dilatancy relations for soil:

$$d'_g = \frac{{}^{RL}_{p'_c}D^{\alpha}_{p'} f(p')}{{}^{RL}_{q}D^{\alpha}_{q_c} f(q)} = \frac{{}^{RL}_{p'}D^{\alpha}_{p'_c} f(p')}{{}^{RL}_{q_c}D^{\alpha}_{q} f(q)} = |M|^{1+\alpha} \frac{\left(p' - p'_c\right) + (2-\alpha)\left(p'_c - p'_0/2\right) + \delta_p}{(q - q_c) + (2-\alpha)q_c + \delta_q} \tag{32}$$

$$d''_g = \frac{{}^{C}_{p'_c}D^{\alpha}_{p'} f(p')}{{}^{C}_{q}D^{\alpha}_{q_c} f(q)} = \frac{{}^{C}_{p'}D^{\alpha}_{p'_c} f(p')}{{}^{C}_{q_c}D^{\alpha}_{q} f(q)} = |M|^{1+\alpha} \frac{(p' - p'_c) + (2-\alpha)(p'_c - p'_0/2)}{(q - q_c) + (2-\alpha)q_c} \tag{33}$$

where $p'_0 = \left[(\eta/M)^2 + 1\right]p'$, is the size of the MCC yielding surface; $\delta_p = \left[p'_0 - \left(p' + p'_c\right)\right](2-\alpha)(1-\alpha)/2$ and $\delta_q = (q + q_c)(2-\alpha)(\alpha-1)/2$. Comparison between Equations (32) and (33) shows that there appears two additional items, i.e., δ_p and δ_q, when using the RL definition. However, further analysis shown in Figure 11a can show that the influence of such two items on soil dialtancy can be compensated by tuning the value of fractional order. A very small difference between d'_g with δ_p and δ_q and d''_g without δ_p and δ_q can be observed in Figure 11b, if a proper fractional order is used. Therefore, for practical application, the contributions from δ_p and δ_q were not considered through the omission of RL derivatives of constants. For the sake of simplicity, a unified d_g is thus suggested, such that:

$$d_g = |M|^{1+\alpha} \frac{(p' - p'_c) + (2-\alpha)(p'_c - p'_0/2)}{(q - q_c) + (2-\alpha)q_c} \tag{34}$$

which also facilitates the calibration of model parameters directly from laboratory test data.

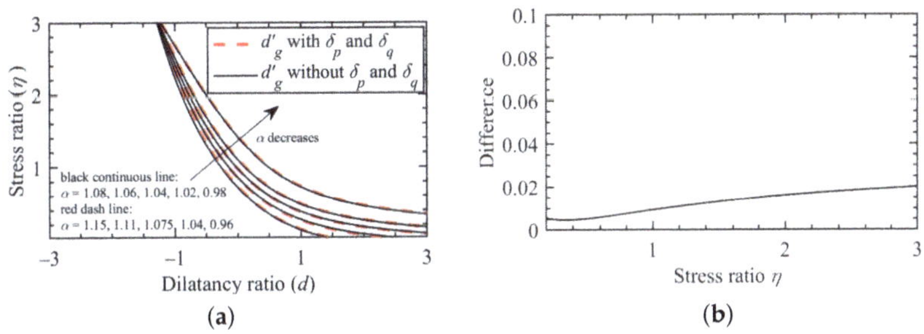

Figure 11. Predicted dilatancy ratios with and without δ_p and δ_q : (**a**) dilatancy line, (**b**) mean difference between predicted $d'g$ with and without δ_p and δ_q.

Moreover, the critical-state deviator stress (q_c) in Equations (32) and (33) can be calculated by checking the geometric position of the current stress and critical-state stress shown in Figure 3, such that:

$$q_c = q + M(p' - p'_c) \tag{35}$$

while the critical-state mean effective pressure can be calculated using the critical state line shown in Figure 3a, such that:

$$p'_c = g(e) \tag{36}$$

where $g(e)$ is a function describing the critical state line of soil in the $e - p'$ plane. $g(e)$ is determined by fitting the critical-state data points. There are different available formulae for $g(e)$, but no matter which formula is used, a unique relation with $p_c^{'}$ can be provided. For example, the $g(e)$ for Toyoura sand [41] shown in Figure 3a can be expressed as:

$$p_c' = p_r \exp\left(\frac{e_\Gamma - e}{\lambda}\right) - p_s \tag{37}$$

where p_r = 1 kPa, is the unit pressure for normalisation; e_Γ and λ are material parameters; p_s is the shift stress, describing the effect of particle breakage on the downward bending of the critical state line shown in Figure 3a.

It can be found from Equation (34) that d_g also has two chances to reach zero: one is at the phase transformation state with a typical value of the fractional, the other is at the critical state with $p^{'} = p_c^{'}$ and $q = q_c$. However, unlike the FP-n based on past stress history and other classic state-dependent models [8,53,54], Equation (34) does not require an additional empirical state parameter, e.g., ψ, to capture the state dependence stress-dilatancy of soil, which is the main advantage of the FP-sn approach.

By considering the effect of the future reference critical state, a series of FP-sn models for modelling the state-dependent strength and deformation behaviour of granular soil and over-consolidated soft soil. Despite the positive model performance, there is still one problem with the FP-sn based on Equation (32): comparatively higher volumetric dilatancy of granular soil could be predicted due to the utilisation of the MCC function. As discussed before, the elastic region of the MCC surface at the 'dry' side of the critical state line is relatively large. A better model prediction can be obtained if using CC-based fractional dilatancy relation. However, it is difficult to analytically solve the fractional differentiations of the CC function, by incorporating the effect of future reference critical state. Further analytical work needs to be conducted.

3. FP-m: The Role of Past and Future Stress States

Development of FP-m

In the previous section, two branches of the FP, i.e., FP-n: the one based on past SLS, and FP-sn: the other based on future reference critical state, were introduced. Even though each branch of the FP can be applied to describe various phenomenological behaviours of geomaterials, a question regarding the further comprehensive development of FP still rises: can one account for the roles of both past and future stress states, since they both can influence the plastic flow of geomaterials? Along with this consideration, we now modify the plastic flow rule by analogy with the FVP [15] to have a third definition of the FP, denoted as FP-m:

$$\mathrm{d}\varepsilon_{ij}^p = \mathrm{d}\lambda \, {}_{\sigma_{ij}'-l_{ij}}^{\mathrm{RC}}\mathrm{D}_{\sigma_{ij}'+\tilde{l}_{ij}}^{\alpha} f\left(\sigma_{ij}'\right) \tag{38}$$

where l_{ij} and $\tilde{l}_{ij}$ are the SLSs along the σ_{ij}'–direction; the Riesz–Caputo fractional operator is adopted, such that

$$ {}_{\sigma_{ij}'-l_{ij}}^{\mathrm{RC}}\mathrm{D}_{\sigma_{ij}'+\tilde{l}_{ij}}^{\alpha} f\left(\sigma_{ij}'\right) = \frac{1}{2}\left[{}_{\sigma_{ij}'-l_{ij}}^{\mathrm{C}}\mathrm{D}_{\sigma_{ij}'}^{\alpha} f\left(\sigma_{ij}'\right) + (-1)^n \, {}_{\sigma_{ij}'}^{\mathrm{C}}\mathrm{D}_{\sigma_{ij}'+\tilde{l}_{ij}}^{\alpha} f\left(\sigma_{ij}'\right)\right] \tag{39}$$

Substituting Equation (39) together with the MCC function into Equation (38), one can obtain the following generalised stress-dilatancy relation:

$$d_g = M^2 \frac{\left[l_p + (2-\alpha)\left(p' - l_p - p_0'/2\right)\right] l_p^{1-\alpha} + (-1)^n \left[\tilde{l}_p - (2-\alpha)\left(p' + \tilde{l}_p - p_0'/2\right)\right] \tilde{l}_p^{1-\alpha}}{\left[l_q + (2-\alpha)\left(q - l_q\right)\right] l_q^{1-\alpha} + (-1)^n \left[\tilde{l}_q - (2-\alpha)\left(q + \tilde{l}_q\right)\right] \tilde{l}_q^{1-\alpha}} \tag{40}$$

where l_p and $\tilde{l}_p$ denote the long SLSs of the past and future stress states, respectively, along the $p^{'}$–axis, while l_q and $\tilde{l}_q$ denote the long SLSs of the past and future stress states, respectively, along with the q–axis. Through parameter analysis, one can find the following specific cases for Equation (40).

- Case A
It can be found that if $\alpha = 1$ in Equation (40), then $n = 1$ and the stress-dilatancy relation reduces to the classic MCC-based one shown below, irrespective of l_p, $\tilde{l}_p$, l_q and $\tilde{l}_q$.

$$d_g = \frac{M^2 - \eta^2}{2\eta} \tag{41}$$

- Case B
If one assumes that the SLSs of past and future stress states are equivalent, i.e., $l_p = \tilde{l}_p$ and $l_q = \tilde{l}_q$, then, the stress-dilatancy relation in Equation (40) can have two possible forms for $\alpha \neq 1$. The first form can be obtained when $\alpha \in (0, 1)$, which also indicates that $n = 1$. Thus, Equation (40) can be derived as:

$$d_g = \left(\frac{l_p}{l_q}\right)^{1-\alpha} \frac{M^2 - \eta^2}{2\eta} \tag{42}$$

where it can be found that the future and past stress states contributes to the dilatancy of geomaterial by multiplying the original MCC-based dilatancy ratio with a factor of $(l_p/l_q)^{1-\alpha}$. Equation (42) can be further simplified by assuming that the SLSs, $l_p = xp'$ and $l_q = y\eta p'$, such that:

$$d_g = d_0 \frac{M^2 - \eta^2}{\eta^{2-\alpha}} \tag{43}$$

where $d_0 = 1/2(x/y)^{1-\alpha}$, is a model parameter, indicating the upward or downward shifting of the dilatancy curve, as shown in Figure 12. With the increase of d_0, the dilatancy ratio at the same stress level increases. As α increases, the dilatancy ratio varies. Note that a similar empirical stress-dilatancy relation was also suggested for modelling crushable soil [55], which can be derived from Equation (42) by assuming a constant value of $(l_p/l_q)^{1-\alpha}$, e.g., $(l_p/l_q)^{1-\alpha} = M^{\alpha-1}$.

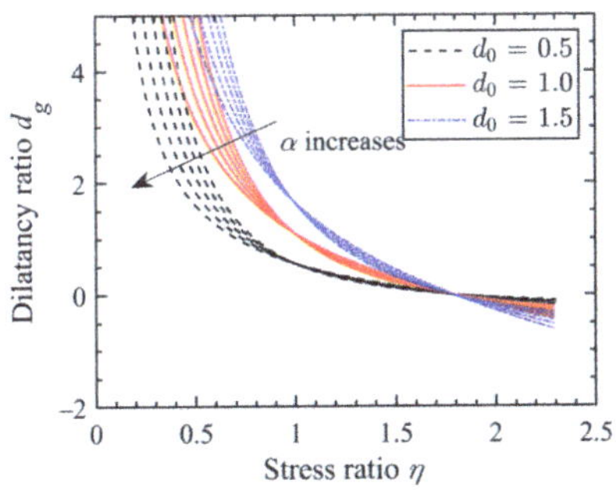

Figure 12. Effects of d_0 and α on the stress-dilatancy response.

4. Conclusions

The FP was developed for modelling the state-dependent nonassociated constitutive behaviour of geomaterials. This study provided a comprehensive introduction and discussion on the development and application of the FP, from the perspective of the role of stress length scale. It can be found that three branches of the FP, i.e., FP-n, FP-sn and FP-m, can be defined, respectively, by considering the effects of past stress state and future reference critical state, or the impact of both past and future stress states. The advantages and disadvantages of each FP approach were discussed. Some main conclusions are summarized as follows:

- Based on the simulation results for geomaterials, the FP-n approach was found to be more effective than the associated flow rule. However, it is difficult for the original FP-n approach to consider state dependence unless an empirical state parameter was introduced. Hence, the FP-sn approach was developed to consider both state dependence and nonassociated plastic flow without using state parameter or additional plastic potential. Moreover, the FP-sn approach can predict a higher volumetric dilatancy of granular soil, due to its large elastic region at the 'dry' side of the critical state line.
- Further analytical work should be needed to propose a modified FP-sn approach by using a yielding surface with a reduced elastic region. Due to the dependence of both past and future stress states on material flow, the FP-m approach was also suggested, where several specific cases of the FP-m based dilatancy relation were discussed, with regard to the role of SLS.
- In future work, the fractional anisotropic damage model can be further studied based on the fractional plastic damage model mentioned in this paper. Moreover, combining with peridynamics and phase field methods, numerical implementation of fractional constitutive model will be an important research direction. By means of a physics-based deep neural network, fractional models can provide a novel sight for challenges faced in multiscale plasticity.

Author Contributions: Conceptualization, writing—original draft preparation, writing-review and editing, design and preparation, P.Q. and Y.S.; writing—review, editing, funding acquisition, W.S. All authors have read and agreed to the published version of the manuscript

Funding: The funding by the Ulam Program (Grant No. PPN/ULM/2020/1/00026) is appreciated.

Data Availability Statement: Part of the data presented in this study are available on request from the corresponding author. The data are not publicly available due to intellectual property.

Conflicts of Interest: The authors declare no conflict of interest.

Appendix A

Caputo's definition of fractional derivatives is [56]:

$$ {}^{C}_{\sigma_c'}D^{\alpha}_{\sigma'} f(\sigma') = \frac{1}{\Gamma(n-\alpha)} \int_{\sigma_c'}^{\sigma'} \frac{f^{(n)}(\chi)\mathrm{d}\chi}{(\sigma'-\chi)^{\alpha+1-n}}, \quad \sigma' > \sigma_c' \tag{A1} $$

$$ {}^{C}_{\sigma'}D^{\alpha}_{\sigma_c'} f(\sigma') = \frac{(-1)^n}{\Gamma(n-\alpha)} \int_{\sigma'}^{\sigma_c'} \frac{f^{(n)}(\chi)\mathrm{d}\chi}{(\chi-\sigma')^{\alpha+1-n}}, \quad \sigma_c' > \sigma' \tag{A2} $$

where Equation (A1) is the left-sided derivative while Equation (A2) is right-sided derivative; D $(= \partial^{\alpha}/\partial\sigma'^{\alpha})$ denotes the partial derivation of function f; $\Gamma(x)$ is the gamma function and $\alpha \in (n-1, n)$, is the fractional order. σ' and σ_c' are the integral limits; χ is the independent variable.

The Riemann–Liouville's fractional derivatives are [56]:

$$ {}^{RL}_{0^+}D^{\alpha}_{x} f(x) = \frac{1}{\Gamma(n-\alpha)} \frac{\mathrm{d}^n}{\mathrm{d}x^n} \int_{0^+}^{x} \frac{f(\tau)\mathrm{d}\tau}{(x-\tau)^{\alpha+1-n}}, \quad x > 0 \tag{A3} $$

$$ {}^{RL}_{x}D^{\alpha}_{0^-} f(x) = \frac{(-1)^n}{\Gamma(n-\alpha)} \frac{\mathrm{d}^n}{\mathrm{d}x^n} \int_{x}^{0^-} \frac{f(\tau)\mathrm{d}\tau}{(\tau-x)^{\alpha+1-n}}, \quad x < 0 \tag{A4} $$

The MCC loading criterion are adopted for modelling. Accordingly, the FP-m stress-dilatancy relation in Equation (40) can be revised as:

$$ d_g = M^2 \frac{\left[l_p + (2-\alpha)(p' - l_p - p_0'/2)\right] l_p^{1-\alpha} - \left[\tilde{l}_p - (2-\alpha)(p' + \tilde{l}_p - p_0'/2)\right] \tilde{l}_p^{1-\alpha}}{\left[l_q + (2-\alpha)(q - l_q)\right] l_q^{1-\alpha} - \left[\tilde{l}_q - (2-\alpha)(q + \tilde{l}_q)\right] \tilde{l}_q^{-\alpha}} \tag{A5} $$

in which the assumptions in Case B are recalled to simplify Equation (A5).

References

1. Gao, Y.; Wu, Y.; Li, D.; Liu, H.; Zhang, N. An improved approximation for the spectral representation method in the simulation of spatially varying ground motions. *Probabilistic Eng. Mech.* **2012**, *29*, 7–15. [CrossRef]
2. Zhang, F.; Gao, Y.; Wu, Y.; Zhang, N. Upper-bound solutions for face stability of circular tunnels in undrained clays. *Géotechnique* **2018**, *68*, 76–85. [CrossRef]
3. Been, K.; Jefferies, M. Stress dilatancy in very loose sand. *Can. Geotech. J.* **2004**, *41*, 972–989. [CrossRef]
4. Been, K.; Jefferies, M.G. A state parameter for sands. *Géotechnique* **1985**, *35*, 99–112. [CrossRef]
5. Van Der Veen, H.; Vuik, C.; De Borst, R. An eigenvalue analysis of nonassociated plasticity. *Comput. Math. Appl.* **1999**, *38*, 107–115. [CrossRef]
6. Lade, P.V.; Nelson, R.B.; Ito, Y.M. Nonassociated flow and stability of granular materials. *J. Eng. Mech.* **1987**, *113*, 1302–1318. [CrossRef]
7. Yu, H.; Khong, C.; Wang, J.; Zhang, G. Experimental evaluation and extension of a simple critical state model for sand. *Granul. Matter* **2005**, *7*, 213–225. [CrossRef]
8. Shi, X.; Zhao, J.; Gao, Y. A homogenization-based state-dependent model for gap-graded granular materials with fine-dominated structure. *Int. J. Numer. Anal. Methods Geomech.* **2021**, *45*, 1007–1028. [CrossRef]
9. Wood, D.M. *Soil Behaviour and Critical State Soil Mechanics*; Cambridge University Press: Cambridge, UK, 1990.
10. Ezzat, M.; El-Bary, A. Unified fractional derivative models of magneto-thermo-viscoelasticity theory. *Arch. Mech.* **2016**, *68*, 285–308.
11. Zenkour, A.; Abouelregal, A. The fractional effects of a two-temperature generalized thermoelastic semi-infinite solid induced by pulsed laser heating. *Arch. Mech.* **2015**, *67*, 53–73.
12. Raslan, W. Application of fractional order theory of thermoelasticity to a 1D problem for a cylindrical cavity. *Arch. Mech.* **2014**, *66*, 257–267.
13. Dinzart, F.; Lipiński, P. Improved five-parameter fractional derivative model for elastomers. *Arch. Mech.* **2009**, *61*, 459–474.
14. Sumelka, W. A note on non-associated Drucker-Prager plastic flow in terms of fractional calculus. *J. Theor. Appl. Mech.* **2014**, *52*, 571–574.
15. Sumelka, W. Fractional viscoplasticity. *Mech. Res. Commun.* **2014**, *56*, 31–36. [CrossRef]
16. Sun, Y.; Sumelka, W. Multiaxial stress-fractional plasticity model for anisotropically overconsolidated clay. *Int. J. Mech. Sci.* **2021**, *205*, 106598. [CrossRef]
17. Lu, D.; Liang, J.; Du, X.; Ma, C.; Gao, Z. Fractional elastoplastic constitutive model for soils based on a novel 3D fractional plastic flow rule. *Comput. Geotech.* **2019**, *105*, 277–290. [CrossRef]
18. Lu, D.; Zhou, X.; Du, X.; Wang, G. A 3D fractional elastoplastic constitutive model for concrete material. *Int. J. Solids Struct.* **2019**, *165*, 160–175. [CrossRef]
19. Qu, P.; Zhu, Q.; Sun, Y. Elastoplastic modelling of mechanical behavior of rocks with fractional-order plastic flow. *Int. J. Mech. Sci.* **2019**, *163*, 105102. [CrossRef]
20. Qu, P.; Zhu, Q.; Zhao, L.; Cao, Y. A micromechanics-based fractional frictional damage model for quasi-brittle rocks. *Comput. Geotech.* **2021**, *139*, 104391. [CrossRef]
21. Sumelka, W.; Nowak, M. Non-normality and induced plastic anisotropy under fractional plastic flow rule: A numerical study. *Int. J. Numer. Anal. Methods Geomech.* **2016**, *40*, 651–675. [CrossRef]
22. Sumelka, W.; Nowak, M. On a general numerical scheme for the fractional plastic flow rule. *Mech. Mater.* **2018**, *116*, 120–129. [CrossRef]
23. Perzyna, P. The constitutive equations for rate sensitive plastic materials. *Q. Appl. Math.* **1963**, *20*, 321–332. [CrossRef]
24. Sun, Y.; Shen, Y. Constitutive model of granular soils using fractional-order plastic-flow rule. *Int. J. Geomech.* **2017**, *17*, 04017025. [CrossRef]
25. Sun, Y.; Gao, Y.; Zhu, Q. Fractional order plasticity modelling of state-dependent behaviour of granular soils without using plastic potential. *Int. J. Plast.* **2018**, *102*, 53–69. [CrossRef]
26. Le, L.A.; Nguyen, G.D.; Bui, H.H.; Andrade, J.E. Modelling the influence of fines content on the instability of silty sands considering grain scale interactions. *Int. J. Plast.* **2021**, *143*, 103020. [CrossRef]
27. Nguyen, H.; Rahman, M.; Fourie, A. Effect of particle shape on constitutive relation: DEM study. *J. Geotech. Geoenvironmental Eng.* **2020**, *146*, 04020058. [CrossRef]
28. Nguyen, H.B.K.; Rahman, M.M.; Fourie, A. The critical state behaviour of granular material in triaxial and direct simple shear condition: A DEM approach. *Comput. Geotech.* **2021**, *138*, 104325. [CrossRef]
29. Sun, Y.; Gao, Y.; Shen, Y. Mathematical aspect of the state-dependent stress–dilatancy of granular soil under triaxial loading. *Géotechnique* **2019**, *69*, 158–165. [CrossRef]
30. Polizzotto, C. Nonlocal elasticity and related variational principles. *Int. J. Solids Struct.* **2001**, *38*, 7359–7380. [CrossRef]
31. Patnaik, S.; Semperlotti, F. A generalized fractional-order elastodynamic theory for non-local attenuating media. *Proc. R. Soc. A* **2020**, *476*, 20200200. [CrossRef]

32. Sun, Y.; Gao, Y.; Song, S. Effect of integrating memory on the performance of the fractional plasticity model for geomaterials. *Acta Mech. Sin.* **2018**, *34*, 896–901. [CrossRef]
33. Sun, Y.; Xiao, Y. Fractional order plasticity model for granular soils subjected to monotonic triaxial compression. *Int. J. Solids Struct.* **2017**, *118*, 224–234. [CrossRef]
34. Wu, L.; Cheng, W.; Zhu, Z. Fractional-Order elastoplastic modeling of sands considering cyclic mobility. *J. Mar. Sci. Eng.* **2021**, *9*, 354. [CrossRef]
35. Zhang, T.; Zhou, X.; Qian, Q. The peridynamic Drucker-Prager plastic model with fractional order derivative for the numerical simulation of tunnel excavation. *Int. J. Numer. Anal. Methods Geomech.* **2022**, *46*, 1620–1659. [CrossRef]
36. Rahman, M.M.; Lo, S.C.; Dafalias, Y. Modelling the static liquefaction of sand with low-plasticity fines. *Géotechnique* **2014**, *64*, 881–894. [CrossRef]
37. Nguyen, G.T.; Chan, E.L.; Tsuji, T.; Tanaka, T.; Washino, K. Resolved CFD–DEM coupling simulation using Volume Penalisation method. *Adv. Powder Technol.* **2021**, *32*, 225–236. [CrossRef]
38. Schofield, A.N.; Wroth, P. *Critical State Soil Mechanics*; McGraw-Hill: London, UK, 1968; Volume 310.
39. Liang, J.; Lu, D.; Zhou, X.; Du, X.; Wu, W. Non-orthogonal elastoplastic constitutive model with the critical state for clay. *Comput. Geotech.* **2019**, *116*, 103200. [CrossRef]
40. Liang, J.; Lu, D.; Du, X.; Wu, W.; Ma, C. Non-orthogonal elastoplastic constitutive model for sand with dilatancy. *Comput. Geotech.* **2020**, *118*, 103329. [CrossRef]
41. Verdugo, R.; Ishihara, K. The steady state of sandy soils. *Soils Found.* **1996**, *36*, 81–91. [CrossRef]
42. Sun, Y.; Sumelka, W.; Gao, Y. Bounding surface plasticity for sand using fractional flow rule and modified critical state line. *Arch. Appl. Mech.* **2020**, *90*, 2561–2577. [CrossRef]
43. Sun, Y.; Nimbalkar, S. Stress-fractional soil model with reduced elastic region. *Soils Found.* **2019**, *59*, 2007–2023. [CrossRef]
44. Sun, Y.; Sumelka, W.; He, S.; Gao, Y. Enhanced Fractional Model for Soil–Structure Interface Considering 3D Stress State and Fabric Effect. *J. Eng. Mech.* **2022**, *148*, 04022054. [CrossRef]
45. Wu, E.; Zhu, J.; Sun, Y.; He, S. A general plastic model for rockfill material developed by using Caputo fractional derivative. *Comput. Geotech.* **2022**, *151*, 104948. [CrossRef]
46. Li, H.; Ma, B.; Zhang, S.; Sheng, D. Mechanical behaviors of soft rocks based on the fractional thermal elastic-plastic theory. *Chin. J. Rock Mech. Eng.* **2020**, *39*, 1311–1320.
47. Shen, W.; Liu, S.; Xu, W.; Shao, J. An elastoplastic damage constitutive model for rock-like materials with a fractional plastic flow rule. *Int. J. Rock Mech. Min. Sci.* **2022**, *156*, 105140. [CrossRef]
48. Qu, P.F.; Zhu, Q.Z. A Novel Fractional Plastic Damage Model for Quasi-brittle Materials. *Acta Mech. Solida Sin.* **2021**, *34*, 706–717. [CrossRef]
49. Zhou, X.; Lu, D.; Du, X.; Wang, G.; Meng, F. A 3D non-orthogonal plastic damage model for concrete. *Comput. Methods Appl. Mech. Eng.* **2020**, *360*, 112716. [CrossRef]
50. Lu, D.; Su, C.; Zhou, X.; Wang, G.; Du, X. A cohesion-friction combined hardening plastic model of concrete with the nonorthogonal flow rule: Theory and numerical implementation. *Constr. Build. Mater.* **2022**, *325*, 126586. [CrossRef]
51. Halilovič, M.; Vrh, M.; Štok, B. NICE—An explicit numerical scheme for efficient integration of nonlinear constitutive equations. *Math. Comput. Simul.* **2009**, *80*, 294–313. [CrossRef]
52. Zhu, Q.; Zhao, L.; Shao, J. Analytical and numerical analysis of frictional damage in quasi brittle materials. *J. Mech. Phys. Solids* **2016**, *92*, 137–163. [CrossRef]
53. Li, X.S.; Dafalias, Y.F. Dilatancy for cohesionless soils. *Géotechnique* **2000**, *50*, 449–460. [CrossRef]
54. Dafalias, Y.F.; Manzari, M.T. Simple plasticity sand model accounting for fabric change effects. *J. Eng. Mech.* **2004**, *130*, 622–634. [CrossRef]
55. Liu, M.; Gao, Y. Constitutive modeling of coarse-grained materials incorporating the effect of particle breakage on critical state behavior in a framework of generalized plasticity. *Int. J. Geomech.* **2017**, *17*, 04016113. [CrossRef]
56. Zhou, Y. Fractional evolution equations and inclusions. In *Analysis and Control*; Elsevier: Amsterdam, The Netherlands, 2015.

materials

Article

Numerical Simulation of the Impact of Water Vapour and Moisture Blockers in Energy Diagnostics of Ventilated Partitions

Barbara Ksit [1], Anna Szymczak-Graczyk [2,*] and Roman Pilch [3]

1 Institute of Building Engineering, Faculty of Civil and Transport Engineering, Poznan University of Technology, Piotrowo 5, 60-965 Poznań, Poland
2 Department of Construction and Geoengineering, Faculty of Environmental and Mechanical Engineering, Poznan University of Life Sciences, Piątkowska 94E, 60-649 Poznań, Poland
3 Department of Architecture and Town Planning, Faculty of Civil, Environmental Engineering and Architecture, Bydgoszcz University of Science and Technology, Prof. S. Kaliskiego 7, 85-796 Bydgoszcz, Poland
* Correspondence: anna.szymczak-graczyk@up.poznan.pl

Abstract: Current trends towards saving energy and designing sustainable buildings result in most designers focusing on achieving the best thermal parameters, thereby neglecting a careful moisture analysis. Excessive moisture content in building partitions degrades the mechanical properties of materials, reduces thermal insulation properties (which leads to an increase in the demand for thermal energy) and worsens the microclimate in rooms. Modern digital solutions help create appropriate models of partitions that work for many years in good environmental conditions. According to the analysis of air parameters, 1 m^3 of air at 20 °C contains approx. 17.3 g of water. When the temperature of the air reaches the dew point temperature, water vapour condenses. The dew point depends on air temperature and relative air humidity; for instance, at the same air temperature of 20 °C, the dew point temperature at 40% relative air humidity is 6 °C, whereas at 90% relative humidity, it is over 18 °C. This means that the higher the value of relative humidity in the room at a certain temperature, the lower the temperature that will cause condensation. The article presents a numerical analysis of the insulation work of flexible materials within the layers of ventilated partitions in an 8-year simulated period of varying environmental conditions. The aim of the article is to analyze different models and variants of ventilated partition operation with respect to the advisability of using a vapour barrier to avoid the problem of destruction of thermal insulation and finishing layers of a ventilated roof.

Keywords: water vapour; operational moisture; numerical analysis; dew point; flexible waterproofing materials

check for updates

Citation: Ksit, B.; Szymczak-Graczyk, A.; Pilch, R. Numerical Simulation of the Impact of Water Vapour and Moisture Blockers in Energy Diagnostics of Ventilated Partitions. *Materials* **2022**, *15*, 8257. https://doi.org/10.3390/ma15228257

Academic Editor: Qing-feng Liu

Received: 10 October 2022
Accepted: 17 November 2022
Published: 21 November 2022

1. Introduction

Water is one of the most important substances for humans, being the source of life, yet it is also the cause of degradation of building partitions. In various forms, it accompanies people every day and wherever they are [1–7]. Renovation processes carried out in multifamily buildings that are subject to municipal management or the exploitation of residential buildings located in areas under conservation protection, where tenants have a varying level of construction knowledge, sometimes contribute to an increase in moisture content in building partitions. Various phenomena observed by researchers lead to an increase in moisture content in walls and ceilings: these include an insufficient number of gravity ventilation chimneys in buildings subject to multiple functional reconstructions; or replacement of window frames (usually wooden) with new ones that no longer allow free air migration and removal of water vapour from rooms. The operating moisture is the water vapour generated by residents and their activities inside a building. Various individual operating habits are usually the key reason for the occurrence of extreme hygrothermal conditions in buildings, which can often significantly exceed the ones specified in building standards [8–10]. The limit value for relative air humidity at the surface of a building is 80%. With ambient air at a temperature of 20 °C and 50% relative humidity, the allowable external wall surface temperature with reference to the protection against fungal coverage is 12.6 °C. The limit value of the dew point temperature on the surface of an internal partition (the temperature at the state of saturation with water vapour, meaning relative air humidity equal to 100%) at 50% relative air humidity is 9.3 °C and the saturated water vapour pressure is 23.40 hPa. As shown in the publications, condensation on building partitions predominantly appears near to critical thermal places, when the internal temperature of external surfaces drops below the dew point [11–17]. The condensate produced in this way contributes to the development of mould fungi, which degrade both living organisms and building elements. The spores produced by their fruiting bodies circulate in the air and are inhaled by people, which make them feel unwell or tired. Also, they may contribute to the development of various diseases or even death. In addition, mould reduces the strength and durability of elements and structures, and the resultant moisture negatively affects the technical parameters of materials and increases the risk of a building catastrophe [5,6,18–22].

Condensation can also occur inside partitions. The transport of water vapour into a partition takes place through the process of diffusion. The course of diffusion depends on the coefficients of water vapour transmission in the materials the partition is made of [15,23,24]. They determine the amount of water vapour, measured in kilograms, which can penetrate through 1 m^2 of a 1-metre thick material in 1 s with a pressure difference between the two sides of the material equal to 1 Pa. On the basis of the amount of the air removed from individual rooms indicated by the standards [25,26], it is possible to estimate the required multiplicity of air exchange inside, which determines how many times the air is changed per hour. For single-family houses, the value of this indicator ranges from 0.5 to 0.8 or even 1.0. The index 1.0 means that all the air in the room will be replaced within one hour. Movement of the air causes a change in heat exchange through convection between a person and their environment, which affects the individual feeling of comfort. Receptors in the skin immediately receive information about the thermal environment in which a human body is located, but when the skin is cooled to approx. 32 °C, it begins to feel cold, and after exceeding 37 °C, it begins to sweat [27–29]. Too low an air velocity in the room prevents proper heat dissipation from a human body, which results in general heat sensation. On the other hand, when the air is exchanged too fast, more heat is given off, which means that a person starts to feel cold. At the comfort temperature, which is from 18 to 21 °C, people do not feel the effect of relative air humidity when it is in the range of 30–60% [28–35].

The basic source of moisture content in buildings is the emission of water vapour related to metabolic changes occurring in their residents and from the activities they carry out inside.

On the subject of diffusion and condensation of moisture in the layers of the building envelope, the literature mandates waterproof layers. Professional practice shows that such layers are standardly adopted by architects for all roof constructions and especially for ventilated roofs. Using the possibilities afforded by numerical analysis, simulation models with different variants of waterproof barrier placement were elaborated for this paper. The operation of such a barrier and the condition of the partition were analyzed over the 8 years of a partition's operation.

The aim of the study was to show that double-sided flat roofs containing air spaces should be treated differently than pitched roofs containing small gaps in the structure of their layers, acting as ventilated spaces. The durability of insulating materials as well as the reliability of their characteristics and parameters depends to a large extent on the environment in which they work. Moreover, deteriorating insulation properties of partitions may lead to degradation and, consequently, to an even greater reduction in their thermal efficiency. Change in the technical parameters of insulating materials caused by excessive moisture sorption over time can involve introducing additional security measures to ensure the safe use of buildings. In extreme cases, it may require serious structural changes or even the repurposing of the the facilities and how they are used. In the case of flat roofs, it is not always necessary to use flexible waterproofing materials in ventilated partitions to ensure the protection of thermal insulation against moisture. The current paper shows that it is worth using multi-variant modelling of building partitions to reduce the energy consumed for heating and to prevent insulating materials from absorbing moisture. The paper presents the results of numerical simulations over the years, forming the basis for our conclusion, namely, that construction solutions that are currently very popular cannot be universally used for all types of roofs. An important aspect of using or not using flexible waterproofing materials is the method of ventilating the space present in the layers of a partition. The results presented in the paper demonstrate the legitimacy of solutions that are less expensive and easier to implement.

2. Materials and Methods

2.1. Operational Concerns Regarding Water Vapour Generation Inside Buildings

The main sources of water vapour in buildings include technological moisture, increasing relative air humidity in the first years after the construction, and operational moisture [9,36]. The people who use the building are themselves the constant generators of a large amount of water vapour. As a result of metabolic changes, humans release much heat, which is emitted to the environment through radiation, convection, and evaporation. Summing up the percentage results of heat dissipation through radiation and convection, 79% of the share is called dry heat (sensible heat), the remaining 21% of heat given off by water evaporation and breathing is called moist heat (latent heat) [24,29,37].

However, a larger amount of water vapour is generated by various activities performed by people, and it has the greatest and fastest impact on the value of relative humidity of the indoor air [10]. On average, it is assumed that the amount of water released by a family of four in metabolic processes is 0.21 L/h i.e., 5 L/day, or 1.25 L/day per person [38]. For two adults, the average emission of water vapour is 6.5 kg/day, and for parents with two children—10.9 kg/day. For one person the emission is assumed to be 4.4 kg/day with the standard deviation 1.73 kg [39]. There is one more source of moisture in the room, which is houseplants that need to be watered regularly. Almost all the water that is provided to plants evaporates, as only 0.2% of it is required for vegetation growth. 5 to 7 potted plants can release approx. 0.5 L of water in 24 h [40,41]. Water vapour is a variable component of the atmospheric air and comes primarily from the process of evaporation of water from the earth's surface and precipitation. The amount of water vapour in a unit of air equal to 1 m^3 decreases or increases depending on surrounding environmental conditions, and its value, referred to as absolute humidity, is expressed in g/m^3 [4,42]. Water vapour content in the air is limited and depends on air temperature. The warmer it is, the more water vapour it can contain. The maximum possible value of filling the air with water vapour is defined as the state of saturation. Exact values of water vapour content in the air with the maximum humidity depending on temperature are compiled in Table 1.

Table 1. Water vapour content in the air with the maximum humidity depending on temperature [42].

Humid Air Temperature RH = 100% [°C]	Water Vapour Content [g/m³]	Humid Air Temperature RH = 100% [°C]	Water Vapour Content [g/m³]	Humid Air Temperature RH = 100% [°C]	Water Vapour Content [g/m³]
−20	0.9	6	7.3	21	18.4
−15	1.4	8	8.3	22	19.5
−10	2.1	10	9.4	23	20.6
−8	2.5	12	10.7	24	21.8
−6	3.0	14	12.1	25	23.1
−4	3.5	16	13.7	26	24.4
−2	4.1	17	14.5	28	27.2
0	4.8	18	15.4	30	30.4
2	5.6	19	16.3	40	51.1
4	6.4	20	17.3	50	82.3

In diagnostics assessing existing building solutions, non-invasive moisture meters (pyrometer Trotec BP25, Heinsberg, Germany) are used, which provide the percentage of relative humidity in analysed rooms (Figure 1).

Figure 1. Examples of air humidity and air temperature tests using a pyrometer and a dew point scanner.

In situ and digital analyses take into account the value of relative humidity i.e., a percentage measure of water vapour content in the air. It is expressed by the ratio of the actual water vapour pressure contained in the air to the maximum water vapour pressure possible at a given temperature (saturated vapour pressure) [4,6,10]. When using the concept of relative humidity, it is always required to provide the air temperature at which it is measured, since the percentage alone says nothing about the actual water vapour content. Research [43] shows that the actual moisture content in the air will remain the same, whereas the saturation level will change: i.e., at 10 °C the value of relative humidity is 100% and at 30 °C it is 28%. In order to eliminate water vapour from rooms, appropriate ventilation systems should be designed.

In the case of gravity ventilation, the occurrence of moisture amplitudes inside rooms is inevitable. To reduce the risk of moisture occurrence in external partitions, ventilation gaps are used [36,44–49]. The dimensions of ventilation gaps must comply with [50] and with the information provided in [51]. Other related standards [52,53] provide parameters that should also be considered in order to correctly model building partitions. Double protection of heterogeneous partitions, with two layers of gaps, is used by introducing an obstacle preventing water vapour from moving from inside a room to the outside, in the form of a layer which prevents its penetration of the partitionThe Polish climate is more hazardous than many for building partitions due to the large amplitude of temperature difference, increasing the number of cycles of condensation formation especially in roofs. As this phenomenon carries the risk of thermal insulation materials becoming damp, it is necessary to minimize the potential damage. Because water vapour in a building penetrates its roof the fastest, in order to ensure this partition provides its protective work, it is most often shaped as a multi-layer structure constructed with the use of flexible waterproofing materials [54–56].

The minimum cross-sections of ventilation gaps in inclined partitions at a roof slope of $\geq$5° and <5° are specified in Tables [50], which are the recommendations of the Association of German Roofers. As early as 1997, the guidelines for roofs with a slope of $\geq$10% [54–56] recommended that the total diffusion resistance should increase with an increase in the length of the ventilating air path, i.e., with an increase in the length of rafters.

A double ventilation gap is required particularly when the roofing is laid on a board (including full boarding), or when a foil with low vapour permeability is used as the initial covering layer (protecting against the penetration of moisture from the outside, but at the same time hindering the diffusion of water vapour from the inside to the outside), and the roofing itself is laid on patches and counter-patches. In addition, it is recommended for roofs with very complex shapes. In such cases, it is necessary to provide separate ventilation for patches and for the main roofing. One ventilation gap should be designed between the thermal insulation material and the layer of initial covering used in a given design (board, full boarding with tar paper, low vapour permeability membrane). In this way, the thermal insulation against moisture is protected (the joint removes the moisture that has penetrated through the vapour barrier to the insulation). It should be a few centimetres wide, with air inlet openings (under the eaves, in the soffit) and outlet openings (under the ridge, ventilation grates in the gable walls). The second gap must be between the initial covering layer and the actual roof covering, thanks to which the roof finishing materials dry faster [45–49].

Due to the various properties of materials related to the transmission of water vapour, the arrangement of layers in external partitions should be carefully selected.

The knowledge of the water vapour permeability coefficient δ or the thickness of the equivalent air layer S_d is needed to calculate the water vapour permeability of the material with thickness d [57–59] calculated with Formula (1):

$$W_p = \frac{\delta}{d} = \frac{\delta_0}{\mu \cdot d} = \frac{2 \cdot 10^{-4}}{S_d} \tag{1}$$

where:

W_p—water vapour permeability $\left[\frac{kg}{m^2 \cdot s \cdot Pa}\right]$

δ—water vapour permeability coefficient, $\left[\frac{kg}{m \cdot s \cdot Pa}\right]$

d—material layer thickness, [m]

δ_0—air vapour permeability ($\delta_0 = 2 \times 10^{-4}$), $\left[\frac{kg}{m \cdot s \cdot Pa}\right]$

μ—diffusion resistance coefficient, [unitless]

Sd—thickness of the equivalent air layer, [m]

The diffusion resistance Z_p is the reciprocal of water vapour permeability and describes the resistance of a building element to water vapour diffusion, according to Formula (2).

$$Z_p = \frac{1}{W_p} = \frac{d}{\delta} = \frac{S_d}{2 \cdot 10^{-4}} \tag{2}$$

In order to determine the value of diffusion resistance of a multilayer element, the values of diffusion resistance of individual material layers should be summed up, as shown by Formula (3):

$$Z_p = \sum_n \frac{S_d}{2 \cdot 10^{-4}} \tag{3}$$

where n is the individual layer of material.

The layers should be arranged in such a manner as to allow free flow of water vapour through the partition and to avoid condensation of water vapour inside it. Table 2 below shows the simulations of the amount of water vapour reaching the attic in 24 h.

Table 2. Simulation of the source of humidity and the amount of water vapour—author's summary [58].

Source of Humidity	Water Vapour Emission in 24 h [g]	Amount of Water Vapour Reaching the Attic in 24 h [g]	
		with 10× Exchange	with 25× Exchange
Respiration and sweat evaporation	5000	3500	2500
Use of a residential building	13,255	9279	9279
Potted plants (5–7 plants per apartment)	500	350	250
Total	18,755 g	13,129 g	12,029 g

Assuming a model building with a cubature of 420 m^3, the air flow is from 4.2 to 10.5 thousand m^3. Thus, in 2940 m^3, there is 40,689.6 g of water vapour, and in 5250 m^3—72,660 g. After adding the appropriate amount of moisture generated in the building, in total, in 2940 m^3, there is 53,818.6 g of water vapour; and in 5250 m^3, as much as 84,689 g (Table 3).

Table 3. Total amount of water vapour reaching the attic depending on the adopted air exchange [58].

	Amount of Water Vapour Reaching the Attic in 24 h at 80% Relative Humidity and Air Temperature 20 °C [g]	Amount of Water Vapour Reaching the Attic in 24 h Generated in a Residential Building [g]	Total [g]
10× exchange	40,689.6	13,129.0	53,818.6
25× exchange	72,660.0	12,029.0	84,689.0

Dividing the obtained amount of water vapour by the volume of air reaching the attic, we obtain the absolute humidity values equal to 18.31 g/m^3 for 10-fold air exchange and 16.13 g/m^3 for 25-fold air exchange.

For 10-fold exchange, we obtain the state of air supersaturation, as the maximum amount of water vapour that the air can hold, according to the guidelines, is 17.3 g/m^3. This result indicates that the value of relative humidity in the rooms is 100%, and 1 g of water drops out of the building partitions from each 1 m^3.

In the case of 25-fold air exchange, the value of relative humidity is at the level of 93% (Table 4) and the amount of water condensed from 1 m^3 of air will need to be calculated.

Table 4. Relative humidity in a model building depending on air exchange [58].

	Absolute Humidity [g/m^3]	Saturation State at 20 °C [g/m^3]	Relative Humidity [%]
10× exchange	18.31	17.30	100% supersaturation state
25× exchange	16.13		93%

In the design of external partitions, due to the diffusion of water vapour, the principle of arranging the layers according to their decreasing diffusion resistance from the inside to the outside is applied [11,12]. As a result, it is more difficult for water vapour flowing through the partition to reach the state of saturation, despite ever decreasing temperature, so that the increased condensation of water vapour inside the partition does not occur. It is important to carry out adequate air ventilation in a given area. Ventilation openings should be designed so that they ensure the required air exchange in the flat roof space and are not blocked by thermal insulation materials.

Designing partitions according to the above-mentioned principle is not always possible, however. Indeed as a rule, it is necessary to analyse the functioning of a particular partition in terms of humidity. Figure 2 shows an example of a geometrically complex roof space in a ventilated roof, analysed here for this article.

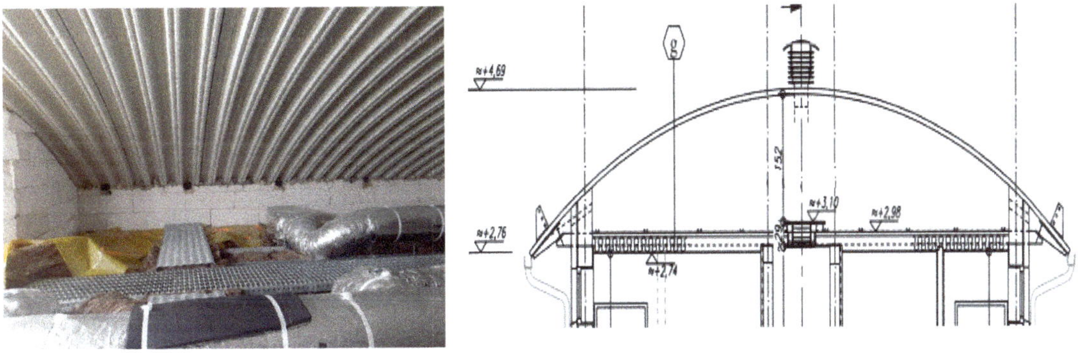

Figure 2. Geometrically complicated roof space in a ventilated roof (analysed here).

2.2. Flexible Waterproofing Materials Used in Ventilated Partitions

These materials are used to protect partitions against water penetration. The division of materials and their names are presented in Table 5 in accordance with the list contained in the Glossary of Roofing Terms and Names [57].

Table 5. Flexible waterproofing products used as initial covering layers [53].

ICF—Initial Covering Foils (Low Vapour-Permeable)		ICM—Initial Covering Membranes (Highly Vapour-Permeable)	
Vapour-tight	Vapour-permeable	Light	Screens

Depending on polymers contained in processed plastics, foils can be divided into thermoplastic (plastomeric) and elastomeric.

Initial covering foils are materials with low vapour-permeability, and are the first plastic products that replaced roof boarding and tar paper over 60 years ago. A roof in which ICF will be used should have two ventilation gaps: one under the covering, and the other with a sublayer of foil. These types of foils are also often called vapour permeable foils.

Initial covering membranes are foils with high vapour-permeability, with $S_d < 0.1$ m (optimally from 0.015 to 0.045 m), and are currently the most common material used for the initial covering of pitched roofs.

In most cases, roof membranes are made of non-woven polypropylene, which is highly vapour-permeable buthas little resistance to the water column, and therefore must be additionally provided with a delicate film, i.e., a functional film and non-woven polypropylene fleece.

Among initial covering membranes, roof screens with increased durability, grammage and strength deserve a special distinction [60–67].

The market of construction materials offers diffusion-active foils, the work of which is to protect structures both in summer and winter. In winter, the average humidity of the vapour barrier environment is approx. 40%. Diffusion is directed outward from the heated interior. The vapour barrier should be highly vapour-tight during this period in order to protect structures from condensation. In summer, the average ambient humidity of the vapour barrier is approx. 80%, and the diffusion flow is reversed. During this period, the vapour barrier should be able to become permeable so as to allow moisture to dry out.

It should also be noted that the vapour barrier foil, which is used in flat roofs with an increased concentration of water vapour (e.g., in rooms with showers), should not be placed on thermal materials. It is also not advisable to wrap the contact points between materials such as between wool and external wall. Moreover, the foil should be attached to the external wall as required, turning it downwards with a slight drip upwards.

2.3. Case Study

In order to analyse the moisture flow of water vapour, the authors modelled a barrier consisting of the following materials: acrylic paint inside, plasterboard, PE foil, mineral wool, glass mineral wool, air space with variable height from 20 to 80 cm, and trapezoidal sheet (Figure 2).

For such a complex case, models based on a system of non-linear partial differential equations describing the non-stationary, coupled transport of heat and moisture in building materials and partitions were used [68]. Simulation calculations were made using the WUFI PRO 6.5 software, used for a one-dimensional analysis of non-stationary processes of heat and moisture flow through building partitions. The software was developed at the Fraunhofer Institute for Building Physics in Holzkirchen (Germany). The simulation calculations included, among others:

- variable properties of the material depending on humidity and temperature;
- additional thermal transport processes, such as latent heat transport by water vapour flows;
- additional heat sources due to solar radiation;
- parameters dependent on the environmental conditions, such as wind and rain action.

In the calculations performed, the parameters of the external climate were adopted on the basis of a "typical meteorological year" included in the software.

Natural gravity ventilation was adopted for the analyses, and the foil parameters took into account possible leaks resulting from the perforation of the foil with mechanical fasteners used for attaching plasterboards or for fixing lamps. A high level of internal air humidity was adopted, and the calculation variants assumed an 8-year simulated period of environmental actions. This is long enough to show a possible increase in moisture content in built-in materials and occurrence of conditions for the development of biological aggression. The transport of moisture in the partition was assumed to be two-way.

2.4. Calculation Variants

Various arrangements were considered, relative to the polyethylene foil applied or its absence, and to ventilation of the space above the thermal insulation materials used. The technical data sheets provided by the manufacturers lacked the information specifying both the material density and diffusion resistance parameters. Therefore, on the basis of their technical knowledgeand after considering various options, the authors chose the most realistic parameters for the declared materials, as adapted to the technological processes taking place during execution and operation.

The analysed variants are presented below:

- Variant 1—PE foil under the grate for fixing plasterboards. Roof ventilation equal to 20 changes per hour (Figure 3).

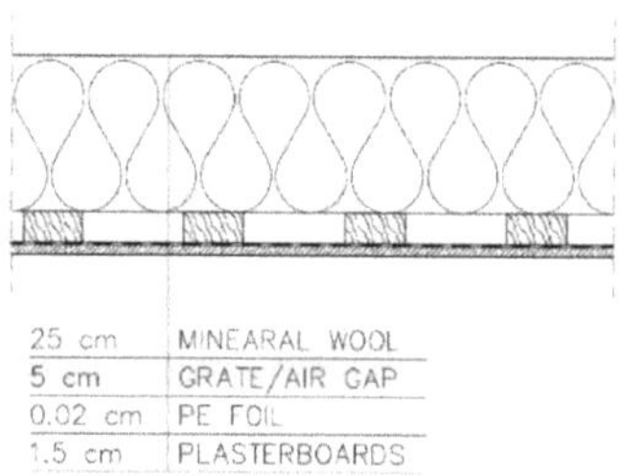

Figure 3. Variant 1: PE foil under the grate for fixing plasterboards. Roof ventilation equal to 20 changes per hour.

- Variant 2—PE foil above the grate for fixing plasterboards. Roof ventilation equal to 20 changes per hour (Figure 4).

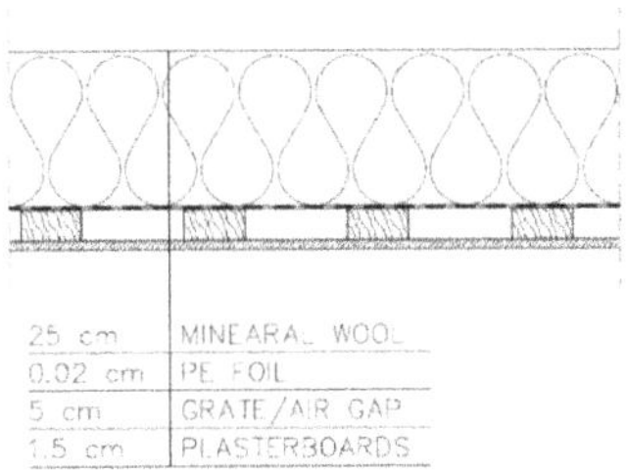

Figure 4. Variant 2: PE foil above the grate for fixing plasterboards. Roof ventilation equal to 20 changes per hour.

– Variant 3—PE foil under the grate for fixing plasterboards. Mineral wool fills the air gap formed by the grate. Roof ventilation equal to 20 changes per hour (Figure 5).

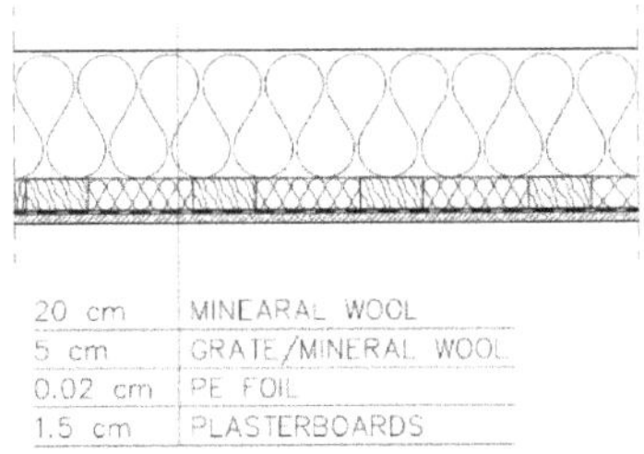

Figure 5. Variant 3: PE foil under the grate for fixing plasterboards. Mineral wool fills the air gap formed by the grate. Roof ventilation equal to 20 changes per hour.

– Variant 4—no PE foil in the arrangement of layers. Roof ventilation equal to 20 changes per hour (Figure 6).

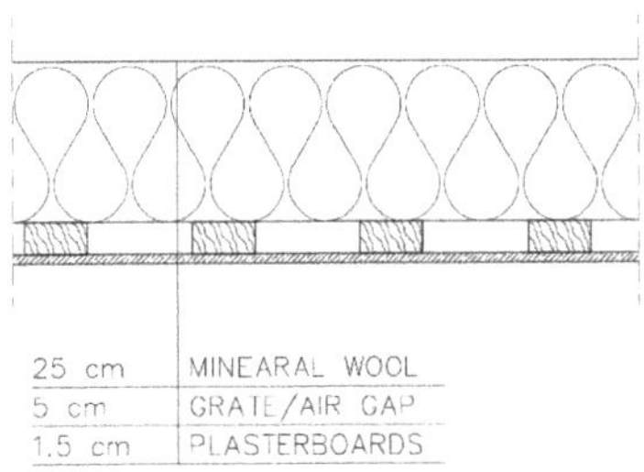

Figure 6. Variant 4: no PE foil in the arrangement of layers. Roof ventilation equal to 20 changes per hour.

– Variant 5—no PE foil in the arrangement of layers. No roof ventilation (Figure 7).

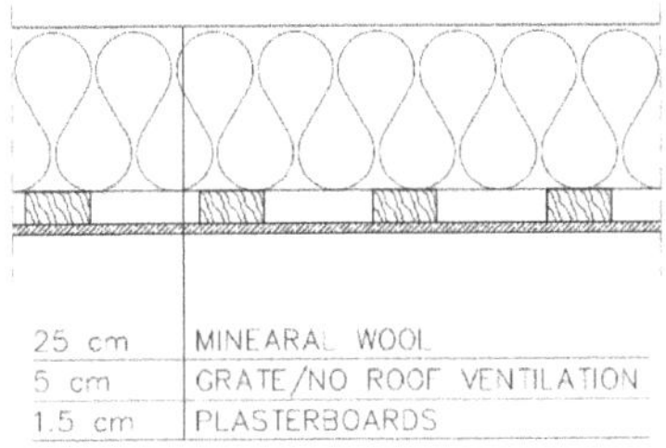

Figure 7. Variant 5: no PE foil in the arrangement of layers, no roof ventilation.

3. Results

The calculations are summarized below, assuming the value of water vapour pressure 2150 Pa and the corresponding values of temperature and relative humidity (Figure 8).

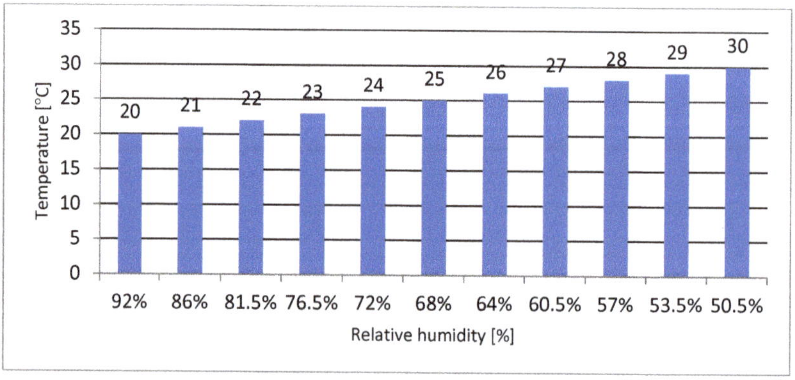

Figure 8. Dependence of temperature [°C] on relative humidity [4].

The highest temperature was 25 °C and corresponded to 70% relative humidity. The corresponding partial pressure of water vapour was 2217 Pa. For individual calculation variants from the data cloud, there are shown changes in moisture content in thermal insulation materials and in plasterboard. In addition, an isopleth of the internal surface is shown (interpretation based on the WUFI software). The WUFI program divides materials into:

- LIM B I: bio-utilizable substrate, i.e., wallpaper, plasterboard, products made from easily degradable material, material for permanently elastic joints, etc.;
- LIM B II: substrates with porous structure, i.e., plasters, mineral building materials, some types of wood, insulating materials not belonging to group I, etc. If strongly contaminated, these materials belong to group I [69].

On the isopleth map, the material curves developed according to the guidelines of the numerical program LIM B I and LIM B II—below which mould development is not normally expected—express the limits for building materials or for the conditions for their incorporation into buildings.

The colour of each point on the isopleth indicates when that point occurred during the computation. The diagrams show the changes in moisture content of individual materials over eight years for variants 1–5. Moisture content [kg/m^3] is marked on the vertical axis and the time period from 01 January 2023 to 01 January 2031 was simulated on the horizontal axis. The isopleth diagrams of the internal surface of the partition show the relationship between relative humidity [%] and temperature for variants 1–5 [70].

For variants 1, 2 and 3, the same analysis results were found for changes in moisture content in both wool and plasterboard. Figure 9 shows the changes in moisture content of glass mineral wool over the analysed period of eight years, while Figure 10 shows the changes in moisture content of plasterboard over the same period.

For variants 1–3 (Figures 3–5), the moisture level for mineral wool during the analysed period ranged from 1.7% to 4%, and for plasterboard from 0.44% to 0.80%. For these same variants 1–3, the humidity—temperature isopleth is presented in Figure 11.

The isopleths of the LIM B I (dashed line) and LIM B II (solid line) curves, i.e., contour charts of the mould growth rate as a function of temperature (horizontal axis) and humidity (vertical axis) as shown in the diagram, do not intersect with the area of points for the analysed partition.

Mould spots appeared at a minimum relative humidity of 41% and a temperature of 18 °C. In the isopleth (Figure 10), there is no critical value, i.e., no point of intersection of the curves with the area of points.

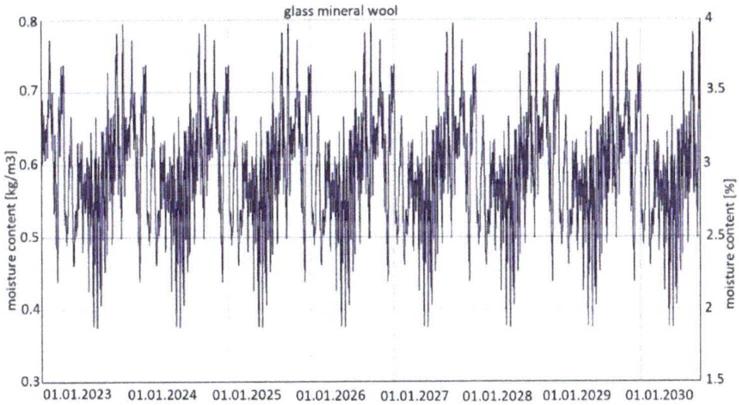

Figure 9. Changes in moisture content in glass mineral wool over eight years, variants 1–3 [70].

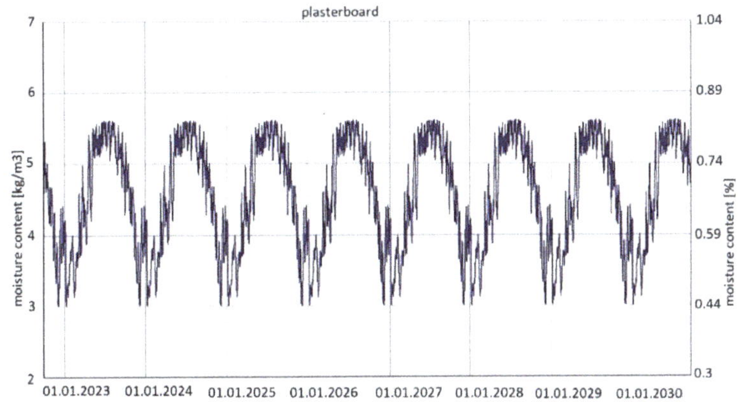

Figure 10. Changes in moisture content in plasterboard over eight years, variants 1–3 [70].

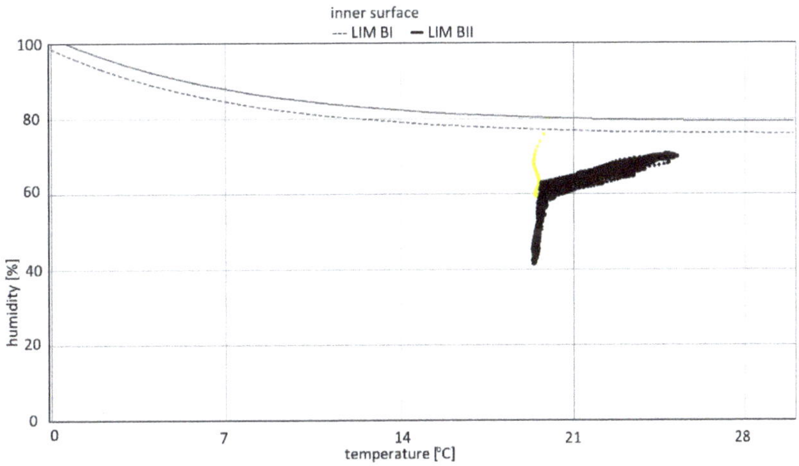

Figure 11. Isopleth of the inner surface of the partition, variants 1–3 [70].

For variant 4 (Figure 6), moisture content depending on the season is shown in the diagrams for mineral wool (Figure 12) and for plasterboard (Figure 13), while the relative humidity—temperature isopleth is shown in Figure 14.

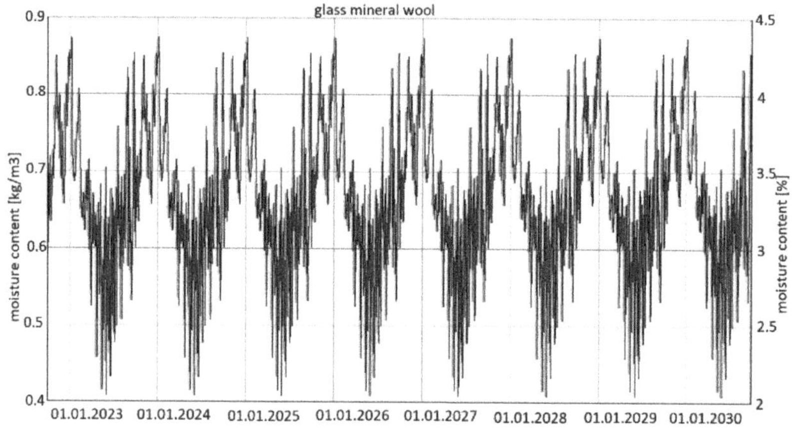

Figure 12. Changes in moisture content in glass mineral wool over eight years, variant 4 [70].

The moisture level for mineral wool during the analysed period ranged from 2.1% to 4.7%; and for plasterboard, depending on the season, it ranged from 0.3% to 0.85%.

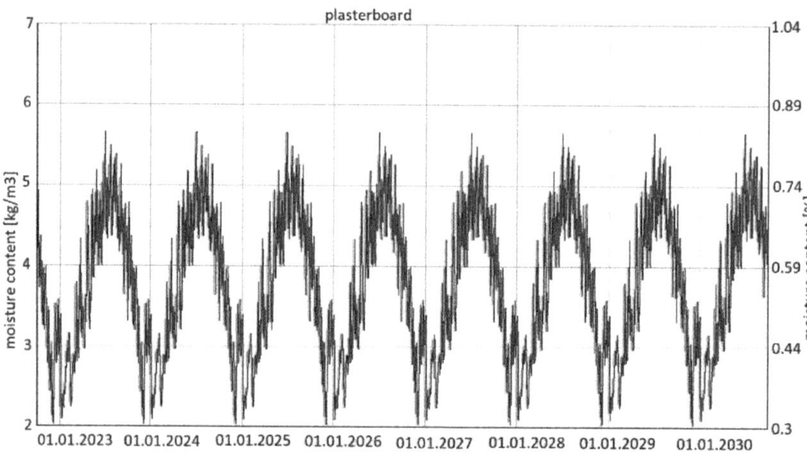

Figure 13. Changes in moisture content in plasterboard over eight years, variant 4 [70].

As previously, the isopleths of the LIM B I (dashed line) and LIM B II (solid line) curves, i.e., contour charts of the mould growth rate as a function of temperature (horizontal axis) and humidity (vertical axis) appearing in the diagram, do not intersect with the area of points for the analysed partition. Mould spots appeared at a minimum relative humidity of 30% and a temperature of 18 °C. There is no critical value in the isopleth (Figure 14), i.e., no point where the curves intersect with the area of points.

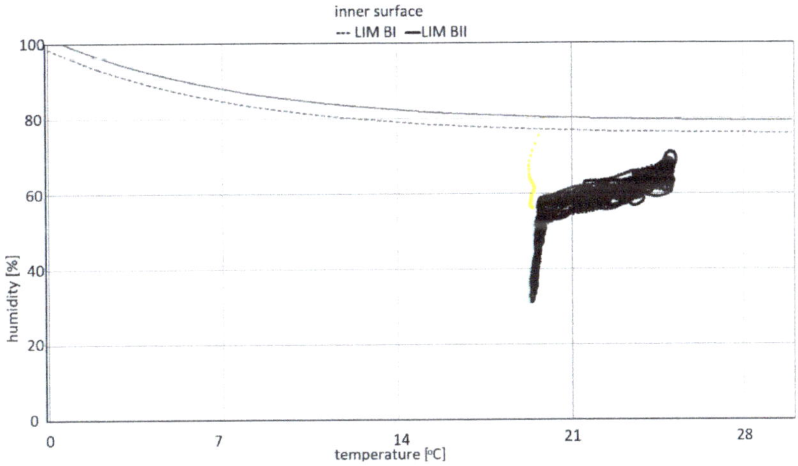

Figure 14. Isopleth of the inner surface of the partition, variant 4 [70].

For the last of the analysed variants, variant 5 (Figure 7), moisture content depending on the season is shown in the diagrams for mineral wool (Figure 15) and for plasterboard (Figure 16), while the relative humidity—temperature isopleth is presented in Figure 17.

Moisture content in variant 5 depending on the season shown in the diagram (Figure 15) for mineral wool increased throughout the research period, starting from 2% in winter to 12% (maximum value last year in summer). The maximum value was approx. 2.4 kg/m^3.

Moisture content in variant 5 depending on the season shown in the diagram (Figure 16) for plasterboard fluctuated throughout the research period from 0.3% in winter to 0.92% (maximum value last year in summer). The maximum value was approx. 6.1 kg/m^3.

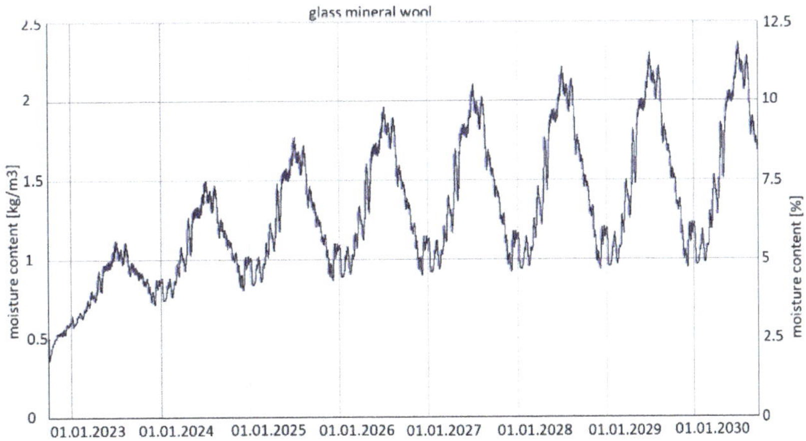

Figure 15. Changes in moisture content in glass mineral wool over eight years, variant 5 [70].

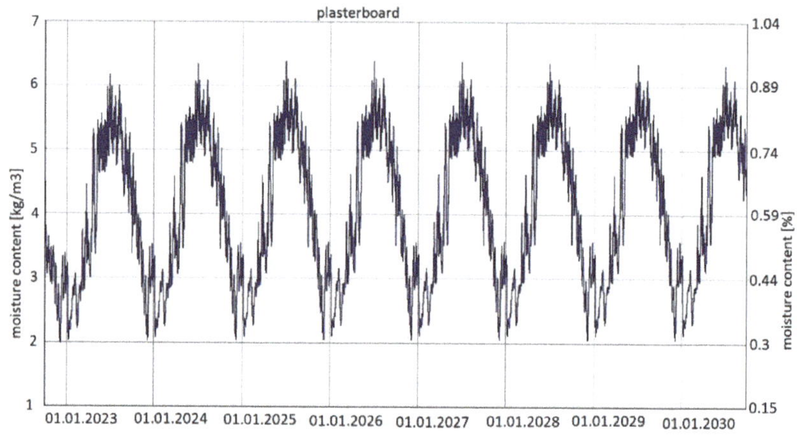

Figure 16. Changes in moisture content in plasterboard over eight years, variant 5 [70].

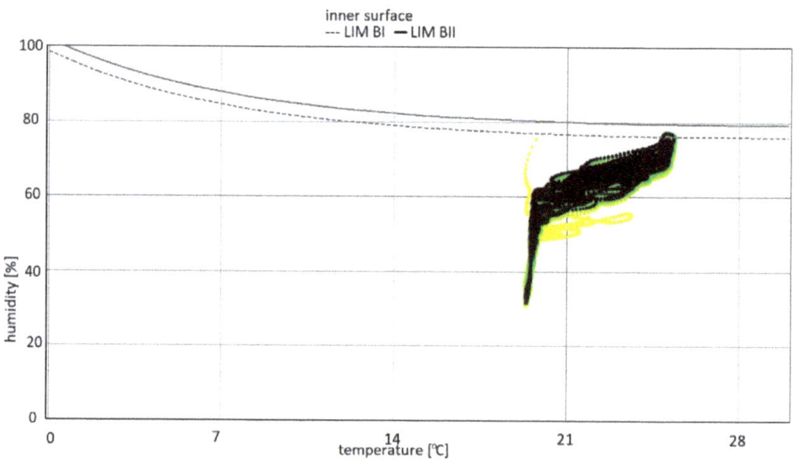

Figure 17. Isopleth of the inner surface of the partition, variant 5 [70].

The isopleths of the LIM B I (dashed line) and LIM B II (solid line) curves i.e., contour charts of the mould growth rate as a function of temperature (horizontal axis) and humidity (vertical axis), do intersect with the area of points for the partition analysed. The risk of mould development appeared at a minimum relative humidity of 78% and a temperature of approx. 24 °C (Figure 17).

By analysing the isopleths for variants 1–5 (Figures 11, 14 and 17), it can be stated that the points from yellow to black show the situation on the inner surface during the simulation period. The coordinates of each point are determined by temperature and relative humidity. If the point is located above the LIM boundary curve as in variant 5, then mould development can be expected; whereas if below the limit as in variants 1–4, then no biological aggression is to be expected. As the analysis showed, the conditions conducive to the development of mould aggression occurred when the flat roof ceased to be well ventilated. The mere arrangement of the moisture blocking layer is irrelevant in variants 1–3.

Summarizing, it can be concluded that the results shown in the Figures regarding thermal insulation materials and plasterboards clearly indicate the trend of fluctuating results depending on the season. Figures 9, 10, 12, 13, 15 and 16 indicate changes in moisture content in thermal insulation materials (Figures 9, 12 and 15) and in plasterboard (Figures 10, 13 and 16), while isopleths (Figures 11, 14 and 17) indicate the possibility of mould development on the inner surface of the material.

Variant 5 shows an upward trend in the moisture content of the thermal material—amounting to 12% in the 8th year of operation, and in the case of variants 1 and 2, the moisture content of mineral wool is at the same level at a max. of 4%. For variant 5, the moisture content peaks at 5% throughout the analyzed period. In the case of moisture content in the gypsum board, the use of a moisture barrier of PE film—whether under the grid, above the grid or by filling the air gap with material—does not affect the amount of moisture appearing in the board, and is 0.44–0.80%. In variants 4 and 5, the amplitude of changes in the moisture content of the gypsum board is greater, namely 0.3–0.85% (variant 4) and 0.3–0.92% (variant 5). As can be seen from the analysis, the moisture content of gypsum plasterboard does not exceed about 0.9%, which does not have a degrading effect on the operation of the partition (except for variant 5). In the case of thermal material, moisture as low as 4% reduces the thermal insulation of the partition by 50%, and at 12%, causes its thermal insulation parameters to drop to 20%. It should be noted here that in variant 5, the moisture content of wool never falls below 5% as early as the second year of operation of the partition. In variants 1, 2, 3 and 4, moisture can build up to 4%, but in summer periods the moisture level falls below 2%. This kind of wetting, especially short-term, does not damage the structure of the wool, and should not result in the permanent deterioration of its thermal or strength properties. In the case of variant 5, the dampness is above 5% from the 2nd year of operation onwards, that is, the wool is soggy and wet, which indicates the possibility of layer degradation and mycological changes.

4. Discussion

Many building are characterized by insufficient air exchange, which may result in the symptoms of sick building syndrome (SBS). A large number of existing buildings are equipped with natural ventilation, but whose effectiveness is reduced by energy-saving activities [71].

Article [72] studies ventilation driven by thermal buoyancy in the air cavity of inclined roofs. The influence of air cavity design and roof inclination on the airflow is investigated. Combinations of different roof inclinations, air cavity heights and applied heating power on the air cavity top surface were examined. The study showed that increased air cavity height led to increased airflow and decreased surface temperatures in the air cavity. Increased roof inclination and heating power applied to the roofing also increased the airflow.

Thanks to the numerical simulation of temperature fields and other parameters of modelled partitions, it is possible to obtain a lot of information about their operation in the long term [62,73]. The analysed partition meets the current thermal standards of buildings. The obtained results of simulation calculations show no influence of the position of PE foil (acting as a vapour barrier, or type of "vapour retarder") against the plasterboard fixing grate on the moisture condition of thermal insulation materials or plasterboard used. This is proven by the results for variants 1 and 2 of the calculations. Moreover, the possible sagging of "rock" mineral wool boards and their contact with the surface of PE foil does not affect moisture content of other built-in materials, as shown in variant 3 of the calculations. When the natural ventilation of roof airspace works (correctly), lack of PE foil does not significantly affect the moisture condition of thermal insulation materials or plasterboards, as shown in variant 4 of the calculation. Only the time of the highest moisture content of glass mineral wool is shifted. Lack of a vapour barrier with simultaneous lack of ventilation of roof airspace creates a risk of a gradual increase in moisture content of thermal insulation materials. In addition, conditions for the development of mould appear on the surface of plasterboards. The occurrence of these phenomena is indicated by variant 5 of the calculations. In double-sided ventilated flat roofs, when there is no water vapour concentration, no vapour barrier is required. Above wet rooms (water vapour pressure above 2150 Pa, i.e., temperature inside $\geq$27 °C and approx. 20 h of hot water evaporation), if it is impossible to use a double-sided ventilated flat roof with a well-ventilated airspace, a solution with an appropriate vapour barrier is used, the type of which depends on the value of diffusion resistance of subsequent layers. For the analysed case, the water vapour pressure did not exceed the given limit at the given temperatures. The highest temperature was 25 °C and corresponded to 70% relative humidity. Even with extreme environmental conditions, with active ventilation (gravitational or mechanical), during the simulated period, the problem of condensation of an excessive amount of water vapour will not occur regardless of whether or not the vapour barrier layers are arranged on the inside of wet rooms.

5. Conclusions

The "sick" building syndrome (SBS) or mycological changes appearing on walls and ceilings are largely affected by the excessive amount of water generated in buildings. Moistened materials contribute to the destruction and faster wear of structures. Properly made ventilation and modelled partition layers protect roof structures against moisture and related further damage.

The numerical design simulation showed that flat roofs containing air spaces should be treated differently to pitched roofs containing small, ventilated spaces in their layers. In the case of flat roofs, it is not always necessary to use flexible waterproofing materials in ventilated partitions to protect their thermal insulation against moisture.

For variants 1–3 (Figures 3–5), the moisture level for mineral wool during the analysed period ranged from 1.7% to 4%; for variant 4 (Figure 6), it ranged from 2.1% to 4.7%; while for variant 5 (Figure 7), moisture content for mineral wool increased throughout the research period, starting from 2% in winter to 12% (maximum value last year in summer).

For variants 1–3 (Figures 3–5), the moisture level for plasterboard ranged from 0.44% to 0.80%; for variant 4 (Figure 6), it ranged from 0.3% to 0.85%; while for variant 5 (Figure 7), moisture content for plasterboard fluctuated throughout the research period from 0.3% in winter to 0.92% (maximum value last year in summer).

For variants 1–3 (Figures 3–5), mould spots appeared at a minimum relative humidity of 41% and a temperature of 18 °C; and for variant 4, mould spots appeared at a minimum relative humidity of 30% and a temperature of 18 °C. For variants 1–4 (Figures 3–6), there are no critical values in the isopleth (Figures 11 and 14), i.e., the point where the curves intersect with the area of points. However, for variant 5 (Figure 7), the risk of mould development appeared at a minimum relative humidity of 78% and a temperature of approx. 24 °C (Figure 17).

The numerical analysis with the assumptions of extreme parameters showed that the problem with moisture will appear in variant 5, i.e., in the absence of PE foil in the system of layers and in the absence of ventilation of the flat roof.

The numerical analysis showed that proper multi-variant modelling reduces the energy loss for heating buildings and eliminates the problem of damage to thermal and finishing materials. Additionally, it should be noted that the use of a flexible waterproofing material in the layers of a double-sided ventilated flat roof is not necessary, even if humidity in the rooms under the roof exceeds 70% and ventilation is carried out by gravity.

The manuscript presents a variety of solutions aimed at designers to protect a ventilated roof from moisture penetrating the thermal insulation and ceiling (gypsum board) layers. An analysis of the validity of the use of a ventilated waterproofing layer in a multilayer partition is presented, with a simulation of the work of such a layer or its absence in the partition over 8 years. In the previously available literature, there has been little research or analysis on the validity of the use of a waterproofing layer placed in different locations of the ventilated ceiling, or evaluation of the variation of moisture in individual layers.

Author Contributions: Conceptualization, B.K. and A.S.-G.; methodology, B.K.; software, B.K.; validation, B.K., A.S.-G. and R.P.; formal analysis, R.P.; investigation, A.S.-G.; resources, B.K.; data curation, B.K.; writing—original draft preparation, B.K. and A.S.-G.; writing—review and editing, B.K., A.S.-G. and R.P.; visualization, R.P.; supervision, B.K., A.S.-G. and R.P.; project administration, A.S.-G.; funding acquisition, R.P. All authors have read and agreed to the published version of the manuscript.

Funding: This research received no external funding.

Institutional Review Board Statement: Not applicable.

Informed Consent Statement: Not applicable.

Data Availability Statement: Not applicable.

Acknowledgments: The publication was co-financed within the framework of the Ministry of Science and Higher Education programme as "Regional Initiative Excellence" in the years 2019–2022, Project No. 005/RID/2018/19.

Conflicts of Interest: The authors declare no conflict of interest.

References

1. Szymczak-Graczyk, A.; Laks, I.; Ksit, B.; Ratajczak, M. Analysis of the Impact of Omitted Accidental Actions and the Method of Land Use on the Number of Construction Disasters (a Case Study of Poland). *Sustainability* **2021**, *13*, 618. [CrossRef]
2. Ksit, B.; Szymczak-Graczyk, A.; Nazarewicz, B. Diagnostics and renovation of moisture affected historic buildings. *Civ. Environ. Eng. Rep.* **2022**, *32*, 59–73. [CrossRef]
3. Karagiannis, N.; Karoglou, M.; Bakolas, A.; Moropoulou, A. Effect of temperature on water capillary rise coefficient of building materials. *Build. Environ.* **2016**, *106*, 402–408. [CrossRef]
4. Frössel, F. *Osuszanie Murów i Renowacja Piwnic [Wall Drainage and Cellar Renovation]*; Polcen Sp. z o.o.: Warsaw, Poland, 2007.
5. Nowogońska, B. Intensity of damage in the aging process of buildings. *Arch. Civ. Eng.* **2020**, *66*, 19–31.
6. Szymczak-Graczyk, A.; Ksit, B.; Laks, I. Operational Problems in Structural Nodes of Reinforced Concrete Constructions. *IOP Conf. Ser. Mater. Sci. Eng.* **2019**, *603*, 032096. [CrossRef]
7. Weber, J. Baudiagnose und Geräte. In *Bauwerksabdichtung in der Altbausanierung: Verfahren und Juristische Betrachtungsweise*; Springer Vieweg: Wiesbaden, Germany, 2012; pp. 99–135.
8. *PN-EN ISO 7730*; Ergonomics of the Thermal Environment. Analytical Determination and Interpretation of Thermal Comfort with the Use of PMV and PPD Indexes and Local Thermal Comfort Criteria. International Organization for Standardization: Geneva, Switzerland, 2005.
9. Chojnacka, A. Wybrane aspekty dotyczące komfortu termicznego w pomieszczeniach [Selected aspects of thermal comfort in rooms]. *Ciepłownictwo Ogrzew. Went.* **2007**, *4*, 8–13.
10. Mijakowski, M. Podstawy opisu zjawisk kształtujących wilgotność powietrza w pomieszczeniach [Basics of description of phenomena shaping indoor air humidity]. *Ciepłownictwo Ogrzew. Went.* **2001**, *10*, 16–19.
11. Pogorzelski, J.A. Fizyka cieplna budowli. In *Thermal Physics of Buildings*; Wydawnictwo Naukowe PWN: Warsaw, Poland, 1976.
12. Pogorzelski, J.A. Zagadnienia cieplno-wilgotnościowe przegród budowlanych [Thermal and humidity issues of building partitions]. In *Budownictwo Ogólne. Fizyka Budowli [General Construction Physics of Structures]*; Klemm, P., Ed.; Arkady: Warsaw, Poland, 2005; Volume 2.
13. Telejko, M.; Zender-Świercz, E. Attempt to improve indoor air quality in kindergartens. *Procedia Eng.* **2016**, *161*, 1704–1709. [CrossRef]
14. Zender-Świercz, E. Analysis of the impact of the parameters of outside air on the condition of indoor air. *Int. J. Environ. Sci. Technol.* **2017**, *14*, 1583–1590. [CrossRef]
15. Szymczak-Graczyk, A.; Gajewska, G.; Laks, I.; Kostrzewski, W. Influence of Variable Moisture Conditions on the Value of the Thermal Conductivity of Selected Insulation Materials Used in Passive Buildings. *Energies* **2022**, *15*, 2626. [CrossRef]
16. Alizamir, M.; Kim, S.; Zounemat-Kermani, M.; Heddam, S.; Kim, N.W.; Singh, V.P. Kernel Extreme Learning Machine: An Efficient Model for Estimating Daily Dew Point Temperature Using Weather Data. *Water* **2020**, *12*, 2600. [CrossRef]
17. Cichoń, A.; Worek, W. Analytical Investigation of a Novel System for Combined Dew Point Cooling and Water Recovery. *Appl. Sci.* **2021**, *11*, 1481. [CrossRef]
18. Heßler, M. Unikanie kondensacji i pleśni [Avoiding Condensation and Mould]. *Świat Szkła* **2013**, *12*, 30–32.
19. Zyska, B. *Zagrożenia Biologiczne w Budynku [Biological Hazards in the Building]*; Arkady: Warsaw, Poland, 1999.
20. Gutarowska, B.; Piotrowska, M.; Koziróg, A. *Grzyby w Budynkach. Zagrożenia, Ochrona, Usuwanie [Fungi in Buildings. Threats, Protection, Deletion]*; Wydawnictwo Naukowe PWN: Warsaw, Poland, 2019.
21. Szymczak-Graczyk, A.; Walczak, Z.; Ksit, B.; Szyguła, Z. Multi criteria diagnostics of historic buildings with the use of 3D laser scanning (a case study). *Bull. Pol. Acad. Sci. Tech. Sci.* **2022**, *70*, e140373. [CrossRef]
22. Muselli, M.; Carvajal, D.; Beysens, D.A. A Study to Explore the Dew Condensation Potential of Cars. *Atmosphere* **2022**, *13*, 65. [CrossRef]
23. Geryło, R. Powierzchniowa kondensacja pary wodnej—Przegrody przeszklone [Surface condensation of water vapour—Glazed partitions]. *Świat Szkła* **2010**, *12*, 23–25.
24. Macieik, L. Zawilgocenia ściany mieszkania na skutek złej eksploatacji [Moisture in the wall of the apartment as a result of poor exploitation]. *Warstwy Dachy Ściany* **2008**, *1*, 90–91.

25. *PN-83/B-03430*; Ventilation in Residential Buildings, Collective Residence, and Public Utility Buildings. Requirements. Polski Komitet Normalizacyjny: Warsaw, Poland, 2000.
26. *EN 15251:2007*; Indoor Environmental Input Parameters for Design and Assessment of Energy Performance of Buildings Addressing Indoor Air Quality, Thermal Environment, Lighting and Acoustics. International Energy Agency: Paris, France, 2007.
27. Śliwowski, L. *Mikroklimat Wnętrz i Komfort Cieplny Ludzi w Pomieszczeniach [Microclimate of Interiors and Thermal Comfort of People Indoors]*; Oficyna Wydawnicza Politechniki Wrocławskiej: Wrocław, Poland, 1999.
28. Baumgarth, S.; Hörner, B.; Reeker, J. *Poradnik Klimatyzacji [Air Conditioning Handbook]*; SYSTHERM: Poznań, Poland, 2010; Volume 1.
29. Traczyk, W.Z.; Trzebski, A. *Fizjologia Człowieka z Elementami Fizjologii Klinicznej i Stosowanej [Human Physiology with Elements of Clinical and Applied Physiology]*; Wydawnictwo Lekarskie PZWL: Warsaw, Poland, 2004.
30. Fischer-Uhlik, H. Mikroklimat pomieszczeń. Cz. 1. [Microclimate of rooms. Part 1]. *Świat Szkła* **2002**, *10*, 32–36.
31. Fischer-Uhlik, H. Mikroklimat pomieszczeń. Cz. 2. [Microclimate of rooms. Part 2]. *Świat Szkła* **2002**, *11*, 36–38.
32. Charkowska, A. Komfort cieplny w środowisku zimnym, termicznie umiarkowanym i gorącym [Thermal comfort in a cold, thermally moderate and hot environment]. *Chłodnictwo Klim.* **2010**, *10*, 48–52.
33. Śliwowski, L.; Lis, A. Wybrane problemy komfortu cieplnego osób w pomieszczeniach [Selected problems of thermal comfort of people indoors]. *Ciepłownictwo Ogrzew. Went.* **2002**, *33*, 45–49.
34. Adrian, Ł. Klimatyzacja i wentylacja budynków użyteczności publicznej w aspekcie komfortu cieplnego [Air conditioning and ventilation of public buildings in terms of thermal comfort]. *Chłodnictwo Klim.* **2010**, *11*, 18–25.
35. Young, T. III An essay on the cohesion of fluids. *Philos. Trans. R. Soc. Lond.* **1805**, *95*, 65–87.
36. Patoka, K. Regulatory pary w przegrodach dachowych [Steam regulators in roof partitions]. *Mater. Bud.* **2021**, *584*, 32–34.
37. Zieliński, J.; Przybylski, J. Ile wody tracimy z oddechem [How much water do we lose with the breath]. *Pneumonol. Alergol. Pol.* **2002**, *80*, 339–342.
38. Styrofoam Manufacturers Association Guide. Available online: http://www.styropiany.pl/poradnik/poradnik_03_fiz_budowli.pdf (accessed on 30 June 2022).
39. Available online: http://ekoenergia.dzien-e-mail.org/wentylacja/22-powietrzew-pomieszczeniach/265-wilgo-w-pomieszczeniach-mieszkalnych.html (accessed on 30 June 2022).
40. Trechsel-Heinz, R. *Moisture Control in Buildings*; ASTM Manual Series: Philadelphia, PA, USA, 1994.
41. Škerget, L.; Tadeu, A.; Almeida, J. Unsteady Coupled Moisture and Heat Energy Transport through an Exterior Wall Covered with Vegetation. *Energies* **2021**, *14*, 4422. [CrossRef]
42. Patoka, K. *Poradnik. Wentylacja Dachów i Stropodachów [Handbook. Roof and Flat Roof Ventilation]*; Dom Wydawniczy MEDIUM: Warsaw, Poland, 2010.
43. Moisture—A Factor Affecting the Building. Available online: http://www.ogrzewnictwo.pl/artykuly/inne/wilgoc-czynnikoddzialywujac-na-budynek (accessed on 30 June 2022).
44. Patoka, K.; Wiluś, S. Zasady doboru warstw wstępnego krycia dla dachów skośnych. Cz. 3. Sposób wykorzystania klas szczelności. Wentylacja dachów [Principles of selecting the initial covering layers for pitched roofs. Part 3. The way of using the tightness classes. Roof ventilation]. *Izolacje* **2018**, *23*, 88–92.
45. Kain, G.; Idam, F.; Federspiel, F.; Réh, R.; Krišťák, Ľ. Suitability of Wooden Shingles for Ventilated Roofs: An Evaluation of Ventilation Efficiency. *Appl. Sci.* **2020**, *10*, 6499. [CrossRef]
46. Wierzelewski, P. Zastosowanie folii w konstrukcjach dachów skośnych [Application of foil in pitched roof structures]. *Warstwy Dachy Ściany* **2004**, *1*, 103–105.
47. Patoka, K. Funkcje i podział folii wstępnego krycia [Functions and division of the initial covering foil]. *Mater. Bud.* **2003**, *5*, 52–53.
48. Patoka, K. Folie wstępnego krycia—Niezbędny element nowoczesnego dachu spadzistego [Initial covering foils—An indispensable element of a modern pitched roof]. *Mater. Bud.* **2000**, *2*, 77–78.
49. Patoka, K. Funkcje warstw paroizolacyjnych w przegrodach dachowych [Functions of vapour barrier layers in roof partitions]. *Mater. Bud.* **2021**, *582*, 32–34.
50. *DIN 4108-3:2001-07*; Thermal Insulation and Energy Saving in Buildings. Part 3. Protection Against Moisture Depending on Climatic Conditions, Requirements, Calculation Method and Guidelines for Design and Execution. Beuth Publishing Din: Berlin, Germany, 2001.
51. *PN-EN ISO 6946:2017*; Building Components and Building Elements. Thermal Resistance and Heat Transfer Coefficient. Calculation Methods. International Organization for Standardization: Geneva, Switzerland, 2017.
52. *PN-EN ISO 13788:2017*; Heat and Humidity Properties of Building Components and Building Elements. Internal Surface Temperature Necessary to Avoid Critical Surface Moisture and Interlayer Condensation. Calculation Methods. International Organization for Standardization: Geneva, Switzerland, 2017.
53. *PN-EN ISO 9346:2009*; Heat and Humidity Performance Properties of Buildings and Construction Materials. Physical Quantities Concerning Mass Transfer. Dictionary. International Organization for Standardization: Geneva, Switzerland, 2009.
54. *Reguły Dekarskie Niemieckiego Związku Dekarzy. Instrukcja. Instrukcja Izolacji Cieplnej Dachów [Roofing Rules of the German Roofers' Association. Instruction. Thermal Insulation of Roofs and Walls]*; Wydawnictwo Rudolf Müller: Cologne, Germany, 1997.
55. *Reguły Dekarskie Niemieckiego Związku Dekarzy. Instrukcja. Izolacja Cieplna Dachów i Ścian [Roofing Rules of the German Roofers' Association. Instruction. Thermal Insulation of Roofs and Walls]*; Wydawnictwo Rudolf Müller: Cologne, Germany, 2004.

56. Patoka, K. Zmiany w zasadach wentylowania dachów w wytycznych Związku Dekarzy Niemieckich [Changes in the rules of roof ventilation in the guidelines of the Association of German Roofers]. *Mater. Bud.* **2018**, *546*, 38–40.
57. Dictionary of Roofing Terms and Names. Available online: https://static.dekarz.com.pl/default/TDFvI9lMPSDSlownikdekar.pdf (accessed on 30 June 2022).
58. Kłopocka, L. Wentylacja Dachu a Analiza Bilansu Pary Wodnej w Budynku Mieszkalnym [Roof Ventilation and the Analysis of Water Vapour Balance in a Residential Building]. Bachelor's Thesis, Politechnika Poznańska, Poznań, Poland, 2015.
59. Orlik-Kożdoń, B. Microclimate Conditions in Rooms: Their Impact on Mold Development in Buildings. *Energies* **2020**, *13*, 4492. [CrossRef]
60. van Hees, R.P.J.; Koek, J.A.G. Treatment of rising damp. A laboratory evaluation method. In Proceedings of the International Colloquium on Methods of Evaluating Products for the Conservation of Porous Building Materials in Monuments, Rome, Italy, 19–21 June 1995; pp. 403–418.
61. Patoka, K. Wentylacja membran wstępnego krycia (MWK) [Ventilation of initial covering membranes (ICM)]. *Warstwy Dachy Ściany* **2008**, *3*, 28–31.
62. Ksit, B.; Szymczak-Graczyk, A. Rare weather phenomena and the work of large-format roof coverings. *Civ. Environ. Eng. Rep.* **2019**, *29*, 123–133. [CrossRef]
63. Weber, C. Membrany dachowe—Kluczowe zagadnienia dotyczące starzenia i trwałości [Roofing Membranes—Key Aging and Durability Issues]. *Warstwy Dachy Ściany* **2008**, *2*, 120–121.
64. Patoka, K. Ocena właściwości membran wstępnego krycia [Assessment of the properties of initial covering membranes]. *Mater. Bud.* **2008**, *6*, 39–41.
65. Horta, A. Sekrety membran dachowych [Secrets of Roofing Membranes]. *Warstwy Dachy Ściany* **2005**, *3*, 68–70.
66. Podeszwa, M. Cele i zasady stosowania folii i membran dachowych [Objectives and principles of using roofing membranes and foils]. *Warstwy Dachy Ściany* **2009**, *2*, 86–87.
67. Olaoye, T.S.; Dewsbury, M.; Künzel, H. Laboratory Measurement and Boundary Conditions for the Water Vapour Resistivity Properties of Typical Australian Impermeable and Smart Pliable Membranes. *Buildings* **2021**, *11*, 509. [CrossRef]
68. Künzel, H.M. One- and Two-Dimensional Calculation Using Simple Parameters. In *Simultaneous Heat and Moisture Transport in Building Components;* Fraunhofer IRB Verlag: Stuttgart, Germany, 1995.
69. Available online: https://wufi.de/download/WUFI41_help_e.pdf (accessed on 30 June 2022).
70. Ksit, B.; Gaczek, M. *Raport z Badań Stropodachu Eliptycznego [Report on the Study of an Elliptical Flat Roof];* Politechnika Poznańska: Poznań, Poland, 2022.
71. Zender-Świercz, E. Microclimate in Rooms Equipped with Decentralized Façade Ventilation Device. *Atmosphere* **2020**, *11*, 800. [CrossRef]
72. Bunkholt, N.S.; Säwén, T.; Stockhaus, M.; Kvande, T.; Gullbrekken, L.; Wahlgren, P.; Lohne, J. Experimental Study of Thermal Buoyancy in the Cavity of Ventilated Roofs. *Buildings* **2020**, *10*, 8. [CrossRef]
73. Geryło, R.; Młoźniak, D.; Pogorzelski, J.A. Wybrane problemy numerycznej symulacji pól temperatury w Zakładzie Fizyki Cieplnej ITB [Selected problems of numerical simulation of temperature fields in the Department of Thermal Physics ITB]. *Work. Inst. Tech. Bud.* **2000**, *29*, 24–37.

materials

MDPI

Article

Sensitivity Analysis of Open-Top Cartons in Terms of Compressive Strength Capacity

Damian Mrówczyński [1], Tomasz Gajewski [2] and Tomasz Garbowski [3,*]

[1] Doctoral School, Department of Biosystems Engineering, Poznan University of Life Sciences, Wojska Polskiego 28, 60-637 Poznań, Poland
[2] Institute of Structural Analysis, Poznan University of Technology, Piotrowo 5, 60-965 Poznań, Poland
[3] Department of Biosystems Engineering, Poznan University of Life Sciences, Wojska Polskiego 50, 60-627 Poznań, Poland
* Correspondence: tomasz.garbowski@up.poznan.pl

Abstract: Trays in which fruit and vegetables are transported over vast distances are not only stored in extreme climatic conditions but are also subjected to long-term loads. Therefore, it is very important to design them correctly and select the optimal raw material for their production. Geometric parameters that define the shape of the packaging may also be optimized in the design process. In this work, in order to select the most important parameters that affect the load capacity of a tray, sensitivity analysis was used. A sensitivity analysis is often the first step in the process of building artificial-intelligence-based surrogates. In the present work, using the example of a specific tray's geometry, the selection of starting parameters was carried out in the first step, based on the Latin hypercube sampling method. In the next step, local sensitivity analyses were performed at twenty selected starting points of the seventeen-dimensional space of the selected parameters. Based on the obtained results, it was possible to select the parameters that have a significant impact on the load capacity of the tray in the box compression test and whose influence is negligible or insignificant.

Keywords: sensitivity analysis; corrugated board; trays; open-top packaging

check for updates

Citation: Mrówczyński, D.; Gajewski, T.; Garbowski, T. Sensitivity Analysis of Open-Top Cartons in Terms of Compressive Strength Capacity. *Materials* **2023**, *16*, 412. https://doi.org/10.3390/ma16010412

Academic Editor: Swarup Roy

Received: 15 December 2022
Revised: 27 December 2022
Accepted: 28 December 2022
Published: 1 January 2023

1. Introduction

Open-top, corrugated board cartons have become one of the most popular ways to transport fruits and vegetables. Long-distance transport applies to raw food products such as tomatoes, peppers, bananas, apples, lemons, and many more. The natural material of paper favors the long-distance transport of such products, especially when moisture and temperature changes occur frequently during transport. Corrugated cardboard packaging is also perceived by consumers as ecologically friendly, and scientific studies have confirmed that customers are willing to pay more for environmentally sustainable packaging [1–3]. For instance, for decades bananas, have been effectively transported in cartoon boxes. In 2020, 21.5 million tons of bananas were exported, from which about 16.5 million tons were exported from Africa. The European Union and the United States are the biggest importers, accounting for 26% and 21% of the share in global imports in 2020 [4]. The ventilation conditions and the maturation of bananas during transport are crucial throughout the production process.

Open-top, corrugated board cartons prevail over other types of packaging if the particular ventilation conditions are ensured. The ventilation holes allow air to circulate inside the packaging, which also enables water to evaporate from the inside of the container, which, in turn, protects the raw food products from undesirable moisture. Superior air flow in the packaging facilitates the disposal of ethylene, which causes the fruit to ripen as an aging-stimulating hormone. Therefore, on the one hand, the better the ventilation, the better the fruit's condition, while on the other hand, the more ventilation holes, the lower the packaging strength (i.e., less material to bear the loads) [5–7]. Both objectives are contradictory, although the appropriate location of the ventilation holes can significantly reduce the loss of the load-bearing capacity of the packaging and, at the same time, maximize air flow. In Figure 1, an example of an open-top carton for the transport of fruits/vegetables is presented. Optimizing the use of material to maximize the efficiency of the packaging (regardless of the measure, e.g., compressive strength, resistance to random vibrations, ventilation of products, etc.) is, and will always be, a significant problem in the field of packaging design [8].

Figure 1. Selected example of an open-top box for the transport of fruits or vegetables.

Open-top, corrugated board cartons constitute a very specific form of packaging with a complex structure, which is significantly different from typical flap packaging such as F0201 according to FEFCO codes [9]. Due to the complexity of this type of packaging, the use of known analytical formulas to determine its compression load capacity is inadequate [10–16]. The popular formulas are adequate for simple flap boxes. Many geometrical parameters for determining the shape of open-top packaging for fruits/vegetables make it difficult, and even impossible, to derive the analytical formula for computing the compressive strength of this type of packaging. Therefore, to avoid relying on the experience of the designer, numerical tools for the estimation of strength should be adopted. However, building a model for complex packaging is not a trivial task; it requires experience in numerical modelling. In this regard, the utilization of artificial neural networks (ANNs) preceded by a numerical study could be the solution, for instance, as performed in [17,18]. Numerical methods and ANNs are used effectively in similar problems in the food/packaging industry [19–22] and others [23,24]. An ANN model must be versatile enough to cover a moderately wide range of possible designs. ANNs can be efficient, even for a large number of parameters; however, many parameters require numerous pre-processing computations to prepare input data for the ANN's training, testing, and validation [25]. The increase in the computational cost with the increasing number of parameters is non-linear.

A good practice in building an artificial neural network is to analyze the sensitivity of potential input parameters to the network so that the output argument of the network depends meaningfully on the input arguments. It is recommended that the ANN inputs (i.e., all the model parameters) that have proven to exert a very low level of influence on the ANN's output (i.e., BCT value) are removed from the model to limit the pre-processing cost. A reliable sensitivity analysis for open-top corrugated board cartons for fruits and vegetables is not available in the scientific literature; therefore, addressing this gap is the main aim of the paper. There are papers that analyze the role of horticultural cartons' vent hole design on cooling efficiency and compression strength [26–28]; however, this constitutes research into a different type of cartoon.

In this paper, we will indicate the sensitivity of an open-top cartoon in respect to its main geometrical parameters. We will point out which of the parameters may be omitted while building the ANN network to estimate the compression strength of the open-top box in order to limit the computational cost of ANN input data. Resultantly, the parameters that seem crucial for the corrugated board packaging industry do not have a significant impact on the compression strength of the packaging.

2. Materials and Methods

2.1. Geometric Parametrization of Open-Top Boxes

There are many types of trays for fruit and vegetables in the corrugated carboard packaging offered by its respective industry, but there are some constructions that have repeatable folding and similar dimensions. Manufacturers often have their own packaging designs selected on the basis of their own experience. As part of this work, a type of tray that is relatively popular in the market was selected. The geometry of such packaging is shown in Figure 2 as 2D drawing and in 3D view. In this type of packaging, the sidewalls fold vertically, and the elements of the shorter walls form rectangular triangles reinforcing the corners. One of the sides of the triangular corners is glued to the longer sidewalls, which ensures the rigidity of the corners. Moreover, on the bottom edges of the tray, there are circular or oval ventilation holes, usually two for each edge. In addition, there are trapezoidal folds at the tops of the sidewalls (one for each wall). These folds are bent down and glued to strengthen the sidewalls (not shown in the drawings).

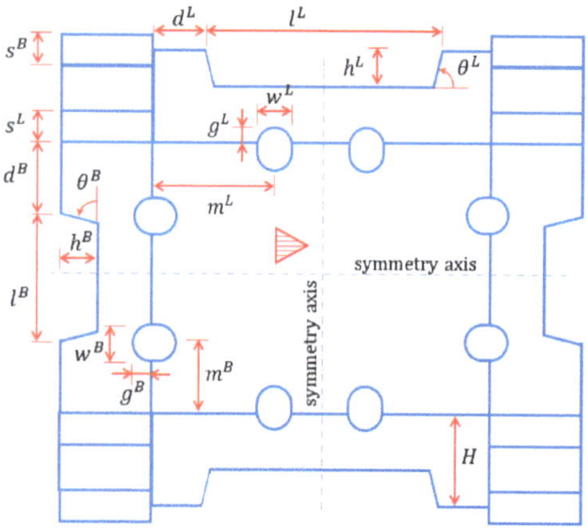

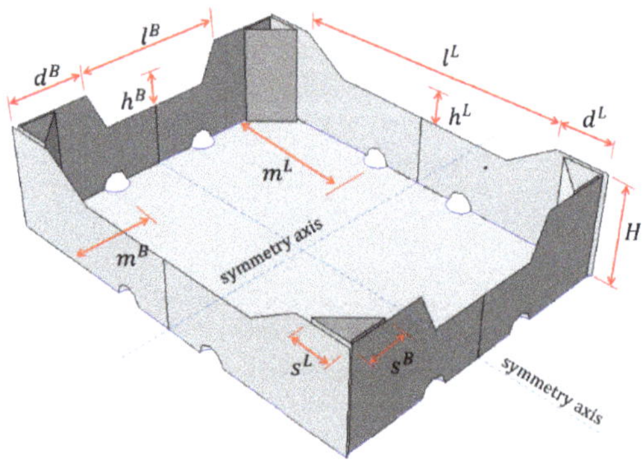

Figure 2. Parametric model used in the study: (**a**) 2D grid and (**b**) 3D scheme (for clarity, w^L, g^L, w^B, g^B, θ^L, and θ^B parameters are not shown here).

Based on the selected packaging design, over a dozen geometric parameters were defined, which are labelled in Figure 2. The selection criteria were as follows: (i) the suspected significance of the impact on the box compression test (BCT) value and (ii) its importance in terms of wicking or other functions of packaging. These selected design parameters are as follows:

- d^L, half of the horizontal length of the non-folded part of the longer sidewalls;
- d^B, half of the horizontal length of the non-folded part of the shorter sidewalls;
- H, the height of the stiffening triangles and the box;
- l^L, the width of the trapezoidal folds on the longer sidewalls;
- h^L, the height of the trapezoidal folds on the longer sidewalls;
- l^B, the width of the trapezoidal folds on the shorter sidewalls;
- h^B, the height of the trapezoidal folds on the shorter sidewalls;
- s^L, the length of the sides of the stiffening triangles on the longer sidewalls;
- s^B, the length of the sides of the stiffening triangles on the shorter sidewalls;
- w^L, the width of the edge holes on the longer sidewalls;
- g^L, the height of the edge holes on the longer sidewalls;
- m^L, the dist. of the edge holes on the longer walls from the shorter walls to its axis;
- w^B, the width of the edge holes on the shorter sidewalls;

- g^B, the height of the edge holes on the shorter sidewalls;
- m^B, the dist. of the edge holes on the shorter walls from the longer walls to its axis;
- θ^L, inclination of the arms of the trapezoidal folds on the longer sidewalls;
- θ^B, inclination of the arms of the trapezoidal folds on the shorter sidewalls.

Note that the d^L and d^B modifications change the in-plane dimensions of the box model (because, in this case, l^L and l^B are constant), while the l^L and l^B modifications do not change these dimensions (due to simultaneous changes of d^L and d^B).

After determining the box's design parameters, listed above, a numerical algorithm was created using MATLAB software in order to automatically generate the FE model in Abaqus FEA, which is capable of simulating a box compression test. After creating the algorithm, the twenty sets of design parameters were determined according to a Latin Hypercube Sampling (LHS) strategy. One of the examples of LHS algorithm was published by Jin et al. [29]. LHS was used to explore the space of seventeen design parameters most effectively in practically applicable ranges. The use of such a strategy was dictated by the need to build a dataset concerning the sensitivity of the system in a global sense. Exploration of the parameter space in selected 20 locations allows one to understand the relationship between the sensitivity of a given parameter and its initial value. The final sets of design parameters are presented by numbers in Table 1 and graphically by box designs in Figure 3. It is worth mentioning that the presented approach allows for a very even sampling of the space of all seventeen parameters of the model, whose sensitivity to minor perturbations in these parameters at different points in the space may be different. Therefore, this approach allows for the acquirement of averaged responses in the form of sensitivities to all model parameters in the full range of parameters' space.

Table 1. Geometric parameters of selected cases of open-top boxes; all dimensions, apart from the last two columns, are shown in mm.

Box Case	d^L	d^B	H	l^L	h^L	l^B	h^B	s^L	s^B	w^L	g^L	m^L	w^B	g^B	m^B	θ^L (°)	θ^B (°)
1	366	282	94	274	30	141	37	36	31.5	25	12.5	134.5	25	12.5	91.5	33.5	65
2	330	282	94	238	30	141	37	36	31.5	25	12.5	134.5	30	15	91.5	33.5	65
3	366	254	94	274	30	113	37	36	31.5	30	15	134.5	25	12.5	91.5	33.5	65
4	366	282	85	274	30	141	37	36	31.5	25	12.5	90	25	12.5	70	33.5	65
5	366	282	94	214	30	141	37	36	31.5	20	10	114.5	20	10	111.5	90	90
6	366	282	94	220	30	141	37	50	40	32	16	134.5	36	18	91.5	33.5	65
7	386	282	94	220	30	141	37	30	40	32	16	134.5	25	12.5	91.5	45	45
8	386	302	94	220	40	141	40	30	40	32	16	134.5	25	12.5	91.5	75	60
9	391	262	104	220	40	141	40	25	25	28	14	125	25	12.5	91.5	75	60
11	396	262	90	220	40	141	40	45	32	35	14	125	30	12.5	91.5	80	85
12	401	265	98	240	20	161	20	38	40	34	17	125	28	14	91.5	55	45
13	386	268	98	186	20	101	20	38	32	20	10	125	20	10	91.5	55	45
14	388	271	98	206	15	121	25	42	45	20	10	155	20	10	111	55	45
15	392	252	94	206	15	121	25	30	27	20	10	92	20	10	85	55	45
16	396	252	81	300	15	170	25	30	27	20	10	135	20	10	111	20	35
17	398	252	81	300	35	170	35	30	27	20	10	135	20	10	111	37	35
18	396	252	83	261	25	150	35	36	36	35	10	135	35	10	85	65	55
19	310	290	85	174	25	150	35	30	28	16	8	95	16	8	85	65	55
20	320	271	81	174	25	160	35	40	28	22	11	115	22	11	95	40	45

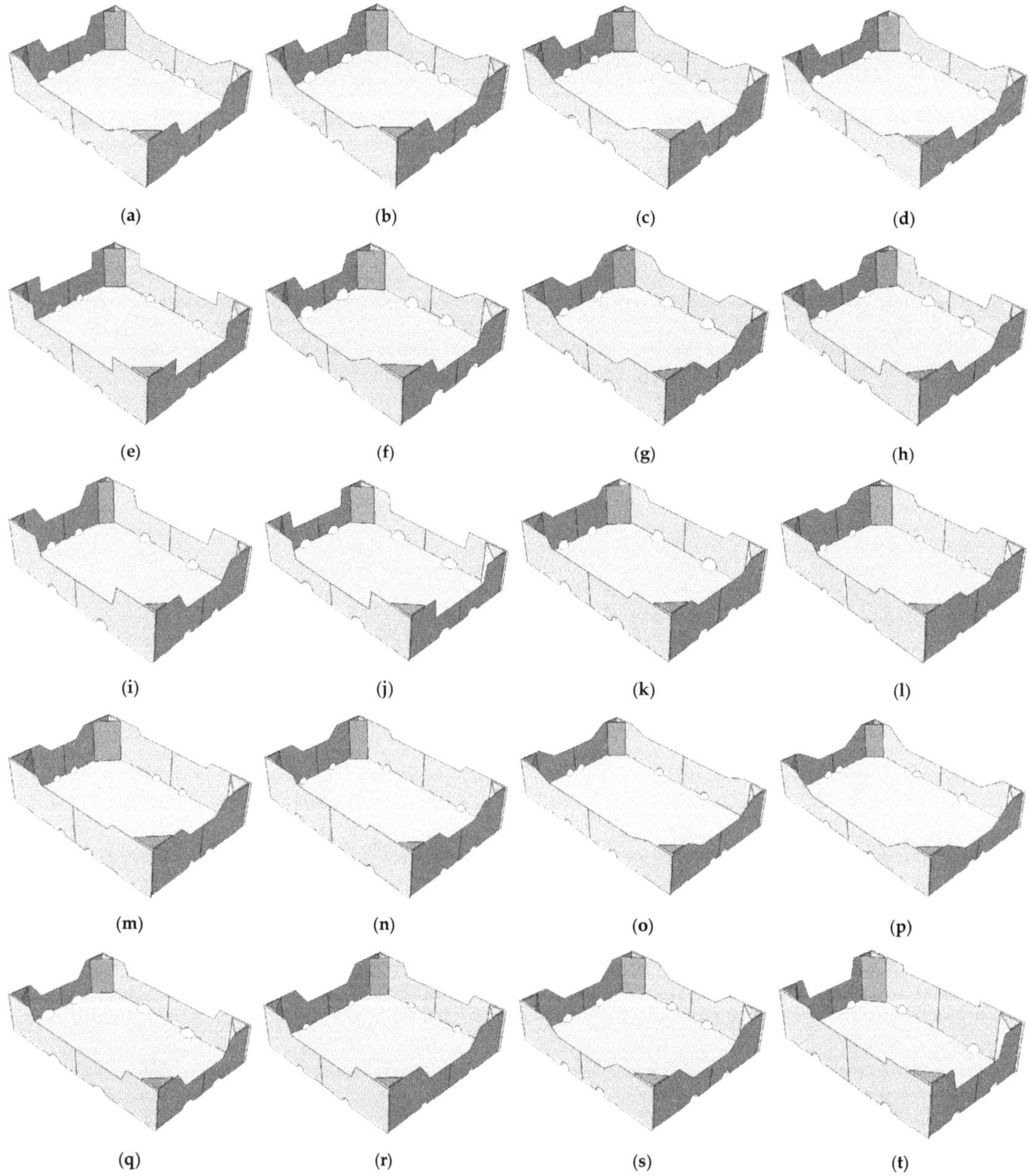

Figure 3. Cases of selected open-top boxes for the storage of fruits. **(a–t)** selected twenty carton geometries.

2.2. Finite Element Model of Open-Top Boxes

Numerical models were created in commercial software FE (Abaqus Unified FEA software [30]) to simulate box compression test (BCT) of open-top boxes. In order to reduce the number of finite elements and shorten computational time, only 1/4 of the box was modeled. The bottom of the packaging was omitted because the bottom does not contribute to the compression-related load-bearing capacity. To ensure robust behavior of the model, the above-mentioned simplifications have been replaced with appropriate boundary conditions (see Figure 4). Symmetrical boundary conditions on the two sidewalls were applied and the vertical displacements of the top edges of the box panels were blocked. Out-of-plane displacements of the bottom and top edges of the panels were also blocked. The analysis consists of two computational steps. In the first step, a buckling analysis was performed in order to compute the first mode, which was applied as imperfection to the model, and in the second nominal step, the packaging was loaded by applying vertical displacement on the top edges. In Figure 4, the boundary conditions for both steps are shown. The figure also presents the corner panels (green color) that are glued to the sidewall fragments (red color) in order to assemble the packaging. In numerical model, this connection of panels has been mapped by special numerical techniques, so-called 'tie connection', which ensures the integrity (continuity of displacements) of the structure between two parts considered.

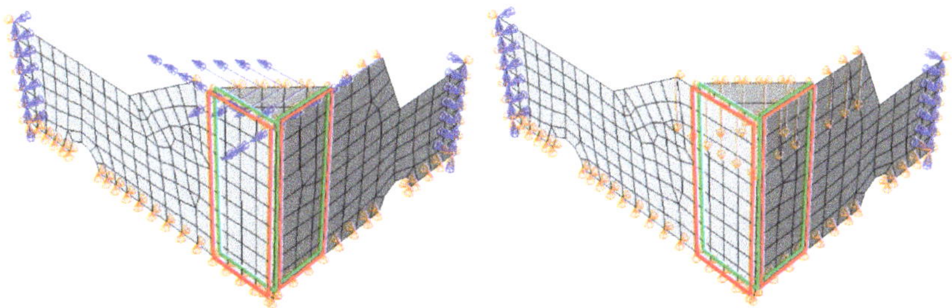

(**a**) (**b**)

Figure 4. Finite element model of open-top boxes for transport of fruits with boundary conditions and mesh (the red and green rectangles indicate where the vertical faces are in contact): (**a**) buckling step; (**b**) compression step.

The analysis was performed for three different corrugated boards that were modeled as linear elastic orthotropic material with Hill plasticity [31]. In Table 2, the parameters of three materials used in the model are given. The material data were determined by the BSE System via FEMAT [32] from mechanical tests of corrugated board samples. Samples were prepared in laboratory and conditioned in a climate chamber. For each test, 10 samples were used to acquire statistically representative material data. In the first column of Table 2, the grade symbol represents the type of the wave and the grammage of the cardboard in g/m^2. Columns 2–7 contain elastic material parameters: E_1 and E_2 are the moduli of elasticity, ν_{12} is the Poisson's ratio, G_{12} is the in-plane shear stiffness, and G_{13} and G_{23} are the transverse shear stiffnesses. The last two columns contain plastic parameters of the material; σ_0 is the initial yield stress and R_{11} is the yield stress ratio in the machine direction of the corrugated cardboard.

Table 2. Material data used in the constitutive models of corrugated boards considered herein.

Grade	E_1 (MPa)	E_2 (MPa)	ν_{12} (–)	G_{12} (MPa)	G_{13} (MPa)	G_{23} (MPa)	σ_0 (MPa)	R_{11} (–)
B-840	2032	1111	0.40	1184	7	11	3.05	0.95
EB-880	1636	907	0.40	963	8	11	3.50	0.65
EB-965	1616	750	0.44	898	7	11	3.01	0.74

For each material, 20 boxes with dimensions shown in Table 1 were analyzed. Each of the parameters from Table 1 for each grade and box design was subjected to a 1% perturbation. This means that 18 analyses were performed for each of the 20 geometries (one reference case and 17 analyses with one parameter changed). In total, this yielded

1080 numerical models (3 materials × 20 packaging designs × 18 analyses). In each model, 4-node quadrilateral shell elements with full integration, named S4 according to [30], were used, and they were completed with 3-node triangular shell elements with full integration, named S3 according to Abaqus FEA. A global mesh size equal to 10 mm was assumed, which resulted in a different number of nodes and elements for each geometry case. For example, for the first case, 373 elements (368 quadrilateral elements and 5 triangular elements) and 437 nodes were obtained, as shown in Figure 4. The choice of such a finite element dimension was based on the observations made in our previous studies [9,33,34], as well as the validation procedure presented in the next section, which was carried out to verify the computational models and commercial tools used.

2.3. Model Validation

The finite element model used in this research (see Section 2.2) was validated through experimental research. Namely, ten samples of open-top cartons for the storage of vegetable or fruits were manufactured and tested in a mechanical press in order to compare the experimental results of BCT with the numerical prediction according to the computational approach used in the study. In the computational model for validation, the boundary conditions, mesh (element size, element type, etc.), constitutive law, and two-step strength analysis (buckling followed by static analysis) were the same as those described in Section 2.2. The sample of open-top carton for validation is shown in Figure 5a and its numerical model geometry is depicted in Figure 5b. In Figure 5b, the deformed box obtained at maximal compression force was confronted with its numerical counterpart (see Figure 5b). It is visible that the deformation modes are in good agreement. Moreover, if the compression strengths obtained are compared, the numerical prediction was burdened with 6.4% error compared to the average strength obtained from tests.

(**a**) (**b**)

Figure 5. Model validation: (**a**) the box sample during the test and (**b**) its numerical counterpart at maximal strength (maximum reaction force value).

Notably, in this study, the sensitivity of the model to the size of the finite element mesh was also ascertained. It was concluded from the analyses that the models in which elements with dimensions of about 15 mm were used were characterized by slightly increased stiffness in the elastic buckling phase; however, in the non-linear phase, the use of smaller elements (e.g., 5 mm) did not increase the precision of the calculations. Therefore, in the model used for the final verification, a grid with elements of about 10 mm was used.

2.4. Sensitivity Analysis

In this paper, the sensitivity analysis of BCT was performed for 20 open-top carton geometries. In each case, the parameters of the model were the dimensions of the packaging (see Figure 2), which were collected in vector $\mathbf{x}$. The BCT value for the selected set of parameters can be denoted as $h(\mathbf{x})$. Then, by small perturbations of the i-th parameter Δx_i, it is possible to calculate the change in the investigated quantity $h(\mathbf{x} \pm \mathbf{e}_i \Delta x_i)$, where $\mathbf{e}_i$ is the unit vector of the i-th parameter in the parameter space. Determination of the numerical gradient using, e.g., the central difference, allows one to obtain the sensitivity of the compressive strength to the change in the considered parameter, according to the following formula:

$$s = \frac{h(\mathbf{x} + \mathbf{e}_i \Delta x_i) - h(\mathbf{x} - \mathbf{e}_i \Delta x_i)}{2\Delta x_i} \frac{x_i}{h(\mathbf{x})}. \quad (1)$$

The performed analysis is a non-local sensitivity analysis. This means that the computations were carried out for many points in the parameter space (20 packaging designs) to build a dataset concerning the entire range of space, and not just locally at one specific point. In this paper, the approach to compute the sensitivity by Equation (1) is similar to the one used in [33,34].

3. Results

First, the material data used to model the corrugated cardboard were acquired. Then, the geometry of the packaging together with the material data were used to build the models in the FE software. Next, for the created models, buckling analyses were performed, from which maps of displacement of individual panels were obtained in order to determine the initial imperfections of the model to be used in further computations. In order to calculate the buckling modes, the vertical displacement on the top edges of the box was used instead of the loads. In Figure 6, the displacements calculated for examples of cases 1, 19, and 20 are presented.

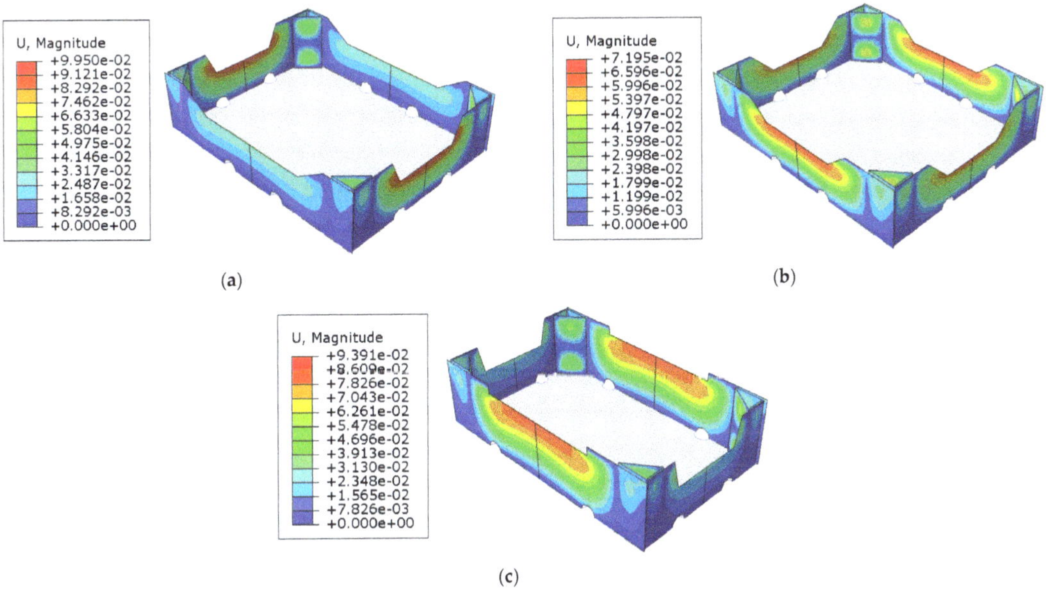

Figure 6. Selected buckling modes used for nominal strength analysis using the finite element method: (**a**) case 1, (**b**) case 19, and (**c**) case 20.

The second step of the numerical analysis was the compression of the packaging. This made it possible to obtain the force–displacement relationship and identify the compression load capacity of the box. In Figure 7, the effective stresses of the Huber–Mises-Hencky distributions for selected cases of open-top boxes for the storage of fruits are shown. The bottom of the packaging was not modeled but was included in the computations using appropriate boundary conditions. In Figures 6 and 7, the bottom of the box is added for visualization purposes.

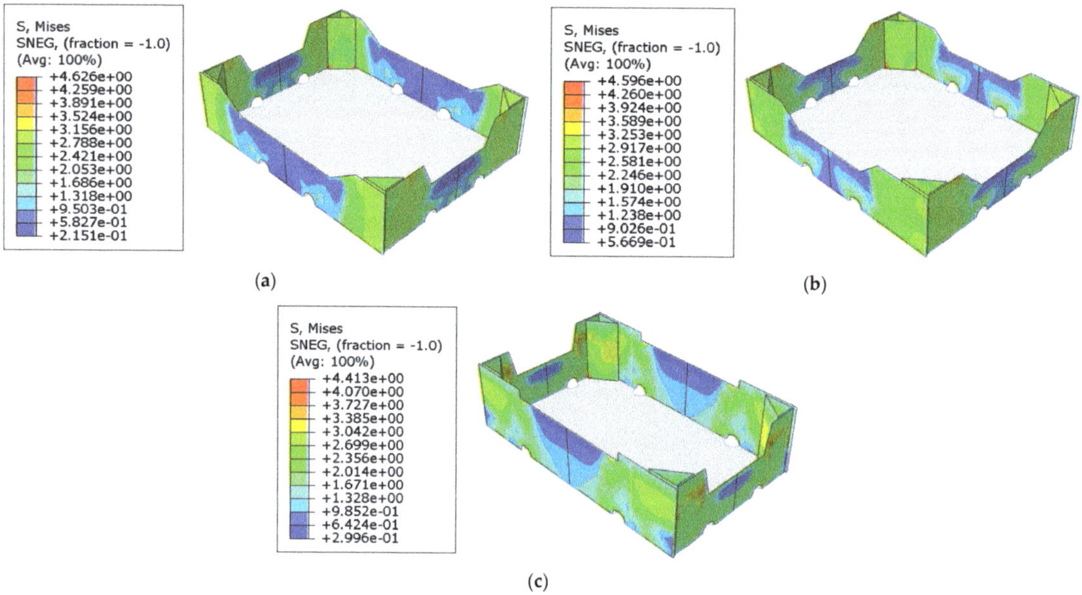

Figure 7. Stress distributions at the moment of reaching the compression load capacity of open-top boxes for storage of fruits: (**a**) case 1, (**b**) case 19, and (**c**) case 20.

As described at the end of Section 2.2, 20 cases of packaging geometry were analyzed (see Table 1). Considering a specific type of geometry, first, the reference value of the compression strength capacity was calculated. Then, each parameter was perturbed individually by 1% and the load capacity of the box was computed. All these computations made it possible to obtain 360 results for each material and a total of 1080 values of the load capacity. Based on the results obtained, the sensitivities of each of the 17 parameters were calculated for 20 packaging geometries and 3 types of corrugated board. All sensitivities were calculated from Equation (1), where $h(\mathbf{x})$ is the load-bearing capacity with respect to top-to-bottom compression. In Table 3, examples of the sensitivities computed for B-840 corrugated cardboard are presented.

To compare which parameters were the most significant in terms of their compressive strength capacity, the average sensitivity of each parameter was determined from 20 geometries. In Figure 8, the averaged values of the sensitivities for three corrugated cardboards are presented and sorted in ascending order. Two horizontal dashed lines indicate levels of 0.05 and 0.10. In addition, in Figure 9, the average sensitivity values of the three corrugated boards used are shown by bar plots. The median of each parameter is also marked with black dots.

Table 3. Sensitivities computed for B-840 corrugated cardboard with the min/max values marked in blue and red for all parameters considered in the study.

Case	d^L	d^B	H	l^L	h^L	l^B	h^B	s^L	s^B	w^L	g^L	m^L	w^B	g^B	m^B	θ^L	θ^B
1	−0.10	0.31	−0.01	−1.23	−0.30	−0.30	0.01	0.14	0.21	0	0	0	0	0	−0.08	−0.25	0.02
2	0.18	0.55	−0.02	−1.17	−0.38	−0.49	−0.17	0.22	0.03	0	−0.04	−0.17	−0.06	−0.06	−0.19	−0.29	−0.04
3	−0.04	0.33	0.12	−1.31	−0.31	−0.25	0.13	0.23	0.22	0	0	0	0.09	0.09	0.03	−0.17	0.15
4	0.22	−0.67	0.20	−0.57	0	−0.39	0.02	0.23	0.21	0	0	0.01	−0.01	−0.01	−0.02	0.08	−1.25
5	0.29	0.29	0.06	−0.45	−0.05	−0.29	−0.01	0.21	0.20	0	0	0	0	0	0	0.12	0.14
6	0.19	0.06	0.15	−0.28	0.01	−0.22	0	0.25	0.18	0	0	0.01	−0.01	−0.02	0.04	0.01	0.03
7	0.25	0.16	0.13	−0.33	0.02	−0.25	0	0.16	0.27	0	0	0.02	0.01	0.01	0.02	0.07	0.02
8	0.27	0.11	0.09	−0.37	−0.01	−0.23	0.01	0.11	0.22	−0.02	−0.02	−0.02	0	0	0.01	0	−0.01
9	0.55	0.31	−0.38	−0.46	−0.02	−0.27	0.06	0.17	0.19	0	0	−0.02	0	0.05	0.05	0	0.07
10	0.36	0.24	0.13	−0.33	0.02	−0.29	0	0.26	0.18	0	0	0	0	0	0.07	0.10	0.11
11	0.29	0.24	0.14	−0.43	0.01	−0.95	0.01	−0.21	0.23	−0.70	0	0	−0.01	−0.01	0.01	−0.06	0.01
12	0.29	0.31	0.19	−0.22	0.05	−0.34	0.02	0.23	0.19	0.01	0	0	0	0	0.01	0.02	0.04
13	0.27	−0.08	0.13	−0.21	0.12	−0.17	−0.06	0.21	0.19	0	0	0	−0.35	0	−0.31	0	0.08
14	0.33	0.25	0.14	−0.33	0.04	−0.26	0.01	0.18	0.17	0	0	0.02	0	0	0	0.03	0.02
15	0.24	0.20	0.63	−0.76	0.12	−0.43	−0.10	0.23	0.26	0	0	0	−0.01	−0.01	0.02	0.12	0.02
16	0.19	0.21	−0.33	−0.76	0	−0.76	0	0.20	−0.03	−0.09	0	0.02	−0.02	−0.02	0.01	−0.06	0.02
17	0.25	0.48	0.35	−0.47	0.08	−0.15	0.27	0.20	0.24	0	0	−0.03	0	0.26	0.27	0.08	0.29
18	0.60	0.29	0.17	−0.24	0.44	−0.24	0.07	0.17	0.20	0.38	0	−0.03	0.07	0	0.07	0.36	0
19	0.23	−0.15	0.35	−0.35	0.18	−0.23	0.11	0.44	0.28	−0.07	0	−0.07	0.09	0.09	0.09	1.82	0.59
20	0.58	0.16	0.02	−0.28	0	−0.29	−0.04	0.28	0.11	0.01	0	−0.07	0	0	0.01	0.04	0.04

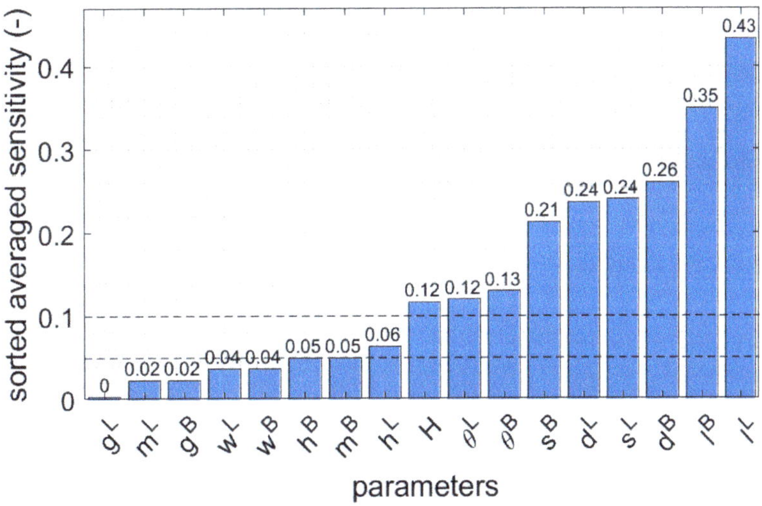

Figure 8. The averaged values of parameter sensitivities for three corrugated cardboards considered (sorted in ascending order).

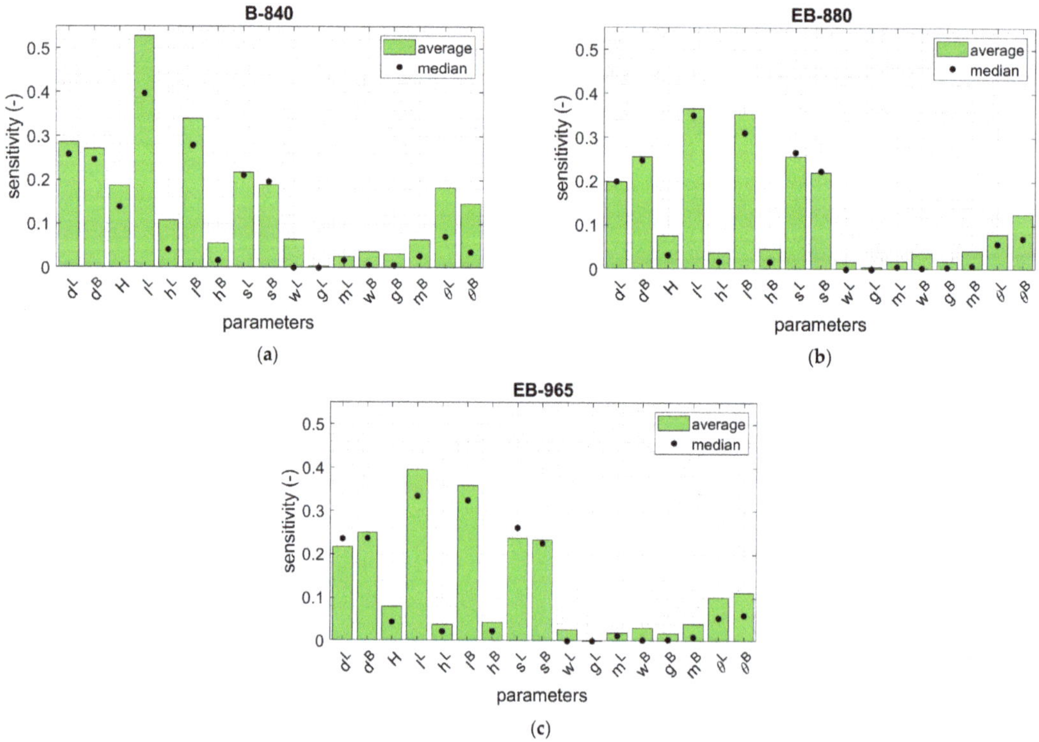

Figure 9. Average and median sensitivity for 20 cases of open-top boxes for the storage of fruits considered for different cardboards: (**a**) B-840, (**b**) EB-880 and (**c**) EB-965.

4. Discussion

All crucial results have been presented in Figures 6–9 and Table 3. Considering Figure 6, in which the maximal displacement values have been presented for the selected cases, it may be observed that the buckling modes are not completely repeatable when compared case to case. Namely, in Figure 6a, the maximal displacements are obtained for shorter sidewalls, while Figure 6b,c shows the opposite situation, i.e., the maximal displacements are acquired for longer sidewalls. It can be concluded that the stiffness of the sidewalls plays a role, which, in a modifiable range of parameters, can switch the maximum buckling displacement between shorter or longer sidewalls. Moreover, regardless of the stiffness ratio of the shorter/longer sidewalls, the buckling mode in the hypotenuse demonstrates two half-waves, which is observed in all cases (see Figure 6). Shorter and longer sidewalls show one half-wave between the boundary conditions. In addition, it is worth noting that the buckling modes were obtained by applying rotations (see Section 2.2); therefore, here, the maximal displacement values are not equal to one.

Considering Figure 7, in which the effective Huber–Mises-Hencky stresses have been presented for the selected cases, it may be observed that the maximal stresses were obtained at the base of the hypotenuse, in which there is a complex stress state. In addition, higher stresses are visible in the upper part of the shorter sidewalls in the gluing zones with stiffening columns, which is particularly visible in Figure 7c. Moreover, all fields of effective stresses show that the highest values are obtained in the stiffening corners, which was expected. However, it seems that the position and shape of the edge holes relative to the arms of the trapezoidal folds affect the stress trajectories of the sidewalls; we acquire different stress zones in the sidewalls depending on the position of the edge holes. This can be seen by comparing the green and blue areas in Figure 7.

In this study, the sensitivities were computed for all the materials considered, namely, B-840, EB-880, and EB-965. However, the detailed tabular values were presented only for the case of B-840 case (see Table 3). The reason for this is the high similarity of the obtained results, as evidenced by the synthetic data shown for all materials in the bar graphs in Figure 9. The results show that d^L and d^B have a large influence on BCT, for which, on average, the sensitivity was 0.25 for the cardboard types (see Figure 8 and compare with Table 3). The longer these dimensions, the stiffer the corners, but the buckling lengths increase. Furthermore, the box height H has lower non-negligible impact because the average sensitivity for all cardboard types was 0.12. The lengths of the sidewall folds, l^L and l^B, have the largest influence on the change in the BCT (0.39, on average, for the cardboard types): their decrease increases the strength of the material that is bearing the compression load in the corners. Two other parameters, the sides of the stiffening triangles, s^L and s^B, play a minor role, but are still very important compared to other parameters. In the case of s^L, the average sensitivity for all cardboards was 0.24, while in the case of s^B, the average sensitivity for all cardboards was 0.21. The larger the triangular sides, the stronger the stiffening corner; its buckling length of the hypotenuse will also increase. The height of the folds of the sidewalls, h^L and h^B, have a minor impact on the BCT, which can be regarded as negligible (average sensitivity for material parameters was 0.055). As suspected by engineering intuition, but now confirmed by the sensitivity study, some parameters related with the dimensions and positions of edge holes can be neglected (see Table 3 and Figure 8); these parameters are w^L, g^L, m^L, w^B, g^B, and m^B. These parameters represent the dimensions of the edge holes (width and height), as well as their positions from the perpendicular sidewalls in both types of sidewalls (shorter and longer). The inclination of the arms of the trapezoidal folds, i.e., θ_L and θ_B, are more important, with an average sensitivity of 0.13. It is worth noting that the conclusion regarding the lack of influence of the size of the ventilation holes on the load capacity of the package is true only when the strength test of the tray is performed using a BCT press. Under natural operating conditions, when a tray is placed on top of another tray, the load on the bottom of the tray becomes significant. This means that the standard for assessing the load capacity of the tray should be different, e.g., considering the load on the bottom of the tray and other boundary conditions (i.e., supports should be applied only at the corners).

The averaged values for all the parameters considered are presented in Figure 8. Seven of the seventeen parameters were observed to have a sensitivity equal to or less than 0.05 (w^L, g^L, m^L, w^B, g^B, m^B, and h^B). These parameters can be assumed as constants by default when making decisions regarding the design of an optimal open-top box for the storage of fruits and vegetables in accordance with the BCT test. If a sensitivity value equal to 0.10 is taken as a threshold, the h^L parameters can also be omitted.

5. Conclusions

In this paper, a parametric model of an open-top, corrugated board carton was built. Several geometrical parameters of box design were specified, and in order to check the impact of each of them on the compression strength of the packaging, a systematic sensitivity analysis was carried out. Several dozen sets of parameters were adopted for the proper exploration of parameter space. In the first step of the simulations, a buckling analysis was performed; then, the actual compression of the box was simulated. The load capacities and the sensitivities to the perturbation of each parameter were obtained from the computations. Based on the analyses carried out, the parameters were ranked in terms of importance, and several of them were deemed to be insignificant in terms of their impact on compressive strength based on the BCT test. In contrast, some of them play a very important role in the construction of the packaging and have a large impact on its load capacity; these parameters should be selected very carefully in the process of the optimal design of trays for containing vegetables or fruit.

In particular, this paper shows that not all geometric features are important from the perspective of compressive strength. The dimensions and location of the ventilation holes on both sidewalls, as well as the height of the trapezoidal folds on the shorter sidewalls, played minor roles in the load-bearing capacity of the packaging in a BCT-based testing protocol. For this reason, the indicated parameters may be assumed as constants and be neglected in the parametric model. Such selection is particularly useful for creating a reduced parametric model for use in an artificial neural network. Limiting the number of parameters allows one to save computational time and more effectively explore the space of crucial parameters in ANN optimization problems.

Author Contributions: Conceptualization, T.G. (Tomasz Garbowski), and T.G. (Tomasz Gajewski); methodology, D.M. and T.G. (Tomasz Garbowski); software, D.M.; validation, D.M. and T.G. (Tomasz Gajewski) and D.M.; formal analysis, D.M.; investigation, D.M. and T.G. (Tomasz Gajewski); writing—original draft preparation, D.M. and T.G. (Tomasz Gajewski); writing—review and editing, D.M., T.G. (Tomasz Gajewski) and T.G. (Tomasz Garbowski); visualization, D.M. and T.G. (Tomasz Gajewski); supervision, T.G. (Tomasz Garbowski) and T.G. (Tomasz Gajewski); project administration, T.G. (Tomasz Garbowski); funding acquisition, T.G. (Tomasz Garbowski). All authors have read and agreed to the published version of the manuscript.

Funding: This research received no external funding.

Institutional Review Board Statement: Not applicable.

Informed Consent Statement: Not applicable.

Data Availability Statement: The data presented in this study are available on request from the corresponding author.

Conflicts of Interest: The authors declare no conflict of interest. The funders had no role in the design of the study; in the collection, analyses, or interpretation of data; in the writing of the manuscript, or in the decision to publish the results.

References

1. Lindh, H.; Olsson, A.; Williams, H. Consumer perceptions of food packaging: Contributing to or counteracting environmentally sustainable development? *Packag. Technol. Sci.* **2016**, *29*, 3–23. [CrossRef]
2. Neill, C.; Williams, R. Consumer Preference For Alternative Milk Packaging: The Case Of An Inferred Environmental Attribute. *J. Agric. Appl. Econ.* **2016**, *48*, 241–256. [CrossRef]
3. Popovic, I.; Bossink, B.A.G.; van der Sijde, P.C. Factors Influencing Consumers' Decision to Purchase Food in Environmentally Friendly Packaging: What Do We Know and Where Do We Go from Here? *Sustainability* **2019**, *11*, 7197. [CrossRef]
4. FAO. Banana Market Review 2020. 2021. Available online: https://agfstorage.blob.core.windows.net/misc/FP_com/2021/10/11/Ban.pdf (accessed on 8 December 2022).
5. Singh, J.; Olsen, E.; Singh, S.P.; Manley, J.; Wallace, F. The Effect of Ventilation and Hand Holes on Loss of Compression Strength in Corrugated Boxes. *J. Appl. Packag. Res.* **2008**, *2*, 227–238.
6. Fadiji, T.; Ambaw, A.; Coetzee, C.J.; Berry, T.M.; Opara, U.L. Application of finite element analysis to predict the mechanical strength of ventilated corrugated paperboard packaging for handling fresh produce. *Biosyst. Eng.* **2018**, *174*, 260–281. [CrossRef]
7. Han, J.; Park, J.M. Finite element analysis of vent/hand hole designs for corrugated fibreboard boxes. *Packag. Technol. Sci.* **2007**, *20*, 39–47. [CrossRef]
8. Frank, B. Corrugated Box Compression—A Literature Survey. *Packag. Technol. Sci.* **2014**, *27*, 105–128. [CrossRef]
9. Mrówczyński, D.; Garbowski, T.; Knitter-Piątkowska, A. Estimation of the compressive strength of corrugated board boxes with shifted creases on the flaps. *Materials* **2021**, *14*, 5181. [CrossRef]
10. McKee, R.C.; Gander, J.W.; Wachuta, J.R. Compression strength formula for corrugated boxes. *Paperboard Packag.* **1963**, *48*, 149–159.
11. Buchanan, J.S.; Draper, J.; Teague, G.W. Combined board characteristics that determine box performance. *Paperboard Packag.* **1964**, *49*, 74–85.
12. Shick, P.E.; Chari, N.C.S. Top-to-bottom compression for double wall corrugated boxes. *Tappi J.* **1965**, *48*, 423–430.
13. Wolf, M. New equation helps pin down box specifications. *Packag. Eng.* **1972**, *17*, 66–67.
14. Kellicutt, K.; Landt, E. Development of design data for corrugated fiberboard shipping containers. *Tappi J.* **1952**, *35*, 398–402.
15. Maltenfort, G. Compression strength of corrugated containers. *Fibre Contain.* **1956**, *41*, 106–121.
16. Garbowski, T.; Gajewski, T.; Grabski, J.K. The Role of Buckling in the Estimation of Compressive Strength of Corrugated Cardboard Boxes. *Materials* **2020**, *13*, 4578. [CrossRef]

17. Archaviboonyobul, T.; Chaveesuk, R.; Singh, J.; Jinkarn, T. An analysis of the influence of hand hole and ventilation hole design on compressive strength of corrugated fiberboard boxes by an artificial neural network model. *Packag. Technol. Sci.* **2020**, *33*, 171–181. [CrossRef]
18. Adamopoulos, S.; Karageorgos, A.; Rapti, E.; Birbilis, D. Predicting the properties of corrugated base papers using multiple linear regression and artificial neural networks. *Drewno* **2016**, *59*, 61–72. [CrossRef]
19. Kumar, A.; Kumar, R.; Subudhi, S. Numerical modeling of forced-air pre-cooling of fruits and vegetables: A review. *Int. J. Refrig.* **2022**, *115*, 217–232. [CrossRef]
20. Jani, D.B.; Mishra, M.; Sahoo, P.K. Application of artificial neural network for predicting performance of solid desiccant cooling systems—A review. *Renew. Sust. Energ. Rev.* **2017**, *80*, 352–366. [CrossRef]
21. Funes, E.; Allouche, Y.; Beltrán, G.; Jiménez, A. A Review: Artificial Neural Networks as Tool for Control Food Industry Process. *J. Sens. Technol.* **2015**, *5*, 28–43. [CrossRef]
22. Frankowski, J.; Zaborowicz, M.; Sieracka, D.; Łochyńska, M.; Czeszak, W. Prediction of the Hemp Yield Using Artificial Intelligence Methods. *J. Nat. Fibers* **2022**, *19*, 13725–13735. [CrossRef]
23. Stręk, A.M.; Dudzik, M.; Kwiecień, A.; Wańczyk, K.; Lipowska, B. Verification of application of ANN modelling in study of compressive behaviour of aluminium sponges. *Eng. Trans.* **2019**, *67*, 271–288. [CrossRef]
24. Araujo, G.; Andrade, F.A.A. Post-Processing Air Temperature Weather Forecast Using Artificial Neural Networks with Measurements from Meteorological Stations. *Appl. Sci.* **2022**, *12*, 7131. [CrossRef]
25. Gajewski, T.; Staszak, N.; Garbowski, T. Parametric Optimization of Thin-Walled 3D Beams with Perforation Based on Homogenization and Soft Computing. *Materials* **2022**, *15*, 2520. [CrossRef] [PubMed]
26. Berry, T.M.; Fadiji, T.S.; Defraeye, T.; Opara, U.L. The role of horticultural carton vent hole design on cooling efficiency and compression strength: A multi-parameter approach. *Postharvest Biol. Technol.* **2017**, *124*, 62–74. [CrossRef]
27. Fadiji, T.; Coetzee, C.J.; Opara, U.L. Compression strength of ventilated corrugated paperboard packages: Numerical modelling, experimental validation and effects of vent geometric design. *Biosyst. Eng.* **2016**, *151*, 231–247. [CrossRef]
28. Fadiji, T.; Coetzee, C.J.; Berry, T.M.; Opara, U.L. Investigating the role of geometrical configurations of ventilated fresh produce packaging to improve the mechanical strength—Experimental and numerical approaches. *Food Packag. Shelf Life.* **2019**, *20*, 100312. [CrossRef]
29. Jin, R.; Chen, W.; Sudjianto, A. An efficient algorithm for constructing optimal design of computer experiments. *J. Stat. Plan. Inference* **2005**, *134*, 268–287. [CrossRef]
30. Abaqus Unified FEA Software. Available online: https://www.3ds.com/products-services/simulia/products/abaqus (accessed on 8 December 2022).
31. Hill, R. A theory of the yielding and plastic flow in anisotropic metals. *Proc. R. Soc. Lond. Ser. A Math. Phys. Sci.* **1948**, *193*, 281–297. [CrossRef]
32. FEMat BSE System. Available online: http://fematsystems.pl/bse-system_en/ (accessed on 8 December 2022).
33. Mrówczyński, D.; Knitter-Piątkowska, A.; Garbowski, T. Non-Local Sensitivity Analysis and Numerical Homogenization in Optimal Design of Single-Wall Corrugated Board Packaging. *Materials* **2022**, *15*, 720. [CrossRef]
34. Mrówczyński, D.; Knitter-Piątkowska, A.; Garbowski, T. Optimal Design of Double-Walled Corrugated Board Packaging. *Materials* **2022**, *15*, 2149. [CrossRef] [PubMed]

materials

Article

Compression Strength Estimation of Corrugated Board Boxes for a Reduction in Sidewall Surface Cutouts—Experimental and Numerical Approaches

Lajos Fehér [1], Renáta Pidl [1] and Péter Böröcz [2,*]

1 Department of Applied Mechanics, Széchenyi István University, Egyetem tér 1, 9026 Győr, Hungary
2 Department of Logistics and Forwarding, Széchenyi István University, Egyetem tér 1, 9026 Győr, Hungary
* Correspondence: boroczp@sze.hu

Abstract: Corrugated cardboard boxes are generally used in modern supply chains for the handling, storage, and distribution of numerous goods. These packages require suitable strength to maintain adequate protection within the package; however, the presence and configuration of any cutouts on the sidewalls significantly influence the packaging costs and secondary paperboard waste. This study aims to evaluate the performance of CCBs by considering the influence of different cutout configurations of sidewalls. The compression strength of various B-flute CCB dimensions (200 mm, 300 mm, 400 mm, 500 m, and 600 mm in length, with the same width and height of 300 mm), each for five cutout areas (0%, 4%, 16%, 36%, and 64%) were experimentally observed, and the results were compared with the McKee formula for estimation. The boxes with cutout areas of 0%, 4%, 16%, 36%, and 64% showed a linear decreasing tendency in compression force. A linear relationship was found between compression strength and an increase in cutout sizes. Packages with 0% and 4% cutouts did not show significant differences in compression strength ($p < 0.05$). Furthermore, this study shows a possible way to modify the McKee estimation for such boxes after obtaining empirical test data since the McKee formula works with a relatively high error rate on corrugated cardboard boxes with sidewall cutouts. Utilizing the numerical and experimental results, a favorable estimation map can be drawn up for packaging engineers to better manage material use and waste. The results of the study showed that the McKee formula does not appropriately estimate the box compression strength for various cutout sizes in itself.

Keywords: paperboard packaging; box compression test; package design; McKee formula; cutout

check for updates

Citation: Fehér, L.; Pidl, R.; Böröcz, P. Compression Strength Estimation of Corrugated Board Boxes for a Reduction in Sidewall Surface Cutouts—Experimental and Numerical Approaches. *Materials* **2023**, *16*, 597. https://doi.org/10.3390/ma16020597

Academic Editors: Tomasz Garbowski and Aleksander Marek

Received: 10 November 2022
Revised: 27 December 2022
Accepted: 4 January 2023
Published: 7 January 2023

1. Introduction

Most of the goods transported use various transport packaging, including the most often used paperboard packaging. This is the leading material to ensure the necessary protection and logistical aims, such as stacking, handling, and forming unit loads during distribution, due to its load-bearing capacity and other beneficial properties [1,2]. The main advantages of paper-based packaging are the reliable protection of products, in addition to relatively low packaging costs, recyclability, and biodegradability [3,4]. Stacking packages on top of each other can cause damage to the packaged product, so before doing this, it is necessary to find an appropriate estimation method or to perform a series of tests to determine the mechanical strength of various constructions of cardboard boxes [4–6]. The situation is more complex if there are some special requirements for the box, such as hand holes, ventilation holes, and openings, respectively [7–11]. Additional considerations involving corrugated cardboard boxes (CCBs) are optimal packaging material costs and sustainable development strategies within their product-packaging range [12]. Basically, the cost of CCB packaging depends on the number of layers and paper/cardboard quality, which are in relation to the mass used [13]. The latter, of course, impacts the cost of packaging solutions and the final mass for destruction or recycling after use. Therefore, it is in the interest of the actors in the industry to find the box with optimal mech properties, which, on the one hand, adequately ensures product protection and, o other hand, leads to acceptable packaging costs and waste savings.

Corrugated cardboard is made of odd layers, usually three or five. In the case three-layered corrugated cardboard, the corrugated layer is placed between the inner an outer flat layers, while five-layered cardboard has two corrugated layers. There are als several types of corrugated layers according to flute height and flute length. The individual layers are bonded together by glue. In the paper industry, the highest flute is denoted by A, followed by flute C, and the lowest normal flute is B. There are so-called micro-flutes with the letters E and F (FEFCO, European Federation of Corrugated Board Manufacturers) [14].

The optimal package design of corrugated boxes for packaging engineers is a major challenge [15]. When trying to ensure product-packaging integrity, the situation is further complicated by environmental conditions that can affect the mechanical behavior of corrugated board, including temperature, relative humidity changes, perforations, prints, etc. [16,17], which can decrease the resilience of the integrity of the packaged product to damages [8–11,18]. However, the most critical issue of compression strength for CCBs is the presence of any openings, handling, or ventilation holes on the walls, and their dimension, orientation, shape, and number [8,19].

One possible way to determine the strength of corrugated boxes is to perform a series of tests in laboratory conditions following standard protocols. In the paper packaging industry, these tests cover the compression, bending, and bursting of corrugated boards. However, the most important and practical tests are the box compression test (BCT) and the edge crush test (ECT). The latter can give information to use as input in an analytical formula for predicting compressive strength. The well-known semi-empirical formula to determine the possible stacking load is McKee's equation [18,20]. The McKee formula is a simple, practical application using parameters of paper, board, and boxes with an arbitrarily chosen constant. The disadvantage of the method is that the formula is applicable to relatively typical box containers without modification of holes, cutouts, and so forth.

In the last 50 years, many researchers have tried to extend the applicability of McKee's formula and presented different approaches. There is a method by Kellicutt and Landt [21] that developed a model for compressive load sizing based on the principle of annular compressive strength. Beldie et al. presented a study in 2001 that modeled the mechanical behavior of corrugated cardboard packages subjected to static compressive loading [22]. Biancolini and Brutti [23] presented a numerical model for splitting the properties of corrugated cardboard boxes with strength calculations. Allerby et al. [24] presented a study in 1985 in which they modified the constants and exponents of McKee's formula. In 1987, Schrampfer et al. extended the applicability of the McKee relationship to a wider range of cutting methods and equipment with a combined board-edge crush technology [25]. Furthermore, the McKee constant was later analyzed in more sophisticated way in complicated cases by Garbowski et al. [26]. There are studies in which the authors have shown that additional tests are required and, therefore, an updated formulation should be used, which was recently modified by Aviles et al. [27] and later by Garbowski et al. [28,29].

As it was mentioned above, the sidewalls of the box are often weakened by cutouts for various purposes. There can be several reasons for this: tab-like cutouts, ventilation holes/openings (mainly for agricultural products), viewing windows, etc. These solutions have a negative influence on the compression strength of the boxes, and this phenomenon was investigated by several authors [8,30–36].

In 2020, Garbowski et al. [35] indicated that a smaller hole on the sidewall would ensure better (greater) compression strength, but it is necessary that it would be located at the center of the wall. On the other hand, the McKee formula cannot give precise results in the aspect of a cutout independent of its position, shape, and size. An additional important issue in the compression strength investigation is the length-to-width ratio of the CCB. Research has shown that if the aspect ratio changes from 1 to 3, then the compression strength increases at first and then decreases; furthermore the maximum compression force can be observed when the aspect ratio was approximately 1.6 [37,38].

It must be mentioned here that after a careful literature review, the authors could not find any published research that measures and/or analyzes the interconnection of these variables on such a wide range of box dimensions and large and growing cutout sizes with a primary focus on material reduction. There is a gap in the literature for a wide variety of experiments in this area. The papers published so far mainly focused on the mechanical properties of individual/special box dimensions, with only a partial understanding of the overall relationships. This study attempts to develop an estimation method for the compression strength of cardboard boxes with various (growing) cutout sizes. Cutout technology is generally used for various purposes such as ventilation, reduction in material, viewing windows, etc. First, empirical box compression tests were performed on a wide range of dimensions, using 250 box samples in total to observe the changes in mechanical strength, and then an analysis was performed comparing the empirical results with the analytical results of the McKee formula. Finally, this paper presents an estimation map for the McKee constant with a given number of cardboard samples along various cutouts. Therefore, this paper can provide novel insights into circumstances for packaging engineers to design boxes based on experimental data using a simple method for compression strength estimation.

2. Materials and Methods

2.1. Samples

For this study, single-wall B-flute corrugated cardboard (Figure 1) boxes with different cutouts were used that were made from the same cardboard material quality. Table 1 shows the mechanical specifications for the tested corrugated cardboard. The corrugated cardboard material composition contained the following:

- Outer liner: 210 GD2 (weight 210 g/m^2, coated white lined chipboard with grey back, quality class 2);
- Fluting medium: 120 HC (weight 120 g/m^2, high compression Wellenstoff);
- Inner liner: 130 TL 3 (weight 130 g/m^2, Testliner, quality class 3).

Figure 1. Structure (**a**) and cross-section (**b**) of corrugated cardboard sample used for this study.

Table 1. Mechanical specification of the B-flute corrugated cardboard tested.

Properties	Specification	Applied Standard
Board Thickness	2.8 mm (±10%)	ISO 3034 (FEFCO No.3)
Grammage	512 g/m^2 (±10%)	ISO 536:1995
Edge crush test (ECT)	5.1 kN/m (±15%)	ISO 3037 (FEFCO No.8)
Bursting strength (BST)	676 kPa (±15%)	ISO 2759 (FEFCO No.4)

Both the assembly of the boxes and the gluing process were carried out by hand. For this study, 5 boxes with different lengths and 5 different cutout areas were used. Table 2 and Figure 2 show the configurations for the samples. Each sample had the same width and height of 300 mm. The ratios of the cutout areas to the sidewall were the following: 0%, 4%, 16%, 36%, and 64%, respectively. The cutouts were cut from the center of the sidewalls along all four sides. In order to evaluate the measurement results, 10 samples were tested for each cutout group for each dimension, for a total of 250 samples.

Table 2. Configuration for dimensions of samples for this study (same width and height of 300 mm).

Length (mm)	Perimeter (mm)	Area without Top and Bottom Flaps (mm^2)	Sizes of Cutout (mm)	Cutout Area (mm^2)	Cutout Ratio (%)
				0	0
			40 × 60/60 × 60	12,000	4
200	1000	300,000	80 × 120/120 × 120	48,000	16
			120 × 180/180 × 180	108,000	36
			160 × 240/240 × 240	192,000	64
				0	0
			60 × 60/60 × 60	14,400	4
300	1200	360,000	120 × 120/120 × 120	57,600	16
			180 × 180/180 × 180	129,600	36
			240 × 240/240 × 240	230,400	64
				0	0
			80 × 60/60 × 60	16,800	4
400	1400	420,000	160 × 120/120 × 120	67,200	16
			240 × 180/180 × 180	151,200	36
			320 × 240/240 × 240	268,800	64
				0	0
			100 × 60/60 × 60	19,200	4
500	1600	480,000	200 × 120/120 × 120	76,800	16
			300 × 180/180 × 180	172,800	36
			400 × 240/240 × 240	307,200	64
				0	0
			120 × 60/60 × 60	21,600	4
600	1800	540,000	240 × 120/120 × 120	86,400	16
			360 × 180/180 × 180	194,400	36
			480 × 240/240 × 240	345,600	64

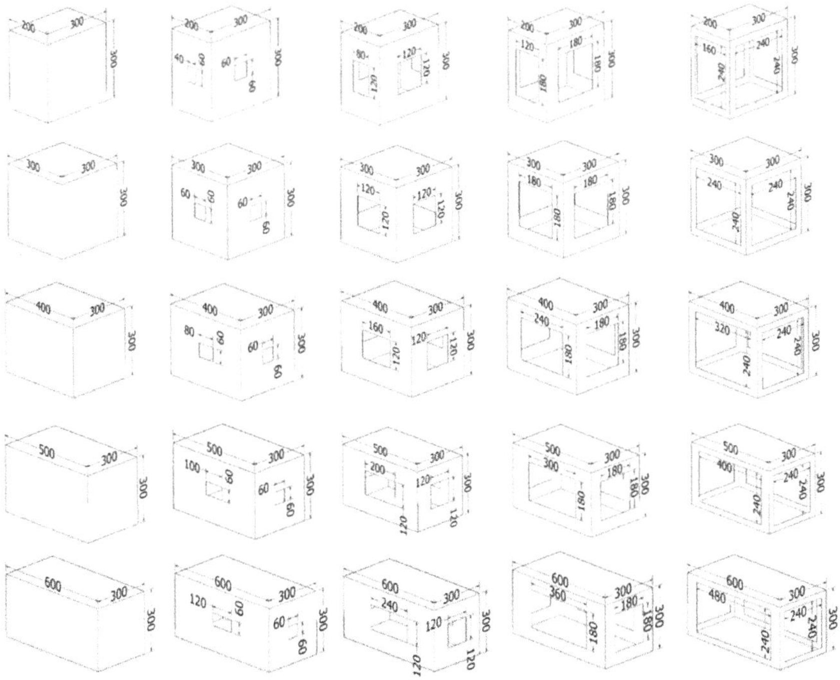

Figure 2. Five types of cutouts used on samples with different dimensions.

2.2. Measurement Setup

To observe and determine the maximum compression force, a BCT (box compression test) was performed on each box. This is a simple load test between a stationary and a moving steel plate. The device used for BCT measurements can be seen in Figure 3. During the tests, the compression force and deformation were continuously recorded. Before the test series, the samples were preconditioned at 30 °C ± 1 °C and 20–30% RH (relative humidity) for 24 h and then conditioned at 23 °C ± 1 °C and 50 ± 2% RH for 24 h in a climate-testing chamber in accordance with the ASTM D4332 standard [39]. Then, BCT tests were promptly executed after conditioning to avoid any additional hygroscopic phenomenon. BCTs were performed according to the ASTM D642 standard [40], so the testing speed of the crosshead was 12.7 mm/min ± 2.5 mm/min until the failure of the box occurred.

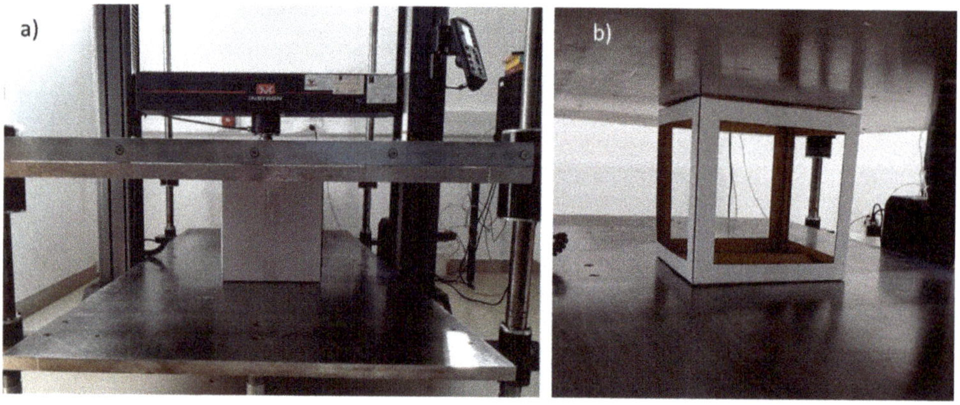

Figure 3. Box compression test (BCT): (**a**) control sample and (**b**) cutout sample (64% cutout area).

2.3. Box Compression Strength—McKee's Formula

The empirical BCT values for various box dimensions (Table 2) were then compared with the BCT values calculated by the McKee formula, so each variant case was compared with the values given by the McKee formula in order to determine which of the variable parameters of the McKee formula could be modified for a unique dimension. The formula, which was created in 1963 by McKee, has already been modified several times since its establishment, and currently, it is widely used in two variants. One version is the full McKee formula (Equation (1)), which is mainly used by researchers and developers and is relatively too complicated for everyday use [20,41]:

$$P = kP_m^b\left(\sqrt{D_x D_y}\right)^{(1-b)} Z^{(2b-1)} \quad (1)$$

The above formula (Equation (1)) gives the BCT value of the box in terms of the compressive force and is obtained using the following corrugated cardboard box parameters: P_m represents the edgewise compression strength of the combined board, and Z is the perimeter of the box. In addition to these two basic parameters, the in-machine flexural stiffness of the combined board and the cross-machine flexural stiffness of the combined board are denoted by D_x and D_y. The latter must be taken into account since the compressive force on the box causes the sidewalls to bulge, which can occur both in the in-machine and cross-machine directions. These values are considered to be non-variable factors for a given box type and size. The context includes the empirical constants denoted by k and b of which k is the multiplier of the whole equation, while b, the empirical constant, is in the exponent, and therefore, the choice of their values will have a major influence on the result obtained for the BCT value. It should be specifically noted here that while the constant is a multiplier without a unit of measure, the constant also modifies the unit of measure in the equation.

The original McKee formula suggests a k constant of 2.028 and a b constant of 0.746. With these data, Equation (1) takes the following form [42]:

$$P = 2.028P_m^{0.746}\left(\sqrt{D_x D_y}\right)^{0.254} Z^{0.492} \quad (2)$$

In practice, it is preferable to use a simplified version of the above formula, which is given by Equation (3) [42]:

$$P = kP_m\left(\sqrt{hZ}\right) \quad (3)$$

The simplified McKee formula uses the cardboard thickness in the equation instead of the bending stiffness, which is denoted by h. The creator of the equation assumes that bending stiffness varies proportionally with the thickness of the cardboard, and although this is a significant mechanical simplification, feedback from the industry confirms this assumption in general practice [43].

In comparing the BCT results, the simplified McKee formula was used, and the aim was to determine whether the practical value of the equation was appropriate or whether another equation should be chosen for the CCB with sidewall cutouts. For our study, the manufacturer gave a value of 5.3 for the material and box type used, and this was used in the simplified calculation. The manufacturer did not reveal the original source of the k value's calculation.

2.4. Data Analysis

The characteristics of the measured datasets can be described with statistical indicators, including the maximum, minimum, average, standard deviation, and standard error for each group. The statistical models were determined using the linear regression method because it was the best fit for the empirical data. With this method, a simple function ($y(x) = ax + b$) was calculated. This method was used in each group where the cutout areas were the predictor variables, and the compression forces were the output values. In linear regression analysis, the R^2 (coefficient of determination) values were used to determine the accuracy of the statistical models. The range of R^2 was between 0 and 1. If the R^2 value equaled one, then this indicated that the model can predict the dependent variable (in our case, the compression force) with 100% accuracy. In order to determine the differences between different the groups, a one-way analysis of variance (ANOVA) was executed with the Tukey post hoc test. The significance level was determined at $p < 0.05$ for the statistical analysis. The following software programs were used for the statistical evaluations: MATLAB R2021b (MathWorks Inc., Natick, MA, USA) and JASP 0.16.3 (the University of Amsterdam, The Netherlands).

3. Results and Discussion

The force–displacement diagrams of the BCT measurements were drawn for all box variants. As a sample, for the boxes of 400 mm × 300 mm × 300 mm, five diagrams are presented for the cut and uncut samples with the ten measurements of each (Figure 4). Figure 4a shows compression force–displacement functions for the 0% cutout, Figure 4b for the 4% cutout, Figure 4c for the 16% cutout, Figure 4d for the 36% cutout, and Figure 4e for the 64% cutout. The numerical data for these measurements are also given in Table 3. For each additional box type, all the datasets are available, which can be requested from the authors.

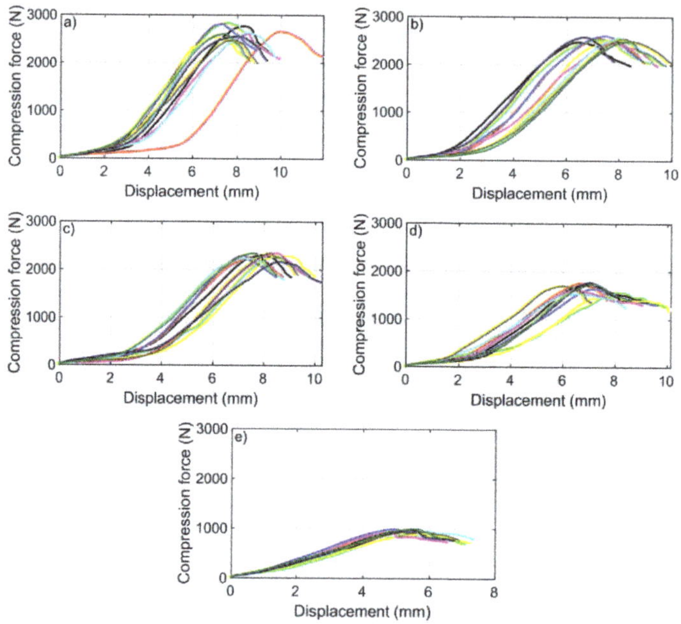

Figure 4. BCT curves: (**a**) with 0% cutout, (**b**) 4% cutout, (**c**) 16% cutout, (**d**) 36% cutout, and (**e**) with 64% cutout (10 samples for box size of 400 × 300 × 300 mm).

Table 3. BCT (N) values of 10 measurements for different cutouts (400 × 300 × 300 mm).

Sample Number	Compression Force (N) for Different Cutout Areas				
	0%	4%	16%	36%	64%
1	2769	2479	2299	1729	937
2	2665	2566	2257	1762	962
3	2843	2528	2295	1581	879
4	2819	2608	2295	1638	988
5	2532	2539	2324	1486	899
6	2606	2465	2342	1616	916
7	2644	2571	2235	1543	993
8	2477	2547	2352	1699	935
9	2555	2582	2170	1772	962
10	2610	2487	2344	1741	992

Table 4 contains the minimum, maximum, mean, fifth highest BCT value (which can be considered as quasi-median), and standard deviation of the BCT results for the different box types, and Figure 5 shows the compression force–displacement diagrams for all box types for all cutout ratios. The diagrams show the fifth highest (quasi-median) maximum compression force value of the ten measurements. By comparing the data in the fifth and sixth columns, it is clear that the quasi-median value was a good approximation of the sample mean, and since these were real measured data, it was reasonable to plot this value. The last column of Table 4 shows the magnitude of the standard deviation of the samples. The 400 mm box size was considered ideal in many respects. The highest measured BCT value was for this box type (2843 N). For box lengths of 500 and 600 mm, the standard deviation increased for cutout areas above 16%. For 500 mm box lengths, it increased from 59 N (36%) to 85 N (64%), and for 600 mm box lengths, it increased from 59 N (16%) to 87 N (36%).

Table 4. BCT results for different box dimensions.

Length (mm)	Cutout Rates (%)	Min. Force (N)	Max. Force (N)	Mean Force (N)	Quasi-Median (N)	Standard Deviation (N)
	0	2026	2462	2261	2261	146
	4	2092	2364	2218	2211	101
200	16	1607	2037	1851	1866	145
	36	1245	1448	1347	1348	64
	64	460	686	615	616	62
	0	2189	2595	2367	2311	136
	4	2117	2382	2275	2284	91
300	16	1841	2198	1982	1973	114
	36	1296	1470	1374	1377	54
	64	690	815	735	731	33
	0	2477	2843	2652	2610	123
	4	2465	2608	2537	2539	47
400	16	2170	2352	2291	2295	57
	36	1486	1772	1657	1638	99
	64	879	993	946	937	40
	0	2040	2667	2402	2356	189
	4	1945	2316	2203	2224	119
500	16	1964	2159	2067	2058	63
	36	1541	1714	1604	1577	59
	64	748	1004	878	899	85
	0	2053	2624	2339	2300	198
	4	2006	2371	2190	2167	108
600	16	1880	2082	1981	1972	59
	36	1476	1745	1591	1598	87
	64	627	915	862	882	85

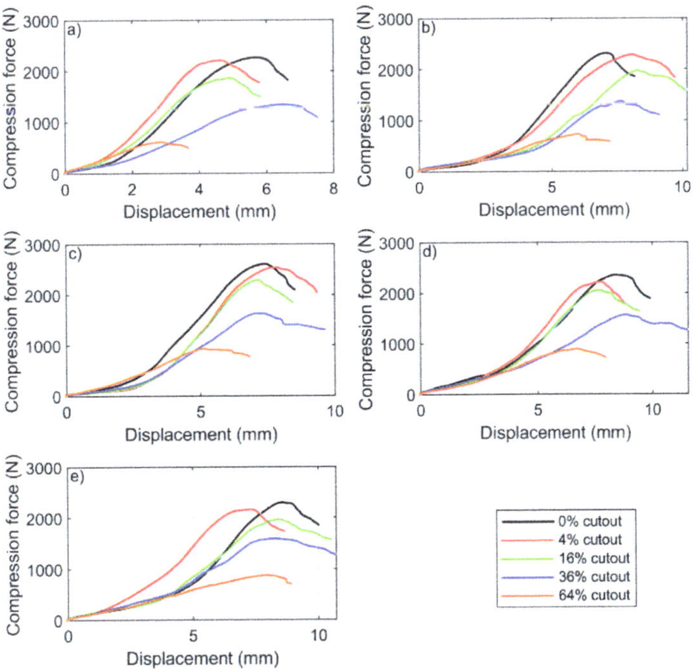

Figure 5. BCT quasi-median values: (**a**) 600 × 300 × 300 mm box, (**b**) 500 × 300 × 300 mm box, (**c**) 400 × 300 × 300 mm box, (**d**) 300 × 300 × 300 mm box, and (**e**) 200 × 300 × 300 mm box.

A cutout of 64% had extremely low values for all box types. The box behaved almost as an edge protector, not capable of bearing the load, even though in principle, the compressive strength of the box is determined by the vertical edges bent into an L shape. Despite this, the fact that the compressive strength of the box was radically reduced demonstrates that the total surface area of the sidewalls played a significant role in the absorption of compressive forces.

3.1. Linear Regression

Using descriptive statistics, the average maximum compression force values were determined in each group (shown in Table 4). Figure 6 shows these average values and the standard errors related to the cutout areas. Based on the data points (Figure 6), it can be identified that there was a decreasing tendency in the average maximum compression forces, as in other studies [8,31,35]. The statistical models were determined in the five box design groups using linear regression. These fitted lines can be seen in Figure 6. There were very high R2 values in each case, between 0.9904 and 0.9988 (Table 5). These indicate that the models described the measured data with very high accuracy. Using these linear functions, the compression forces could be predicted in each box dimension group separately if the box was made from single-wall B-flute corrugated cardboard. However, it has to be noted that in the model of 500 mm and 600 mm groups, there were residuals, so if the cutout rate increased up to 100%, there were residual compression force values that were not possible. The 400 × 300 × 300 group had the steepest regression slope and the highest y-intercept.

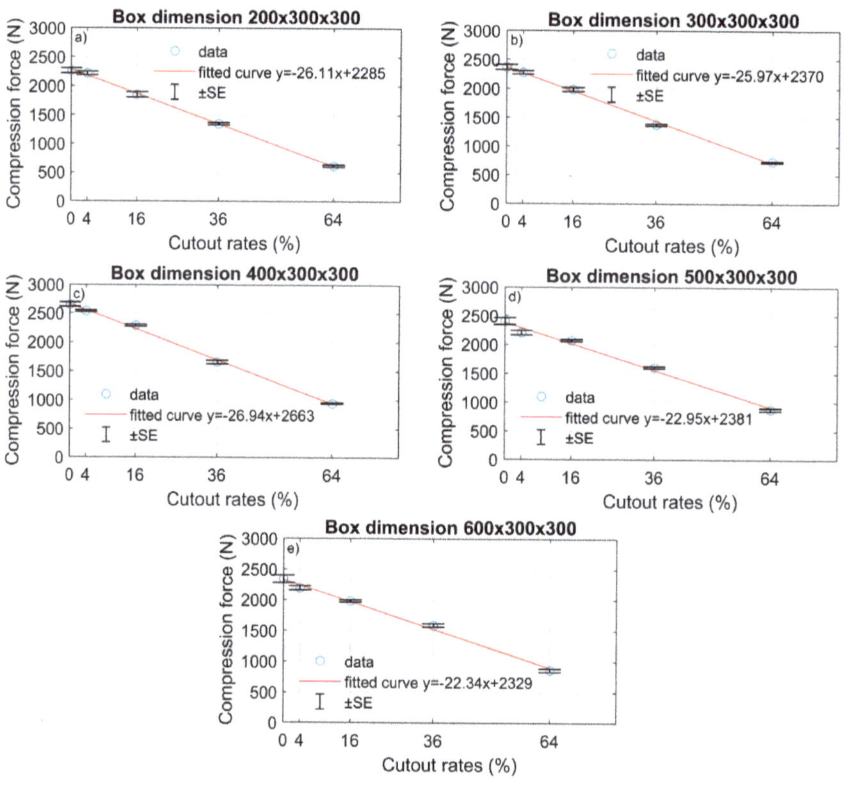

Figure 6. Averages of maximum compression forces and its standard errors (SE): (**a**) 200 × 300 × 300 mm box, (**b**) 300 × 300 × 300 mm box, (**c**) 400 × 300 × 300 mm box (**d**) 500 × 300 × 300 mm box, and (**e**) 600 × 300 × 300 mm box.

Table 5. Statistical model of the BCT values for sample boxes.

Length (mm)	Fitted Linear Curve	R^2
200	$-26.11x + 2285$	0.9988
300	$-25.97x + 2370$	0.9972
400	$-26.94x + 2663$	0.9974
500	$-22.95x + 2381$	0.9904
600	$-22.34x + 2329$	0.994

Figure 7 shows the box weight–average maximum compression force diagrams. Data points with different colors represent the results of different cutout areas. The lines with different colors are the fitted curves of the data points that represent the different box sizes. The data points on the yellow line are above all other measured data, so the 400 × 300 × 300 sized boxes had the best average compression force results. This result shows a good correlation with a previous study [37], where the authors showed that the compressive strength increased at first and then decreased, and in [37], the maximum compression strength appeared when the length-to-width ratio was about 1.6. In our study, the optimal length-to-width ratio was 1.33. It should be highlighted that in [37], the material of the tested corrugated box was different (BC flute corrugated cardboard with five layers). In this study, the highest average compression force was 2651 N ± 39 N for the 400 × 300 × 300 box with 0% cutout area. The lowest average compression force was 614 N ±19 N in the 200 × 300 × 300 box size group with 64% cutout area. In almost each cutout area group (except the 4%), the 200 × 300 × 300 boxes were shown to have the least stiff behavior. In the 4% cutout area group, the 600 × 300 × 300 box had the minimum average compression force (Table 4). The 64% cutout group produced the lowest average compression forces in each box size group. This kind of weight reduction significantly reduced the compressive strength of the boxes. Figure 7 also shows that the slopes of the fitted curves decrease when the size of the box increases. That means the cutouts had more impact related to the weight reduction when the size of the box was bigger.

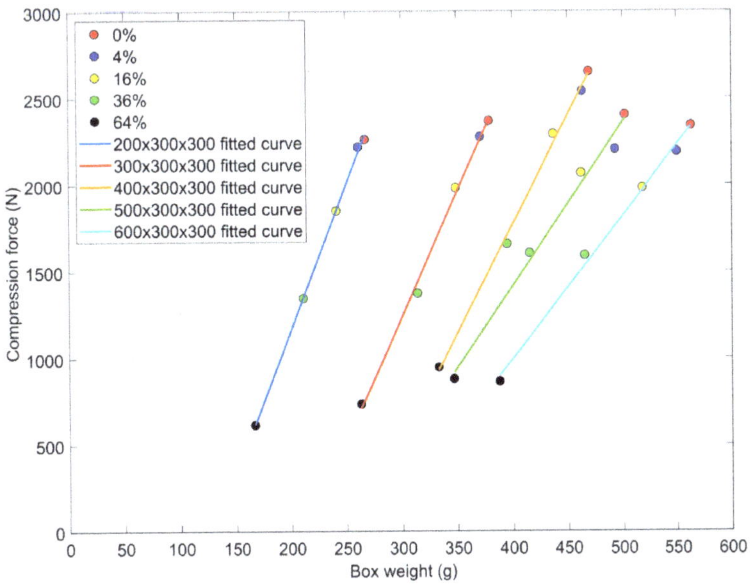

Figure 7. Box weights and averages of maximum compression forces diagram along the various cutouts.

3.2. McKee Comparison

The BCT value, calculated with a coefficient of 5.3 given by the manufacturer, was significantly lower than the measured BCT values for all uncut boxes. In order to calculate the BCT values using the simplified McKee equation more consistently with the measured BCT values, a first approximation of the coefficient k of 7 was chosen. The correlation of the BCT values calculated with the coefficient $k = 5.3$ with the measured BCT values is shown in Figure 8a. The relation of the BCT values calculated with coefficient $k = 7$ to the measured BCT values is shown in Figure 8b. Figure 9 shows the relative error calculated from the results for the different sizes of uncut boxes.

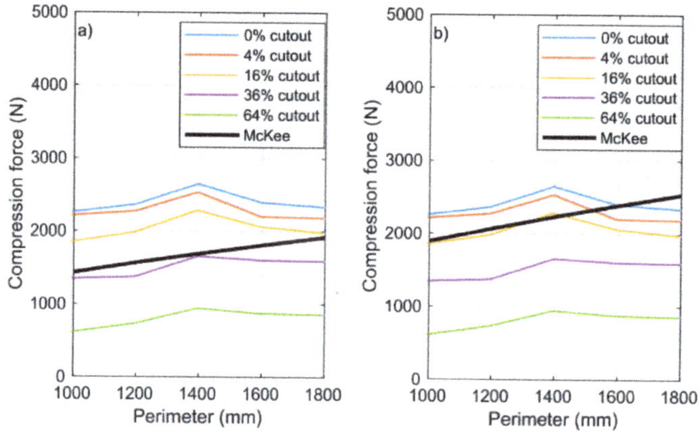

Figure 8. Summary data of BCT values calculated from the McKee equation and measured average BCT values for different cutout rates: (**a**) $k = 5.3$ and (**b**) $k = 7$.

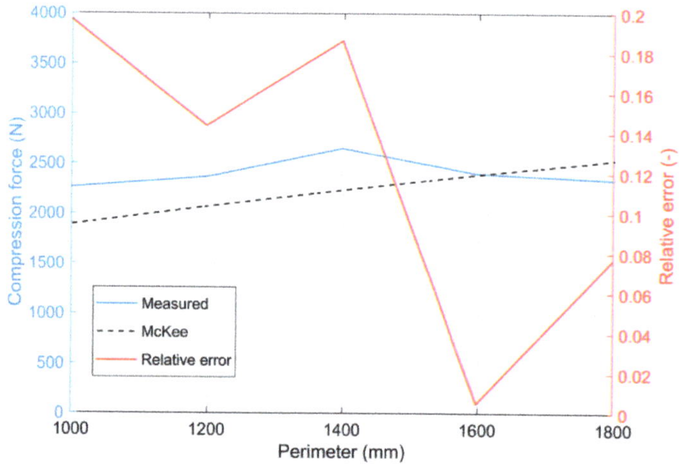

Figure 9. The relative error of the BCT results obtained with the simplified McKee formula ($k = 7$) for the uncut versions of each box type.

From the diagram, it appears that there was rather an optimum value (1400 mm perimeter), below and above which the measured BCT value decreased. However, in both the simplified and the full McKee context, the box perimeter is a multiplier, and hence, as the perimeter increases, the BCT values should also increase. Practically, it can be seen that an unlimited increase in the perimeter simply could not result in a linear increase in the BCT value.

As shown in Figure 10, the value of factor *k* varied between 6 and 8.8 based on the perimeter linear growth of the simplified McKee relation, with a step size of 0.4. The averages of the measured BCT values for each box variation are shown in this interval. From the diagram, the coefficients *k*, applicable to each box size included in the measurement, can be assigned to that box size. This also implies that applying the same *k* coefficient to boxes of different geometric sizes would lead to an error in the estimation of the BCT value. For the cutout samples, the discrepancies were even more significant and could not be handled using the simplified McKee formula.

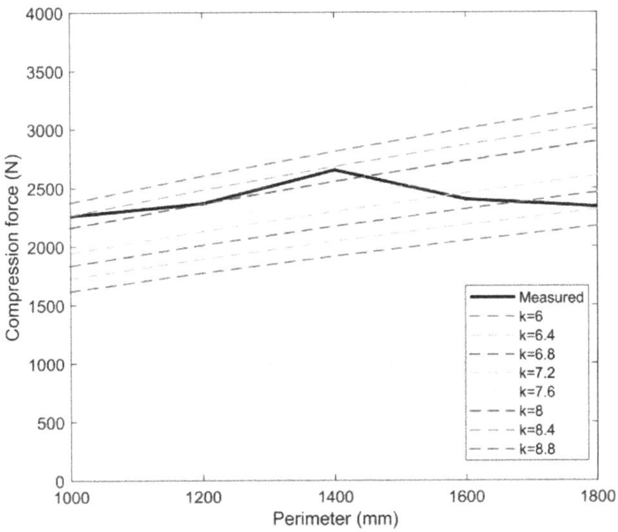

Figure 10. Iteration approximation of coefficient *k* for averages of BCT test results for uncut boxes of different sizes.

In a similar study, Garbowski et al. [35] calculated BCT values using the full McKee formula and compared them with their measured results. In their study, the authors performed the calculations and measurements for several different types of corrugated cardboard, but the geometric dimensions of the box were in a relatively narrow range, only for 300 × 200 × 200 and 300 × 200 × 300 mm (length × width × height). Garbowski et al. calculated the expected BCT value with a relative error of 15.5% for their samples.

In our own measurements, the geometric size of the boxes varied over a wider range, but for the 400 mm boxes with the best BCT results, the error of the simplified McKee formula was extremely high (around 18%), as the measured BCT value was significantly higher than the theoretical value calculated with the simplified McKee formula. For boxes with a length of 500 mm, the measured value and the value obtained from the McKee formula were almost coincident, and the error rate was below 1%. For box lengths of 600 mm, the BCT value given by the McKee relation fell below the measured BCT value; here, the error was almost 8%. The measured BCT values up to a box length of 500 mm exceeded the BCT values calculated from the McKee function for uncut boxes everywhere. It is particularly noticeable that for boxes of 400 mm in length, which gave the highest BCT values in the measurements, the McKee correlation led to much lower BCT values. The linear relationship between the perimeter and the BCT value in the simple McKee formula seems to be falsified based on our actual measurements. This suggests that the McKee relationship is limited as the box perimeter increases, presumably due to the change in the length–width ratio.

3.3. Limitations for Practice

1. The experimental method in this study used only the B-flute corrugated cardboard, so the results fell within a narrow range. In reality, boxes made of corrugated cardboard have an extremely wide range and variation, such as A, B, C, and other flutes with different numbers of layers, etc.; therefore, the results of this study may be limited for general use, but they do cover an important issue for a possible reduction in material and packaging engineering design.
2. It should also be noted here that there are some environmental circumstances that significantly affect box compression strength, including the changes in temperature and relative humidity and the difference between dynamic and static load. This study did not observe these conditions.

4. Conclusions

- The boxes with cutout areas of 0%, 4%, 16%, 36%, and 64% showed a linear decreasing tendency in compression force. The linear regression model described the measured data with very high accuracy.
- The 400 × 300 × 300 mm sized boxes showed the best average compression force results when the length-to-width ratio was 1.33.
- The 64% cutout group produced the lowest average compression forces in each box size group, which means the 64% cutout significantly reduced the compressive strength of the boxes.
- For cutout corrugated cardboard boxes, the discrepancies in mechanical strength are more significant and cannot be handled with the simplified McKee formula. In some cases, the error of the simplified McKee formula was extremely high.
- The McKee relationship is limited as the box perimeter increases, presumably due to the change in the length–width ratio.
- An optimum perimeter value (1400 mm) could be found, below and above which the measured BCT value decreased, as opposed to the McKee formula in which the perimeter is a multiplier.
- The McKee formula works with relatively high error on corrugated cardboard boxes with sidewall cutouts and cannot follow the tendency of compressive forces along various perimeters.

Author Contributions: Conceptualization, R.P. and L.F.; formal analysis, R.P. and L.F.; funding acquisition, R.P. and P.B.; investigation, R.P., L.F. and P.B.; methodology, R.P. and L.F; supervision, P.B.; visualization, R.P. and L.F.; writing—original draft preparation, L.F.; writing—review and editing, L.F. and P.B. All authors have read and agreed to the published version of the manuscript.

Funding: This research received no external funding.

Institutional Review Board Statement: Not applicable.

Informed Consent Statement: Not applicable.

Data Availability Statement: The data published in this research are available on request from the first author and corresponding authors.

Acknowledgments: We would like to acknowledge DS SMITH for making the measurement samples available free of charge for this study. We would also like to thank Zoltán Pánczél for his personal support and guidance in this study.

Conflicts of Interest: The authors declare no conflict of interest.

References

1. Fadiji, T.; Berry, T.; Coetzee, C.; Opara, L. Investigating the Mechanical Properties of Paperboard Packaging Material for Handling Fresh Produce Under Different Environmental Conditions: Experimental Analysis and Finite Element Modelling. *J. Appl. Packag. Res.* **2017**, *9*, 3.
2. Frank, B. Corrugated Box Compression—A Literature Survey. *Packag. Technol. Sci.* **2014**, *27*, 105–128. [CrossRef]

3. Fadiji, T.; Berry, T.M.; Coetzee, C.J.; Opara, U.L. Mechanical Design and Performance Testing of Corrugated Paperboard Packaging for the Postharvest Handling of Horticultural Produce. *Biosyst. Eng.* **2018**, *171*, 220–244. [CrossRef]
4. Talbi, N.; Batti, A.; Ayad, R.; Guo, Y.Q. An Analytical Homogenization Model for Finite Element Modelling of Corrugated Cardboard. *Compos. Struct.* **2009**, *88*, 280–289. [CrossRef]
5. Gong, G.; Liu, Y.; Fan, B.; Sun, D. Deformation and Compressive Strength of Corrugated Cartons under Different Indentation Shapes: Experimental and Simulation Study. *Packag. Technol. Sci.* **2020**, *33*, 215–226. [CrossRef]
6. Quesenberry, C.; Horvath, L.; Bouldin, J.; White, M.S. The Effect of Pallet Top Deck Stiffness on the Compression Strength of Asymmetrically Supported Corrugated Boxes. *Packag. Technol. Sci.* **2020**, *33*, 547–558. [CrossRef]
7. Li, X.; Wang, J.; Gao, D.; Lv, W. A Theoretical and Experimental Study on Corrugated Paperboard Crushing under Quasi-Static Loadings. *Packag. Technol. Sci.* **2018**, *31*, 641–651. [CrossRef]
8. Fadiji, T.; Coetzee, C.; Opara, U.L. Compression Strength of Ventilated Corrugated Paperboard Packages: Numerical Modelling, Experimental Validation and Effects of Vent Geometric Design. *Biosyst. Eng.* **2016**, *151*, 231–247. [CrossRef]
9. Opara, U.L.; Fadiji, T. Compression Damage Susceptibility of Apple Fruit Packed inside Ventilated Corrugated Paperboard Package. *Sci. Hortic.* **2018**, *227*, 154–161. [CrossRef]
10. Fadiji, T.; Coetzee, C.; Chen, L.; Chukwu, O.; Opara, U.L. Susceptibility of Apples to Bruising inside Ventilated Corrugated Paperboard Packages during Simulated Transport Damage. *Postharvest Biol. Technol.* **2016**, *118*, 111–119. [CrossRef]
11. Fadiji, T.; Coetzee, C.; Pathare, P.; Opara, U.L. Susceptibility to Impact Damage of Apples inside Ventilated Corrugated Paperboard Packages: Effects of Package Design. *Postharvest Biol. Technol.* **2016**, *111*, 286–296. [CrossRef]
12. Marsh, K.; Bugusu, B. Food Packaging—Roles, Materials, and Environmental Issues. *J. Food Sci.* **2007**, *72*, R39–R55. [CrossRef]
13. Mrówczyński, D.; Knitter-Piątkowska, A.; Garbowski, T. Optimal Design of Double-Walled Corrugated Board Packaging. *Materials* **2022**, *15*, 2149. [CrossRef]
14. 2022-04-21_FEFCO Code_WEB_unprotected.Pdf. Available online: https://www.fefco.org/sites/default/files/files/2022-04-21_FEFCO%20Code_WEB_unprotected.pdf (accessed on 10 November 2022).
15. Bartolozzi, G.; Pierini, M.; Orrenius, U.; Baldanzini, N. An Equivalent Material Formulation for Sinusoidal Corrugated Cores of Structural Sandwich Panels. *Compos. Struct.* **2013**, *100*, 173–185. [CrossRef]
16. Defraeye, T.; Cronjé, P.; Berry, T.; Opara, U.L.; East, A.; Hertog, M.; Verboven, P.; Nicolai, B. Towards Integrated Performance Evaluation of Future Packaging for Fresh Produce in the Cold Chain. *Trends Food Sci. Technol.* **2015**, *44*, 201–225. [CrossRef]
17. Zaghloul, M.M.Y.; Mohamed, Y.S.; El-Gamal, H. Fatigue and Tensile Behaviors of Fiber-Reinforced Thermosetting Composites Embedded with Nanoparticles. *J. Compos. Mater.* **2019**, *53*, 709–718. [CrossRef]
18. Pathare, P.B.; Opara, U.L. Structural Design of Corrugated Boxes for Horticultural Produce: A Review. *Biosyst. Eng.* **2014**, *125*, 128–140. [CrossRef]
19. Singh, J.; Olsen, E.; Singh, S.P.; Manley, J.; Wallace, F. The Effect of Ventilation and Hand Holes on Loss of Compression Strength in Corrugated Boxes. *J. Appl. Packag. Res.* **2008**, *2*, 227.
20. McKee, R.C.; Gander, J.W.; Wachuta, J.R. Compression Strength Formula for Corrugated Boxes. *Paperboard Packag.* **1963**, *48*, 149–159.
21. Kellicutt, K.; Landt, E. Development of Design Data for Corrugated Fiberboard Shipping Containers. *Tappi* **1952**, *35*, 398–402.
22. Beldie, L.; Sandberg, G.; Sandberg, L. Paperboard Packages Exposed to Static Loads-Finite Element Modelling and Experiments. *Packag. Technol. Sci.* **2001**, *14*, 171–178. [CrossRef]
23. Biancolini, M.E.; Brutti, C. Numerical and Experimental Investigation of the Strength of Corrugated Board Packages. *Packag. Technol. Sci.* **2003**, *16*, 47–60. [CrossRef]
24. Allerby, I.M.; Laing, G.N.; Cardwell, R.D. Compressive Strength—From Components to Corrugated Containers. *Appita Conf. Notes* **1985**, 1–11.
25. Schrampfer, K.E.; Whitsitt, W.J.; Baum, G.A. Combined Board Edge Crush (ECT) Technology. In *Project 2695-24, Report One: A Progress Report to the Fourdrinier Kraft Board Group of the American Paper Institute*; Institute of Paper Chemistry: Appleton, WI, USA, 1987.
26. Garbowski, T.; Gajewski, T.; Grabski, J.K. The Role of Buckling in the Estimation of Compressive Strength of Corrugated Cardboard Boxes. *Materials* **2020**, *13*, 4578. [CrossRef] [PubMed]
27. Avilés, F.; Carlsson, L.A.; May-Pat, A. A Shear-Corrected Formulation for the Sandwich Twist Specimen. *Exp. Mech.* **2012**, *52*, 17–23. [CrossRef]
28. Garbowski, T.; Gajewski, T.; Grabski, J.K. Role of Transverse Shear Modulus in the Performance of Corrugated Materials. *Materials* **2020**, *13*, 3791. [CrossRef]
29. Garbowski, T.; Gajewski, T.; Grabski, J.K. Torsional and Transversal Stiffness of Orthotropic Sandwich Panels. *Materials* **2020**, *13*, 5016. [CrossRef]
30. Han, J.; Park, J.M. Finite Element Analysis of Vent/Hand Hole Designs for Corrugated Fibreboard Boxes. *Packag. Technol. Sci.* **2007**, *20*, 39–47. [CrossRef]
31. Fadiji, T.; Coetzee, C.J.; Berry, T.M.; Opara, U.L. Investigating the Role of Geometrical Configurations of Ventilated Fresh Produce Packaging to Improve the Mechanical Strength—Experimental and Numerical Approaches. *Food Packag. Shelf Life* **2019**, *20*, 100312. [CrossRef]

32. Fadiji, T.; Coetzee, C.J.; Opara, U.L. Analysis of the Creep Behaviour of Ventilated Corrugated Paperboard Packaging for Handling Fresh Produce—An Experimental Study. *Food Bioprod. Process.* **2019**, *117*, 126–137. [CrossRef]
33. Berry, T.M.; Defraeye, T.; Shrivastava, C.; Ambaw, A.; Coetzee, C.; Opara, U.L. Designing Ventilated Packaging for the Fresh Produce Cold Chain. *Food Bioprod. Process.* **2022**, *134*, 121–149. [CrossRef]
34. Mrówczyński, D.; Garbowski, T.; Knitter-Piątkowska, A. Estimation of the Compressive Strength of Corrugated Board Boxes with Shifted Creases on the Flaps. *Materials* **2021**, *14*, 5181. [CrossRef]
35. Garbowski, T.; Gajewski, T.; Grabski, J.K. Estimation of the Compressive Strength of Corrugated Cardboard Boxes with Various Openings. *Energies* **2021**, *14*, 155. [CrossRef]
36. Huang, T.-C.; Wang, T.-F.; Liao, K.-C. Investigations of Structure Strength and Ventilation Performance for Agriproduct Corrugated Cartons under Long-Term Transportation Trip. *Packag. Technol. Sci.* **2022**, *35*, 821–832. [CrossRef]
37. Wei, Z.; Hua, G.J.; Zhao, D.J. Testing Research of Aspect Ratio on Corrugated Box Compression Strength. *Appl. Mech. Mater.* **2011**, *48–49*, 1213–1216. [CrossRef]
38. Ambaw, A.; Fadiji, T.; Opara, U.L. Thermo-Mechanical Analysis in the Fresh Fruit Cold Chain: A Review on Recent Advances. *Foods* **2021**, *10*, 1357. [CrossRef]
39. *ASTM D4332-01*; Standard Practice for Conditioning Containers, Packages, or Packaging Components for Testing. ASTM International: West Conshohocken, PA, USA, 2006.
40. *ASTM D642*; Standard Test Method for Determining Compressive Resistance of Shipping Containers, Components, and Unit Loads. ASTM International: West Conshohocken, PA, USA, 2010.
41. McKee, R.C.; Gander, J.W.; Wachuta, J.R. Edgewise Compression Strength of Corrugated Board. 1961. Available online: https://smartech.gatech.edu/bitstream/handle/1853/1660/1108-4_000_07111961.pdf (accessed on 10 November 2022).
42. Popil, R. Predicting Box Compression Strength. 2021. Available online: https://www.researchgate.net/publication/355201289_Predicting_Box_Compression_Strength (accessed on 10 November 2022).
43. Garbowski, T.; Knitter-Piątkowska, A. Analytical Determination of the Bending Stiffness of a Five-Layer Corrugated Cardboard with Imperfections. *Materials* **2022**, *15*, 663. [CrossRef] [PubMed]

materials

Article

Influence of Imperfections on the Effective Stiffness of Multilayer Corrugated Board

Damian Mrówczyński [1] and Tomasz Garbowski [2,*]

[1] Doctoral School, Department of Biosystems Engineering, Poznan University of Life Sciences, Wojska Polskiego 28, 60-637 Poznań, Poland
[2] Department of Biosystems Engineering, Poznan University of Life Sciences, Wojska Polskiego 50, 60-627 Poznań, Poland
* Correspondence: tomasz.garbowski@up.poznan.pl

Abstract: There are many possible sources of potential geometrical inaccuracies in each layer of corrugated board during its manufacture. These include, among others, the processes of wetting the corrugated layers during profiling, the process of accelerated drying, the gluing process, and any mechanical impact of the pressure rollers on the cardboard. Work taking into account all the above effects in numerical modeling is not well described in the literature. Therefore, this article presents a simple and practical procedure that allows us to easily account for geometric imperfections in the calculation of the effective stiffness of corrugated board. As a main tool, the numerical homogenization based on the finite element method (FE) was used here. In the proposed procedure, a 3D model of a representative volumetric element (RVE) of a corrugated board is first built. The numerical model can include all kinds of geometrical imperfections and is used to calculate the equivalent tensile and bending stiffnesses. These imperfections were included in the 3D numerical model by appropriate modeling of individual layers, taking into account their distorted shape, which was obtained on the basis of a priori buckling analysis. This paper analyzes different types of buckling in order to find the most representative one. The proposed procedure is easy to implement and fully scalable.

Keywords: multilayer corrugated board; numerical homogenization; effective stiffness; geometrical imperfections; buckling

check for updates

Citation: Mrówczyński, D.; Garbowski, T. Influence of Imperfections on the Effective Stiffness of Multilayer Corrugated Board. *Materials* **2023**, *16*, 1295. https://doi.org/10.3390/ma16031295

Academic Editor: Gaetano Giunta

Received: 31 December 2022
Revised: 24 January 2023
Accepted: 1 February 2023
Published: 2 February 2023

1. Introduction

Currently, the growing consumption and the increase in the number of individual shipments make it necessary to ensure the safe storage and transport of goods. At the same time, more emphasis is placed on ecology, which is why companies quickly began to abandon plastic, which not only harms the environment, but also has a negative impact on the reputation of enterprises. For these reasons, corrugated cardboard has begun to conquer the packaging market. It owes its huge popularity to numerous advantages that largely meet the needs of today. Cardboard packaging is recyclable, easy to dispose of, biodegradable, durable under appropriate conditions, and takes up little space before folding. In addition to the positives for the environment, there are also useful properties for companies. Corrugated cardboard products can be easily shaped by adding holes, ventilation holes, or by printing a brand logo on them. It is also possible to create boxes with perforations, which are often used in the food industry. Such packages are opened by tearing off a part of the material along the previously designed perforations, and then immediately put on the shelf. Such a procedure allows for significant time savings, which brings profits for large companies.

In addition to all the above-mentioned advantages, packaging should, above all, effectively protect the goods. For this reason, the load-bearing capacity of packaging has been

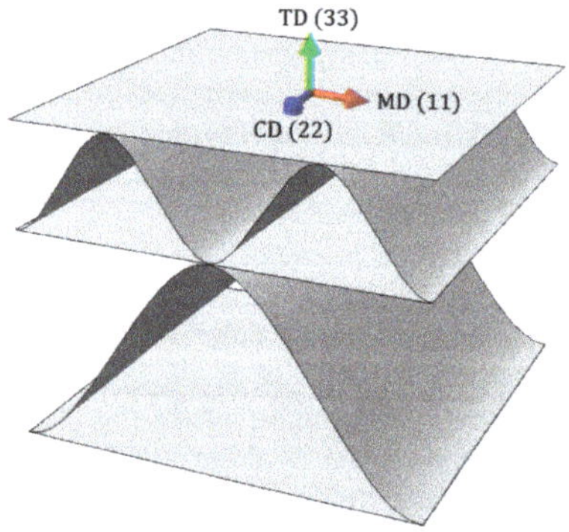

Figure 1. Corrugated cardboard orientation.

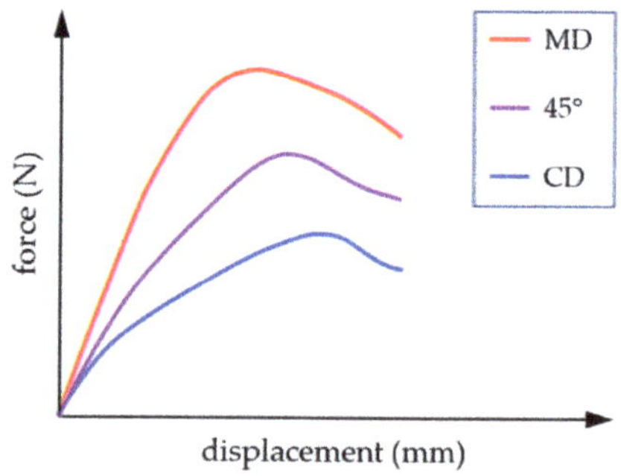

Figure 2. Mechanical behavior of cardboard in different material directions.

Linear elastic orthotropic material can be represented by the stress–strain relationship, as follows:

$$\begin{bmatrix} \varepsilon_{11} \\ \varepsilon_{22} \\ 2\varepsilon_{12} \\ 2\varepsilon_{13} \\ 2\varepsilon_{23} \end{bmatrix} = \begin{bmatrix} 1/E_1 & -\nu_{21}/E_2 & 0 & 0 & 0 \\ -\nu_{12}/E_1 & 1/E_2 & 0 & 0 & 0 \\ 0 & 0 & 1/G_{12} & 0 & 0 \\ 0 & 0 & 0 & 1/G_{13} & 0 \\ 0 & 0 & 0 & 0 & 1/G_{23} \end{bmatrix} \begin{bmatrix} \sigma_{11} \\ \sigma_{22} \\ \sigma_{12} \\ \sigma_{13} \\ \sigma_{23} \end{bmatrix}, \quad (1)$$

where ε_{ij} and σ_{ij} are the strain and stress vector components, E_1 and E_2 are the Young's moduli in the machine and cross directions, respectively, ν_{12} and ν_{21} are the Poisson's coefficients, G_{12} is the in-plane shear modulus, G_{13} and G_{23} are the transverse shear moduli. The compliance/stiffness matrix is symmetrical, so the relationship between Poisson's ratios can be written as follows:

$$\frac{\nu_{12}}{E_1} = \frac{\nu_{21}}{E_2}. \quad (2)$$

In this work, the paper layers were modeled using the linear elastic classical orthotropy described above. The material data were taken from the literature [28] and are presented in Table 1. The thickness of both flat and corrugated layers was assumed to be 0.30 mm.

Table 1. Material data for orthotropic constitutive model of corrugated cardboard layers.

Layers	E_1 (MPa)	E_2 (MPa)	ν_{12} (-)	G_{12} (MPa)	G_{13} (MPa)	G_{23} (MPa)
Liners	3326	1694	0.34	859	429.5	429.5
Fluting	2611	1532	0.32	721	362	362

2.2. *Imperfections–Numerical Study*

The main goal of the work is the numerical analysis of many cases of a corrugated board samples with imperfections and their impact on the stiffness values. In Figure 3, two imperfection shapes are presented; these were considered in this study. The first shape is the most common and corresponds to compression in the machine direction (see Figure 3a). It is formed as a result of thermal and moisture processes during the production of cardboard. The second imperfection mode (see Figure 3b) corresponds to the compression of the sample in the cross direction. This buckling shape was selected in [45] as the most representative for single-walled corrugated board. In both cases, it was assumed that only liners are involved in buckling analysis, and flutes (waves) only support the liners in the right position.

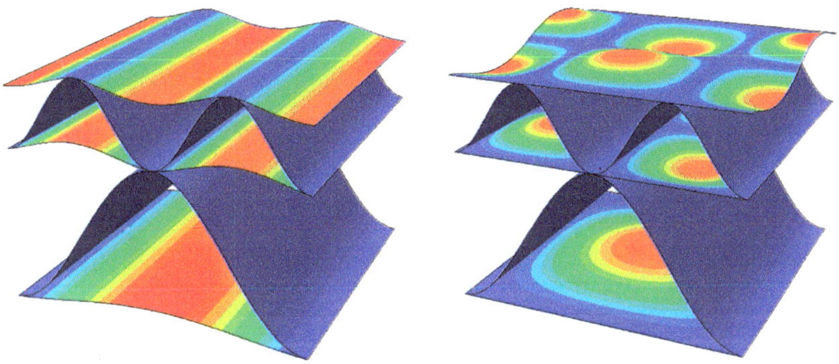

(**a**) (**b**)

Figure 3. Imperfection shapes of double-walled corrugated board corresponding to (**a**) compression in the MD; (**b**) compression in the CD.

All calculations were made for double-walled cardboard composed of high B flute and low E flute (EB cardboard). In Table 2, the geometry of the corrugated layers is presented. The variants also include four imperfection levels, namely 0, 1, 2, and 3% for both imperfection shapes. The amount of the imperfection was taken as a certain part of the buckling length of the liners. Three different buckling lengths can be specified for the analyzed double-walled corrugated board (see Figure 4). The lengths L_1 and L_3 correspond to the periods of the B and E waves, respectively. The buckling length L_2' can take different values depending on the relative position of the crests of E and B waves. The maximum possible value of L_2 is equal to the period of the lower wave, but most often every second segment is divided into two parts by the top of the B wave. For this reason, the buckling length of the middle liner was assumed to be 2/3 of the E wave period.

Table 2. Geometric parameters of cardboard flutes.

Wave (Flute)	Wavelength (mm)	Height (mm)	Take-Up Factor (-)
B	6.5	2.46	1.32
E	3.5	1.15	1.27

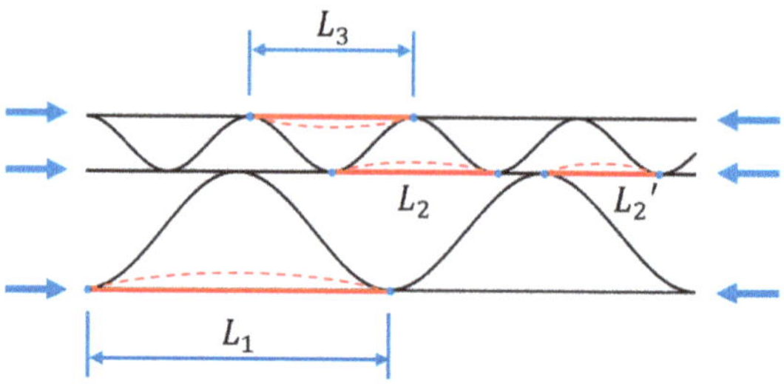

Figure 4. Buckling lengths of corrugated board liners.

Additionally, due to the irregular arrangement of the B and E flutes, the effect of the flute offset on the reduction in cardboard stiffness was verified. For this purpose, for each set of shape and level of imperfection, 10 cases of shifting the lower wave relative to the higher one, ranging from 0% to 90% in increments of 10%, were considered. In Figure 5, all simulated cross-sections are shown.

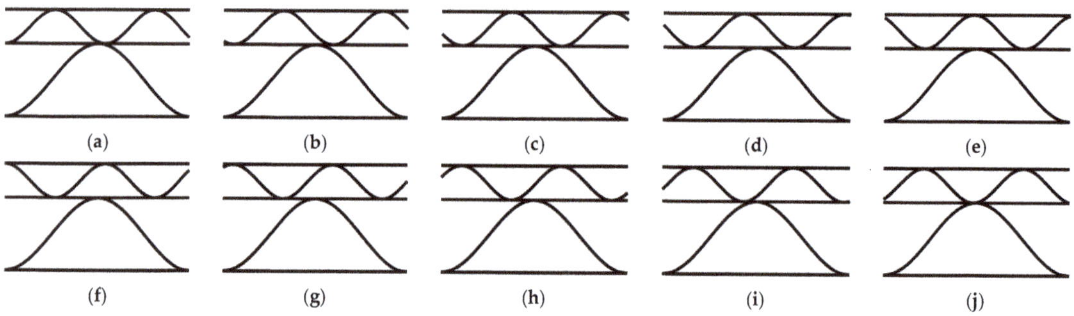

Figure 5. Cross-sections of corrugated board along the wave with offsets of (**a**) 0%, (**b**) 10%, (**c**) 20%, (**d**) 30%, (**e**) 40%, (**f**) 50%, (**g**) 60%, (**h**) 70%, (**i**) 80%, and (**j**) 90%.

Taking into account all the variants described above, each case can be marked with the symbol XX-Y-ZZ. The first two letters XX can be "MD" or "CD", which indicates the imperfection shape corresponding to compression in the MD and CD, respectively. The third sign Y represents the imperfection value and can be 0, 1, 2, or 3, and ZZ is the offset of the lower wave relative to the higher flute and can be 00, 10, 20, 30, 40, 50, 60, 70, 80, or 90.

2.3. Homogenization Procedure

The effect of imperfections on the effective stiffnesses of corrugated board was investigated with the numerical homogenization method proposed by Biancolini in 2005 and extended by Garbowski and Gajewski in 2021. It consists of ensuring the equivalence of the strain energy between the full 3D model of a representative volume element (RVE) and a simplified model of a flat 2D plate. The RVE is a small repeatable fragment of the full 3D model, which in this case is one period of the higher flute of the corrugated board. In the paper, only the main assumptions of the homogenization method used were presented. Detailed information and full derivation can be found in [29].

The homogenization procedure is based on the finite element method. Displacements from linear analysis can be represented by the following basic formula:

$$\mathbf{K}_e\, \mathbf{u}_e = \mathbf{F}_e, \tag{3}$$

where $\mathbf{K}_e$ is the stiffness matrix of the RVE after static condensation, $\mathbf{u}_e$ is a displacement vector of the external nodes, and $\mathbf{F}_e$ is the vector of external forces applied to the considered nodes. The finite element mesh and external nodes are presented in Figure 6.

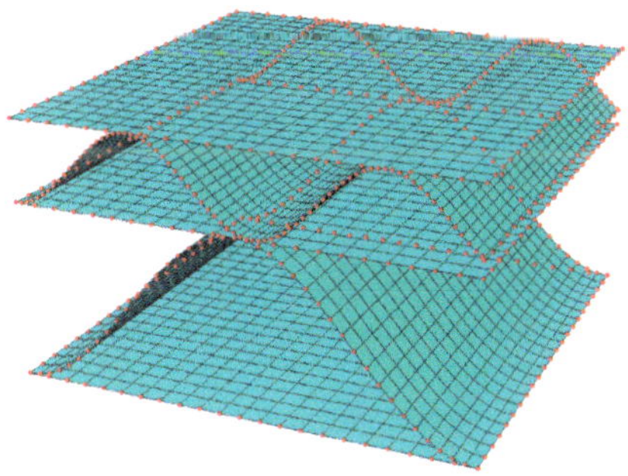

Figure 6. Finite element mesh and external nodes (in red color).

The static condensation procedure is based on the removal of unnecessary unknown degrees of freedom (DOF) and leaving the considered degrees of freedom (called principal DOFs or primary unknowns). In the case of cardboard, the primary unknowns are external RVE nodes. This results in bringing the stiffness matrix only to the external nodes of the model. The FE stiffness matrix after static condensation can be calculated from the following equation:

$$\mathbf{K}_e = \mathbf{K}_{ee} - \mathbf{K}_{ei}\,\mathbf{K}_{ii}^{-1}\mathbf{K}_{ie}, \tag{4}$$

where the components of the stiffness matrix are written for internal (subscript i) and external (subscript e) nodes, as follows:

$$\begin{bmatrix} \mathbf{K}_{ee} & \mathbf{K}_{ei} \\ \mathbf{K}_{ie} & \mathbf{K}_{ii} \end{bmatrix} \begin{bmatrix} \mathbf{u}_e \\ \mathbf{u}_i \end{bmatrix} = \begin{bmatrix} \mathbf{F}_e \\ \mathbf{0} \end{bmatrix}. \tag{5}$$

By the static condensation, the equation of the total elastic strain energy is simplified to the multiplication of the nodal displacements and the external forces acting on these nodes, as follows:

$$E = \frac{1}{2}\mathbf{u}_e^T\,\mathbf{F}_e. \tag{6}$$

Maintaining the appropriate properties of the simplified model is ensured by the balance of the total energy between the 3D model of corrugated board and the 2D plate model. The appropriate definition of displacements at the external edges of the RVE and allowing for bending and membrane behaviors works to ensure the energy balance. The relationship between generalized displacements and strains can be represented by the transformation matrix $\mathbf{H}_j$, as follows:

$$\mathbf{u}_j = \mathbf{H}_j\,\boldsymbol{\epsilon}_j. \tag{7}$$

and considering a single node ($x_j = x$, $y_j = y$, $z_j = z$) the above equation can be written in matrix form, as follows:

$$\begin{bmatrix} u_x \\ u_y \\ u_z \\ \theta_x \\ \theta_y \end{bmatrix}_j = \begin{bmatrix} x & 0 & y/2 & xz & 0 & yz/2 & z/2 & 0 \\ 0 & y & x/2 & 0 & yz & xz/2 & 0 & z/2 \\ 0 & 0 & 0 & -x^2/2 & -y^2/2 & -xy/2 & x/2 & y/2 \\ 0 & 0 & 0 & 0 & -y & -x/2 & 0 & 0 \\ 0 & 0 & 0 & x & 0 & y/2 & 0 & 0 \end{bmatrix}_j \begin{bmatrix} \varepsilon_x \\ \varepsilon_y \\ \gamma_{xy} \\ \kappa_x \\ \kappa_y \\ \kappa_{xy} \\ \gamma_{xz} \\ \gamma_{yz} \end{bmatrix}_j . \tag{8}$$

Substituting the basic FEM equation and the relationship between generalized displacements and strains into Equation (6) results in the following:

$$E = \frac{1}{2}\mathbf{u}_e^T\,\mathbf{K}\,\mathbf{u}_e = \frac{1}{2}\boldsymbol{\epsilon}_e^T\,\mathbf{H}_e^T\,\mathbf{K}\,\mathbf{H}_e\,\boldsymbol{\epsilon}_e \tag{9}$$

and considering a finite element subjected to standard load conditions, such as bending, stretching, shear, elastic internal energy, this can be represented by the following equation:

$$E = \frac{1}{2}\boldsymbol{\epsilon}_e^T\,\mathbf{H}_k\,\boldsymbol{\epsilon}_e\{area\}, \tag{10}$$

where the matrix $\mathbf{H}_k$ is the RVE stiffness matrix, which can be determined from the following:

$$\mathbf{H}_k = \frac{\mathbf{H}_e^T\,\mathbf{K}\,\mathbf{H}_e}{area}. \tag{11}$$

The stiffness matrix $\mathbf{H}_k$ consists of several submatrices corresponding to the basic load conditions, as follows:

$$\mathbf{H}_k = \begin{bmatrix} \mathbf{A}_{3\times3} & \mathbf{B}_{3\times3} & \\ \mathbf{B}_{3\times3} & \mathbf{D}_{3\times3} & \\ & & \mathbf{R}_{2\times2} \end{bmatrix}, \tag{12}$$

where **A** represents the tensile and in-plane shear stiffnesses, **D** represents the bending and torsional stiffnesses, **B** is the coupling subarray of the tension and bending stiffnesses, and **R** is the transversal shear stiffnesses.

The stiffness values in subarray **A** do not depend on the position of a neutral axis. This is not true in the case of subarray **B**, in which the values change depending on the position of the neutral axis. In most symmetrical cases, the stiffness matrix **B** is a zero matrix, unlike for asymmetric cross-sections (e.g., multi-walled corrugated cardboard), for which some components of the matrix **B** are not zero, which affects the stiffness in the subarray **D**. This effect can be suppressed in two ways. The first is to choose the position of the neutral axis so as to minimize the values in the subarray **B**. An alternative solution is to calculate the uncoupled matrix **D** from the following formula:

$$\mathbf{D} = \mathbf{D}' - \mathbf{B}\mathbf{A}^{-1}\mathbf{B}, \tag{13}$$

where $\mathbf{D}'$ contains the original (coupled) bending and torsional stiffnesses for the non-zero matrix **B**.

Numerical computations consisted in applying the selected buckling shape and imperfection value, and then calculating the stiffness matrix of the RVE. In all analyses, the four-node quadrilaterals shell elements with full integration (S4 elements) were used [47]. All models were built in FE commercial software, from where the stiffnesses in all nodes were generated. Next, the effective stiffness matrix was determined using the homogenization method described above. An approximate global size equal to 0.1625 mm was assumed. Due to the different variants of offset, the number of elements changed. For the MD-0-00 case, the model consisted of 9922 nodes, 9760 elements, and 59,532 degrees of freedom. The sensitivity study of the mesh size was already studied in our previous works; therefore, here this part of the research is omitted. It is worth noting that in the presented homogenization method, the use of the finite element method is limited only to building a global stiffness matrix, and no formal loads or boundary conditions are defined. They are directly implemented through the $\mathbf{H}_e$ matrix in Equation (11).

3. Results

The first step was to properly create the corrugated cardboard RVE model in FE software. In the cases of samples with imperfections, the geometry of the liners was changed and replaced with the correct buckling shape (see Figure 3). Then, the material parameters contained in Table 1 were assigned to both liners and waves. Following the homogenization procedure described in Section 2.3, stiffness matrices were determined for all analyzed cases. In Table 3, an example of the stiffness matrix $\mathbf{H}_k$ is presented for a model without imperfections and with 0% offset. No buckling mode is assigned to this case, because the imperfection value is equal to zero; therefore, it was marked with the XX-0-00 symbol.

Table 3. Constitutive stiffness matrix $\mathbf{H}_k$ for the XX-0-00 model.

		A and B			**B and D**			**R**	
		1	2	3	1	2	3	4	5
A and B	1	3460	630	0	741	128	0		
	2	630	2806	0	128	732	0		
	3	0	0	1122	0	0	296		
B and D	1	741	128	0	7692	1357	0		
	2	128	732	0	1357	5375	0		
	3	0	0	296	0	0	2231		
R	4							206	0
	5							0	167

As expected, non-zero components of the **B** matrix appeared in the stiffness matrix $\mathbf{H}_k$, which results from the asymmetry of the double-walled corrugated board sample. Tables 4–10 show the main diagonal components of the stiffness matrix for all analyzed cases. The components $(*)_{12}$ are omitted, but this does not create an error in the analysis because these elements are related to the components $(*)_{11}$ and $(*)_{12}$ in each stiffness matrix. The following tables do not list the elements of matrix **B**, but its effect on matrix **D** has been taken into account by applying Equation (13). The components of the **R** matrix were also omitted. The data contained in Tables 4–10 are also presented in the form of graphs in Figure 7.

Table 4. The selected stiffnesses in models without imperfections and with different offsets.

	XX-0-00	**XX-0-10**	**XX-0-20**	**XX-0-30**	**XX-0-40**	**XX-0-50**	**XX-0-60**	**XX-0-70**	**XX-0-80**	**XX-0-90**
A_{11} (N/mm)	3460	3459	3451	3441	3424	3427	3447	3454	3461	3468
A_{22} (N/mm)	2806	2796	2793	2798	2805	2803	2795	2793	2800	2809
A_{33} (N/mm)	1122	1125	1126	1124	1122	1122	1125	1126	1124	1122
D_{11} (N·mm)	7533	7538	7549	7543	7510	7514	7552	7544	7536	7533
D_{22} (N·mm)	5183	5167	5137	5111	5106	5106	5120	5151	5176	5186
D_{33} (N·mm)	2153	2152	2146	2134	2124	2127	2140	2149	2153	2154

Table 5. The selected stiffnesses with 1% imperfections of compression in the MD shape and with different offsets.

	MD-1-00	MD-1-10	MD-1-20	MD-1-30	MD-1-40	MD-1-50	MD-1-60	MD-1-70	MD-1-80	MD-1-90
A_{11} (N/mm)	3437	3446	3420	3408	3388	3382	3409	3425	3413	3460
A_{22} (N/mm)	2806	2797	2792	2798	2804	2801	2794	2793	2799	2809
A_{33} (N/mm)	1122	1125	1125	1124	1121	1122	1125	1125	1124	1122
D_{11} (N·mm)	7259	7283	7314	7307	7238	7237	7263	7307	7247	7297
D_{22} (N·mm)	5094	5085	5066	5041	5019	5023	5026	5079	5085	5108
D_{33} (N·mm)	2111	2113	2113	2101	2084	2088	2096	2115	2110	2117

Table 6. The selected stiffnesses with 2% imperfections of compression in the MD shape and with different offsets.

	MD-2-00	MD-2-10	MD-2-20	MD-2-30	MD-2-40	MD-2-50	MD-2-60	MD-2-70	MD-2-80	MD-2-90
A_{11} (N/mm)	3361	3375	3332	3314	3291	3266	3305	3339	3313	3406
A_{22} (N/mm)	2804	2796	2790	2796	2802	2799	2793	2791	2796	2809
A_{33} (N/mm)	1122	1124	1125	1123	1121	1122	1124	1125	1123	1121
D_{11} (N·mm)	6799	6848	6902	6896	6798	6772	6787	6893	6790	6884
D_{22} (N·mm)	5002	5000	4993	4968	4931	4937	4931	5004	4992	5029
D_{33} (N·mm)	2069	2074	2080	2068	2043	2049	2051	2082	2068	2081

Table 7. The selected stiffnesses with 3% imperfections of compression in the MD shape and with different offsets.

	MD-3-00	MD-3-10	MD-3-20	MD-3-30	MD-3-40	MD-3-50	MD-3-60	MD-3-70	MD-3-80	MD-3-90
A_{11} (N/mm)	3252	3266	3210	3185	3158	3105	3158	3220	3181	3327
A_{22} (N/mm)	2802	2794	2788	2793	2799	2796	2790	2788	2793	2808
A_{33} (N/mm)	1121	1124	1125	1123	1120	1121	1123	1125	1123	1121
D_{11} (N·mm)	6243	6320	6399	6393	6276	6213	6216	6387	6250	6377
D_{22} (N·mm)	4911	4916	4921	4896	4844	4852	4837	4930	4900	4950
D_{33} (N·mm)	2027	2036	2047	2036	2004	2010	2007	2048	2026	2045

Table 8. The selected stiffnesses with 1% imperfections of compression in the CD shape and with different offsets.

	CD-1-00	CD-1-10	CD-1-20	CD-1-30	CD-1-40	CD-1-50	CD-1-60	CD-1-70	CD-1-80	CD-1-90
A_{11} (N/mm)	3427	3429	3423	3411	3395	3395	3419	3426	3429	3439
A_{22} (N/mm)	2795	2787	2784	2790	2797	2794	2787	2785	2792	2800
A_{33} (N/mm)	1121	1124	1125	1123	1121	1122	1124	1125	1123	1121
D_{11} (N·mm)	7318	7329	7342	7335	7310	7302	7348	7342	7323	7324
D_{22} (N·mm)	5100	5082	5057	5027	5028	5026	5042	5069	5095	5100
D_{33} (N·mm)	2127	2125	2121	2107	2099	2101	2115	2122	2127	2126

Table 9. The selected stiffnesses with 2% imperfections of compression in the CD shape and with different offsets.

	CD-2-00	CD-2-10	CD-2-20	CD-2-30	CD-2-40	CD-2-50	CD-2-60	CD-2-70	CD-2-80	CD-2-90
A_{11} (N/mm)	3342	3351	3346	3335	3319	3313	3347	3351	3348	3364
A_{22} (N/mm)	2767	2763	2761	2767	2776	2772	2767	2764	2769	2776
A_{33} (N/mm)	1119	1122	1123	1121	1119	1120	1122	1123	1121	1119
D_{11} (N·mm)	6922	6956	6969	6965	6954	6914	6985	6990	6932	6952
D_{22} (N·mm)	4967	4947	4933	4895	4908	4900	4921	4943	4967	4966
D_{33} (N·mm)	2097	2094	2092	2077	2070	2073	2086	2093	2098	2096

Table 10. The selected stiffnesses with 3% imperfections of compression in the CD shape and with different offsets.

	CD-3-00	CD-3-10	CD-3-20	CD-3-30	CD-3-40	CD-3-50	CD-3-60	CD-3-70	CD-3-80	CD-3-90
A_{11} (N/mm)	3232	3249	3246	3234	3223	3208	3256	3253	3244	3267
A_{22} (N/mm)	2728	2730	2730	2736	2749	2742	2740	2736	2739	2744
A_{33} (N/mm)	1115	1119	1119	1119	1115	1117	1119	1120	1118	1116
D_{11} (N·mm)	6455	6514	6532	6528	6540	6458	6561	6574	6473	6511
D_{22} (N·mm)	4814	4791	4790	4742	4771	4755	4783	4799	4819	4810
D_{33} (N·mm)	2064	2060	2059	2044	2038	2041	2055	2060	2066	2062

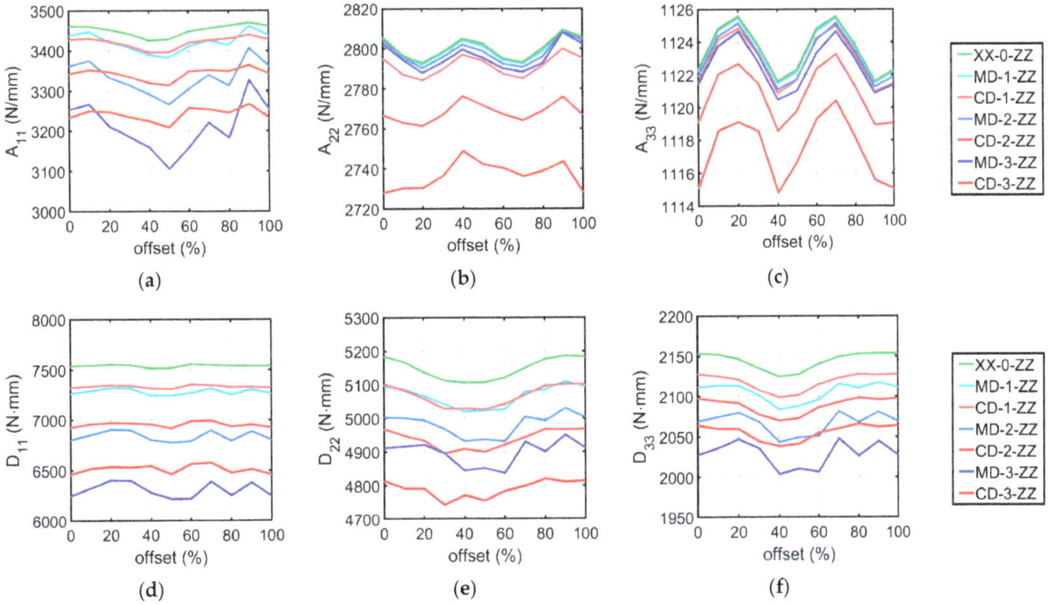

Figure 7. Influence of imperfections and offset on the cardboard stiffness of (**a**) A_{11}, (**b**) A_{22}, (**c**) A_{33}, (**d**) D_{11}, (**e**) D_{22}, and (**f**) D_{33}.

For each shape and level of imperfection, the average stiffness values from all offset cases were calculated. Then, the obtained values for models with imperfections were compared to the average stiffnesses of perfectly shaped cardboard. In Figure 8, the tensile and bending stiffness reduction due to geometrical imperfections is shown.

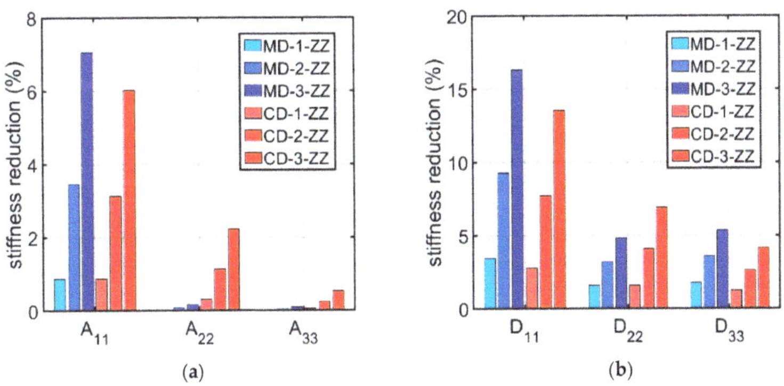

Figure 8. Average percentage stiffness reduction in (**a**) tensile stiffnesses and (**b**) bending stiffnesses.

As can be seen in Tables 4–10 and Figure 7, the stiffnesses vary depending on the offset. Considering each set of shape and level of imperfection, the difference between the maximum and minimum stiffness values was calculated. Then, the obtained difference was related to the average stiffness obtained from all 10 offset cases, obtaining a percentage relative spread. In Table 11, the relative spread for all analyzed sets of imperfection is presented.

Table 11. Relative spread of the extreme stiffnesses for the selected shape and level of imperfection.

Case	Relative Spread (%)					
	A_{11}	A_{22}	A_{33}	D_{11}	D_{22}	D_{33}
MD-3-ZZ	6.9	0.7	0.4	2.9	2.3	2.2
MD-2-ZZ	4.2	0.7	0.4	1.9	2.0	1.9
MD-1-ZZ	2.3	0.6	0.4	1.1	1.8	1.6
XX-0-ZZ	1.3	0.6	0.4	0.6	1.6	1.4
CD-1-ZZ	1.3	0.6	0.4	0.6	1.5	1.4
CD-2-ZZ	1.5	0.5	0.4	1.1	1.5	1.3
CD-3-ZZ	1.8	0.8	0.5	1.8	1.6	1.3

4. Discussion

All crucial results have been shown in Tables 3–11 and Figures 7 and 8. In Table 3, an example of a stiffness matrix determined for the XX-0-00 model is presented, in which the submatrices **A**, **B**, **D**, and **R** are distinguished. The **B** matrix contains non-zero components because the XX-0-00 case cross-section, as in all other models, is asymmetric. For this reason, Equation (13) was used in all corrugated board cases to calculate the uncoupled matrix **D**, whose values are gathered in Tables 4–10. Comparing the stiffness values D_{11}, D_{22}, and D_{33} from Tables 3 and 4, it can be seen that the values in Table 4 are smaller, which results from the uncoupling of the **D** matrix.

Considering Tables 4–10 and Figure 7, where the stiffness results for all models are presented, it can be seen that as the level of imperfection increases, the cardboard stiffness decreases, as expected. The line diagrams also show that depending on the applied wave offset, the stiffnesses change their value. This is shown in more detail in Table 11, which shows the relative spread of the extreme stiffnesses. The greatest fluctuations occur for the A_{11} stiffnesses, are slightly smaller for all bending stiffnesses, and are negligibly small for the A_{22} and A_{33} stiffnesses. In most cases, it is clear that the relative spread increases as the amount of imperfection increases. In addition, the data in this table show that the stiffnesses of the models with the MD imperfection shape are more sensitive to offset changes than the cases with the CD buckling mode.

Based on Tables 4–10 and Figures 7 and 8, it can be concluded that geometric imperfections have a huge influence on the bending stiffness D_{11}. Large decreases were also noted for the stiffnesses A_{11}, D_{22}, and D_{33}. The effect of imperfections on the stiffness A_{22} is small, and the effect is negligible on the A_{33} stiffness. These results are very similar to those obtained from the analysis of the effect of imperfections on the stiffness of single-walled corrugated board conducted by Mrówczyński et al. [45], where the largest decreases were also obtained for D_{11}, and the smallest were obtained for A_{33}. The implementation of the MD buckling mode results in a greater average reduction in stiffnesses A_{11}, D_{11}, and D_{33} compared to the CD imperfection shape, for which a higher average reduction occur for the stiffnesses of A_{22}, A_{33}, and D_{22}. These differences are not that significant and amount to a maximum of 2.75 % (for the stiffness D_{11} between the MD-3-ZZ and CD-3-ZZ models).

To simplify computations and save computational time, it is desirable to select one case that should be computed as representative instead of carrying out an entire numerical study. To determine this, several observations from the conducted analyses should be considered. The imperfection shape of the compression in the CD has smaller relative spread values than the buckling mode of compression in the MD, so the change in the offset does not affect the stiffnesses so significantly. In addition, considering the MD shape does not significantly reduce the stiffnesses A_{22} and A_{33}, which seems to be an unfavorable effect. Based on the calculations, it can be concluded that a reasonable solution is to choose the CD imperfection shape as a representative model. Due to the quite small relative spread, the choice of a representative offset value is not so important, so its value can be chosen arbitrarily. The conclusions drawn from the presented numerical study are consistent with the work of Mrówczyński et al. [45], in which the authors showed that the CD imperfection shape is also the most representative buckling mode for single-walled corrugated cardboard.

All the observations discussed above prove that there is a direct influence of imperfections on the stiffnesses **A**, **B**, and **D**. However, the basic question arises whether it is possible to verify the presented theoretical approach with experimental data. Although testing corrugated board is not very laborious or particularly expensive, nevertheless validating the theoretical analysis presented here is not an easy task, because, firstly, it is practically impossible to produce corrugated board with a given level of imperfection and, secondly, it is also very laborious to check the actual level of imperfection in the cardboard produced. Nevertheless, below, we present a comparison (validation) of the results obtained using the method presented here and the method presented in paper [44] in comparison with the results of laboratory tests conducted by Czechowski et al. and presented in paper [48].

All experimental details are described in [48], while analytical assumptions, in which imperfections are also included, are given in [44]. Here, only the results of the experimental campaign (namely, results of 4-point bending tests of various asymmetric double-walled boards) are shown together with the results obtained using numerical and analytical models. All these results are presented in Table 12, together with the results obtained using the methods presented here with the same amount of imperfection of flat layers as assumed in [44]. It is worth noting that only one of all stiffnesses is checked, namely the bending stiffness in MD. The obtained results do not differ by more than 5% from the results obtained using the analytical model, while the average absolute error (compared to the experimental data) was 7.5%.

Table 12. Bending stiffness in MD for all considered models.

Board ID	Face-Up	EXP (Mean) [48] (Nm)	FEM [48] (Nm)	Analytical [44] (Nm)	Present Model (Nm)
Board 1	EB	8.32	7.62	7.13	7.54
	BE	8.47	7.58	7.84	8.34
Board 2	EB	10.97	9.88	11.15	11.78
	BE	11.58	9.81	11.65	12.20
Board 3	EB	7.25	7.61	7.15	7.54
	BE	9.50	7.53	7.85	8.16
Board 4	EB	9.10	7.53	7.24	7.69
	BE	11.10	7.45	7.98	8.26
Board 5	EB	11.46	10.42	10.89	11.33
	BE	12.97	10.37	11.52	12.02
Board 6	EB	8.20	8.45	8.86	9.01
	BE	9.12	8.40	9.27	9.88

5. Conclusions

The article presents an approach of numerical homogenization of double-walled corrugated board, taking into account the initial imperfections of its layers. The creation of many cardboard models in various configurations made it possible to study various aspects of the considered problem. In the study, corrugated cardboard was homogenized using the numerical method based on the strain energy equivalence between a representative 3D model and a simplified plate. The analysis of five-layered cardboard with imperfections is not such an easy task. The cross-sectional shape of the cardboard depends on the storage and load conditions. For this reason, it is important to indicate a representative case that adequately reflects the properties of the material.

Based on the conducted analyses, it can be seen that the bending stiffnesses and the tensile stiffness along the wave are the most sensitive to the imperfect shape of the cardboard. Calculations were made for a wide range of cases, which allowed us to choose a representative imperfection model. The conclusions drawn from the numerical analyses are consistent with the research that was carried out in the previous work for single-walled corrugated board.

Author Contributions: Conceptualization, D.M. and T.G.; methodology, D.M. and T.G.; software, T.G. and D.M.; validation, D.M. and T.G.; formal analysis, D.M.; investigation, D.M.; resources, D.M.; data curation, D.M.; writing—original draft preparation, D.M.; writing—review and editing, T.G.; visualization, D.M.; supervision, T.G.; project administration, T.G.; funding acquisition, T.G. All authors have read and agreed to the published version of the manuscript.

Funding: This research received no external funding.

Institutional Review Board Statement: Not applicable.

Informed Consent Statement: Not applicable.

Data Availability Statement: The data presented in this study are available on request from the corresponding author.

Conflicts of Interest: The authors declare no conflict of interest. The funders had no role in the design of the study; in the collection, analyses, or interpretation of data, in the writing of the manuscript, or in the decision to publish the results.

References

1. Kellicutt, K.; Landt, E. Development of design data for corrugated fiberboard shipping containers. *TAPPI J.* **1952**, *35*, 398–402.
2. Maltenfort, G. Compression strength of corrugated containers. *Fibre Contain.* **1956**, *41*, 106–121.
3. McKee, R.C.; Gander, J.W.; Wachuta, J.R. Compression strength formula for corrugated boxes. *Paperboard Packag.* **1963**, *48*, 149–159.
4. Allerby, I.M.; Laing, G.N.; Cardwell, R.D. Compressive strength—From components to corrugated containers. *Appita Conf. Notes* **1985**, 1–11.
5. Schrampfer, K.E.; Whitsitt, W.J.; Baum, G.A. *Combined Board Edge Crush (ECT) Technology*; Institute of Paper Chemistry: Appleton, WI, USA, 1987.
6. Batelka, J.J.; Smith, C.N. *Package Compression Model*; Institute of Paper Science and Technology, Georgia Institute of Technology: Atlanta, GA, USA, 1993.
7. Frank, B. Corrugated Box Compression—A Literature Survey. *Packag. Technol. Sci.* **2014**, *27*, 105–128. [CrossRef]
8. Garbowski, T.; Gajewski, T.; Grabski, J.K. Estimation of the compressive strength of corrugated cardboard boxes with various openings. *Energies* **2021**, *14*, 155. [CrossRef]
9. Archaviboonyobul, T.; Chaveesuk, R.; Singh, J.; Jinkarn, T. An analysis of the influence of hand hole and ventilation hole design on compressive strength of corrugated fiberboard boxes by an artificial neural network model. *Packag. Technol. Sci.* **2020**, *33*, 171–181. [CrossRef]
10. Fadiji, T.; Coetzee, C.J.; Opara, U.L. Compression strength of ventilated corrugated paperboard packages: Numerical modelling, experimental validation and effects of vent geometric design. *Biosyst. Eng.* **2016**, *151*, 231–247. [CrossRef]
11. Garbowski, T.; Gajewski, T.; Grabski, J.K. Estimation of the compressive strength of corrugated cardboard boxes with various perforations. *Energies* **2021**, *14*, 1095. [CrossRef]
12. Mrówczyński, D.; Garbowski, T.; Knitter-Piątkowska, A. Estimation of the Compressive Strength of Corrugated Board Boxes with Shifted Creases on the Flaps. *Materials* **2021**, *14*, 5181. [CrossRef] [PubMed]
13. Stott, R.A. Compression and stacking strength of corrugated fibreboard containers. *Appita J.* **2017**, *70*, 76–82.
14. Junli, W.; Quancheng, Z. Effect of moisture content of corrugated box on mechanical properties. *J. Lanzhou Jiaotong Univ.* **2006**, *25*, 134–136.
15. Gallo, J.; Cortés, F.; Alberdi, E.; Goti, A. Mechanical behavior modeling of containers and octabins made of corrugated cardboard subjected to vertical stacking loads. *Materials* **2021**, *14*, 2392. [CrossRef]
16. Zhang, Y.-L.; Chen, J.; Wu, Y.; Sun, J. Analysis of hazard factors of the use of corrugated carton in packaging low-temperature yogurt during logistics. *Procedia Environ. Sci.* **2011**, *10*, 968–973. [CrossRef]
17. Urbanik, T.J.; Frank, B. Box compression analysis of world-wide data spanning 46 years. *Wood Fiber Sci.* **2006**, *38*, 399–416.

18. Nordstrand, T.; Carlsson, L. Evaluation of transverse shear stiffness of structural core sandwich plates. *Compos. Struct.* **1997**, *37*, 145–153. [CrossRef]
19. Nordstrand, T. Basic Testing and Strength Design of Corrugated Board and Containers. Ph.D. Thesis, Lund University, Lund, Sweden, 2003.
20. Avilés, F.; Carlsson, L.A.; May-Pat, A. A shear-corrected formulation of the sandwich twist specimen. *Exp. Mech.* **2012**, *52*, 17–23. [CrossRef]
21. Garbowski, T.; Gajewski, T.; Grabski, J.K. Torsional and transversal stiffness of orthotropic sandwich panels. *Materials* **2020**, *13*, 5016. [CrossRef]
22. Garbowski, T.; Gajewski, T.; Grabski, J.K. Role of transverse shear modulus in the performance of corrugated materials. *Materials* **2020**, *13*, 3791. [CrossRef]
23. Urbanik, T.J.; Saliklis, E.P. Finite element corroboration of buckling phenomena observed in corrugated boxes. *Wood Fiber Sci.* **2003**, *35*, 322–333.
24. Maneengam, A.; Siddique, M.J.; Selvaraj, R.; Kakaravada, I.; Arumugam, A.B.; Singh, L.K.; Kumar, N. Influence of multi-walled carbon nanotubes reinforced honeycomb core on vibration and damping responses of carbon fiber composite sandwich shell structures. *Polym. Compos.* **2022**, *43*, 2073. [CrossRef]
25. Gu, F.; Chen, L.; Zhu, X.; Lu, X.; Fang, D. Fabrication and uniaxial compression mechanical behavior of composite corrugated-core sandwich cylinder with thin-wall metal liner. *Mech. Adv. Mater. Struct.* **2022**, *29*, 949–962. [CrossRef]
26. Sohrabpour, V.; Hellström, D. Models and software for corrugated board and box design. In Proceedings of the 18th International Conference on Engineering Design (ICED 11), Copenhagen, Denmark, 15–18 October 2011.
27. Allaoui, S.; Benzeggagh, M.L.; Aboura, Z.; Talbi, N. Elastic behaviour of corrugated cardboard: Experiments and modeling. *Compos. Struct.* **2004**, *63*, 53–62.
28. Biancolini, M.E. Evaluation of equivalent stiffness properties of corrugated board. *Compos. Struct.* **2005**, *69*, 322–328. [CrossRef]
29. Garbowski, T.; Gajewski, T. Determination of transverse shear stiffness of sandwich panels with a corrugated core by numerical homogenization. *Materials* **2021**, *14*, 1976. [CrossRef]
30. Hohe, J. A direct homogenization approach for determination of the stiffness matrix for microheterogeneous plates with application to sandwich panels. *Compos. Part B* **2003**, *34*, 615–626. [CrossRef]
31. Buannic, N.; Cartraud, P.; Quesnel, T. Homogenization of corrugated core sandwich panels. *Compos. Struct.* **2003**, *59*, 299–312. [CrossRef]
32. Abbès, B.; Guo, Y.Q. Analytic homogenization for torsion of orthotropic sandwich plates. *Appl. Compos. Struct.* **2010**, *92*, 699–706. [CrossRef]
33. Ramírez-Torres, A.; Penta, R.; Rodríguez-Ramos, R.; Merodio, J.; Sabina, F.J.; Bravo-Castillero, J.; Guinovart-Díaz, R.; Preziosi, L.; Grillo, A. Three scales asymptotic homogenization and its application to layered hierarchical hard tissues. *Int. J. Solids Struct.* **2018**, *130–131*, 190–198. [CrossRef]
34. Ramírez-Torres, A.; Di Stefano, S.; Grillo, A.; Rodríguez-Ramos, R.; Merodio, J.; Penta, R. An asymptotic homogenization approach to the microstructural evolution of heterogeneous media. *Int. J. Non Linear Mech.* **2018**, *106*, 245–257. [CrossRef]
35. Suarez, B.; Muneta, M.L.M.; Sanz-Bobi, J.D.; Romero, G. Application of homogenization approaches to the numerical analysis of seating made of multi-wall corrugated cardboard. *Compos. Struct.* **2021**, *262*, 113642. [CrossRef]
36. Garbowski, T.; Knitter-Piątkowska, A.; Mrówczyński, D. Numerical homogenization of multi-layered corrugated cardboard with creasing or perforation. *Materials* **2021**, *14*, 3786. [CrossRef]
37. Mrówczyński, D.; Knitter-Piątkowska, A.; Garbowski, T. Non-Local Sensitivity Analysis and Numerical Homogenization in Optimal Design of Single-Wall Corrugated Board Packaging. *Materials* **2022**, *15*, 720. [CrossRef] [PubMed]
38. Mrówczyński, D.; Knitter-Piątkowska, A.; Garbowski, T. Optimal Design of Double-Walled Corrugated Board Packaging. *Materials* **2022**, *15*, 2149. [CrossRef]
39. Nguyen-Minh, N.; Tran-Van, N.; Bui-Xuan, T.; Nguyen-Thoi, T. Static analysis of corrugated panels using homogenization models and a cell-based smoothed mindlin plate element (CS-MIN3). *Front. Struct. Civ. Eng.* **2019**, *13*, 251–272. [CrossRef]
40. Beck, M.; Fischerauer, G. Modeling Warp in Corrugated Cardboard Based on Homogenization Techniques for In-Process Measurement Applications. *Appl. Sci.* **2022**, *12*, 1684. [CrossRef]
41. Nordstrand, T.M. Parametric study of the post-buckling strength of structural core sandwich panels. *Compos. Struct.* **1995**, *30*, 441–451. [CrossRef]
42. Nordstrand, T. Analysis and testing of corrugated board panels into the post-buckling regime. *Compos. Struct.* **2004**, *63*, 189–199. [CrossRef]
43. Lu, T.J.; Chen, C.; Zhu, G. Compressive behaviour of corrugated board panels. *J. Compos. Mater.* **2001**, *35*, 2098–2126. [CrossRef]
44. Garbowski, T.; Knitter-Piątkowska, A. Analytical Determination of the Bending Stiffness of a Five-Layer Corrugated Cardboard with Imperfections. *Materials* **2022**, *15*, 663. [CrossRef]
45. Mrówczyński, D.; Knitter-Piątkowska, A.; Garbowski, T. Numerical Homogenization of Single-Walled Corrugated Board with Imperfections. *Appl. Sci.* **2022**, *12*, 9632. [CrossRef]
46. Cillie, J.; Coetzee, C. Experimental and Numerical Investigation of the In-Plane Compression of Corrugated Paperboard Panels. *Math. Comput. Appl.* **2022**, *27*, 108. [CrossRef]

47. Abaqus Unified FEA Software. Available online: https://www.3ds.com/products-services/simulia/products/abaqus (accessed on 29 December 2022).
48. Czechowski, L.; Kmita-Fudalej, G.; Szewczyk, W.; Gralewski, J.; Bienkowska, M. Numerical and Experimental Study of Five-Layer Non-Symmetrical Paperboard Panel Stiffness. *Materials* **2021**, *14*, 7453. [CrossRef] [PubMed]

materials

MDPI

Article

Compressive Strength of Corrugated Paperboard Packages with Low and High Cutout Rates: Numerical Modelling and Experimental Validation

Lajos Fehér [1], Damian Mrówczyński [2], Renáta Pidl [1] and Péter Böröcz [3,*]

[1] Department of Applied Mechanics, Széchenyi István University, Egyetem tér 1, 9026 Győr, Hungary
[2] Research and Development Department, Femat Sp. z o.o., Wagrowska 2, 61-369 Poznań, Poland
[3] Department of Logistics and Forwarding, Széchenyi István University, Egyetem tér 1, 9026 Győr, Hungary
* Correspondence: boroczp@sze.hu

Abstract: The finite element method is a widely used numerical method to analyze structures in virtual space. This method can be used in the packaging industry to determine the mechanical properties of corrugated boxes. This study aims to create and validate a numerical model to predict the compression force of corrugated cardboard boxes by considering the influence of different cutout configurations of sidewalls. The types of investigated boxes are the following: the width and height of the boxes are 300 mm in each case and the length dimension of the boxes varied from 200 mm to 600 mm with a 100 mm increment. The cutout rates were 0%, 4%, 16%, 36%, and 64% with respect to the total surface area of sidewalls of the boxes. For the finite element analysis, a homogenized linear elastic orthotropic material model with Hill plasticity was used. The results of linear regressions show very good estimations to the numerical and experimental box compression test (BCT) values in each tested box group. Therefore, the numerical model can give a good prediction for the BCT force values from 0% cutout to 64% cutout rates. The accuracy of the numerical model decreases a little when the cutout rates are high. Based on the results, this paper presents a numerical model that can be used in the packaging design to estimate the compression strength of corrugated cardboard boxes.

Keywords: paperboard packaging; finite element method; box compression test; numerical model; cutout; compression force

Citation: Fehér, L.; Mrówczyński, D.; Pidl, R.; Böröcz, P. Compressive Strength of Corrugated Paperboard Packages with Low and High Cutout Rates: Numerical Modelling and Experimental Validation. *Materials* **2023**, *16*, 2360. https://doi.org/10.3390/ma16062360

Academic Editor: Nicholas Fantuzzi

Received: 23 February 2023
Revised: 11 March 2023
Accepted: 14 March 2023
Published: 15 March 2023

1. Introduction

In logistics, the packaging of products is essential [1–5]. The basic function of packaging is to protect packaged products against the effects occurring in logistics. The typical loading conditions in logistics during transport, storage, and loading processes can be divided into two groups: mechanical and climatic loads [6–8]. External loads can cause damage to packaging, which can interrupt logistics chains [9–11]. Much of the packaging used in logistics is made of corrugated board and can be closed boxes, open boxes on the top, or even sidewall-less boxes. The main advantages of corrugated packaging are its economy, reliable protection of products, relatively low specific weight, low packaging costs, and the recyclability and biodegradability of the paper [12,13]. Corrugated board is always made up of odd layers. The number of layers in practice is 3, 5, or 7. The corrugated layers can be A, B, C as normal corrugated; and D, E, F as micro-corrugated according to the corrugation height [14]. The flat layers are bonded to the corrugated layer with a water-soluble adhesive. Corrugated product manufacturers sell their products in sheet form. From these flat sheets, the packaging manufacturers cut out the expanded form of the packaging material. These cutout flat sheets are delivered to the users. To make the final packaging, the producers fold them in the appropriate way to produce the finished box and fix the sleeves and the top and bottom sheets. As can be seen from the foregoing, the corrugated board packaging is delivered to the place of use in a flat condition, so that it is transported with good volume utilization of the vehicles.

For the production of corrugated sheets, there are several quality grades of both flat and corrugated sheets, basically classified by their fiber content:

Flat sheets can be: kraftliner, which contains only primary pulp fibers; bodyliner (duplex), which contains primary fibers in one part and recycled fibers in the other part; and srenc, which are made by processing mixed waste paper. The papers forming the corrugated layer can be divided into two categories: fluting (semi-chemical), which contains primary fibers and is made suitable for retaining the fluting by special chemical treatment; wellenstoff, which is virtually identical to the base material for flat layers; and srenc.

The paper material is highly sensitive to moisture and all its strength properties are significantly reduced by high moisture content. High fiber content materials (kraftliner, fluting) are less sensitive to moisture than paper materials containing recycled fibers. For this reason, when corrugated board packaging is tested, the type of paper used should always be specified for each layer.

For both papers and corrugated sheets made from them, the direction of manufacture should be interpreted. In the paper industry, the direction of manufacture is abbreviated as MD, whereas the direction perpendicular to the manufacture is abbreviated as CD. The mechanical properties (tensile strength, bending stiffness, compressive strength) of the paper in the production direction are significantly higher than those of the CD. Similarly, the thickness direction (ZD) that can be interpreted for corrugated sheets is important because it determines the second moment of area of the corrugated sheet under compressive loading [15–17].

In box manufacturing, FEFCO designates each box type by a numerical code, the simplest and most commonly used box variant being the so-called slotted box, designated by the code FEFCO0201 [14].

In logistics, corrugated boxes are stacked for both storage and transport. The determination of the stacking load capacity for corrugated boxes is still a problem for designers [18].

The best-known empirical formula for sizing for stack loading is the McKee equation, which attempts to determine the compressive strength of a corrugated box based on the Edge Crush Test (ECT) of the corrugated sheet [19]. The ECT of a corrugated board can be estimated by artificial intelligence [20]. Several authors have modified the original equation, and Kellicut and Landt attempted to use the ring crush test (RCT) value [21]. Beldie and co-workers developed a mechanical model for corrugated cardboard boxes subjected to static compressive loading [8]. The authors modelled the corrugated cardboard box as an orthotropic, linearly elastic-plastic laminate. Nowadays, FEM (finite element method) techniques are continuously improving, and several software tools have become available for the strength analysis of statically indeterminate structures. There are several studies in the literature that investigate the numerical analysis of the transverse shear stiffness of corrugated paper sheets [22–26]. The applicable model can, in principle, be simplified by homogenizing the material [27–31]. The corrugated paper sheet is assumed to be a composite material, neglecting that it is made up of multiple different layers. The method of homogenization was demonstrated by Hohe for sandwich panels by basing the approach on strain energy [32]. To investigate the effect of wrinkling on the local strength of corrugated paper sheets, a comparison of laboratory experimental and FEM results was performed by Thakkar et al. [33]. Beex and Peerlings [34] also performed similar experiments. Leminen et al. performed experimental, as well as numerical, studies on the effect of compression creasing on the mechanical properties of corrugated sheet [35].

In practice, it is often necessary to make holes and cutouts of various sizes in the sidewalls of the box for various purposes. These can have several purposes:

- Hand holes for carrying [18];
- Ventilation openings [36];
- Products requiring cold storage;
- Window-like cutouts for reading product identifiers or codes;
- Window-like cutouts to reduce the amount of corrugated board used.

The effect of sidewall cutouts in reducing the compressive strength of the box has been investigated by several authors, but good estimation for BCT value on various sizes and locations of the cutouts has not yet been published. Experiments and modelling have generally been carried out on specific box types used in practice by the authors, and very often, conflicting data have been obtained [7,36].

It would be advisable to measure BCT (box compression test) values for each variant under laboratory conditions in a methodical way, by gradually reducing the surface area of the sidewalls, and to develop a FEM parameterization based on this, which represents the measured BCT values with a good approximation. This procedure could also help designers to place a cut of any shape in the actual cut location, and the model could be used to numerically determine the reduction in BCT value.

2. Materials and Methods

2.1. Samples

The box samples were made from single-wall B-flute corrugated cardboard material. The material properties of the tested corrugated cardboard are shown in Table 1. The corrugated cardboard contained the following components:

- Outer liner: 210 GD2 (weight 210 g/m^2, coated white lined chipboard with grey back, quality class 2);
- Fluting medium: 120 HC (weight 120 g/m^2, high compression Wellenstoff);
- Inner liner: 130 TL 3 (weight 130 g/m^2, Testliner, quality class 3).

Figure 1 shows the tested box samples. Five different box lengths with 5 different cutout rates were tested; therefore, 25 types of samples were analyzed for this study. The widths and heights of the samples were the same, 300 mm in each case. The length dimension of the boxes varied from 200 mm to 600 mm with a 100 mm increment. The cutout rates were 0%, 4%, 16%, 36%, and 64% with respect to the total surface area of the sidewalls of the boxes. These cutouts were positioned in the middle of the sidewalls of the boxes along all four sides, as shown in Figure 1.

Table 1. Material properties of corrugated cardboard used in the study.

Properties	Specification	Standard
board thickness	2.8 mm	ISO 3034 (FEFCO no. 3)
grammage	512 g/m^2	ISO 536:1995
edge crush test (ECT)	5.1 kN/m	ISO 3037 (FEFCO no. 8)
bending stiffness (BS)	4.23 Nm (MD) 2.90 Nm (CD)	ISO 5628:1990
bursting strength (BST)	676 kPa	ISO 2759 (FEFCO no. 4)

Figure 2 shows a few examples of the tested box samples. In Figure 2, each size and cutout group are presented with one or two examples. The exact sizes of the tested box samples and the sizes of the cutouts are shown in Figure 1.

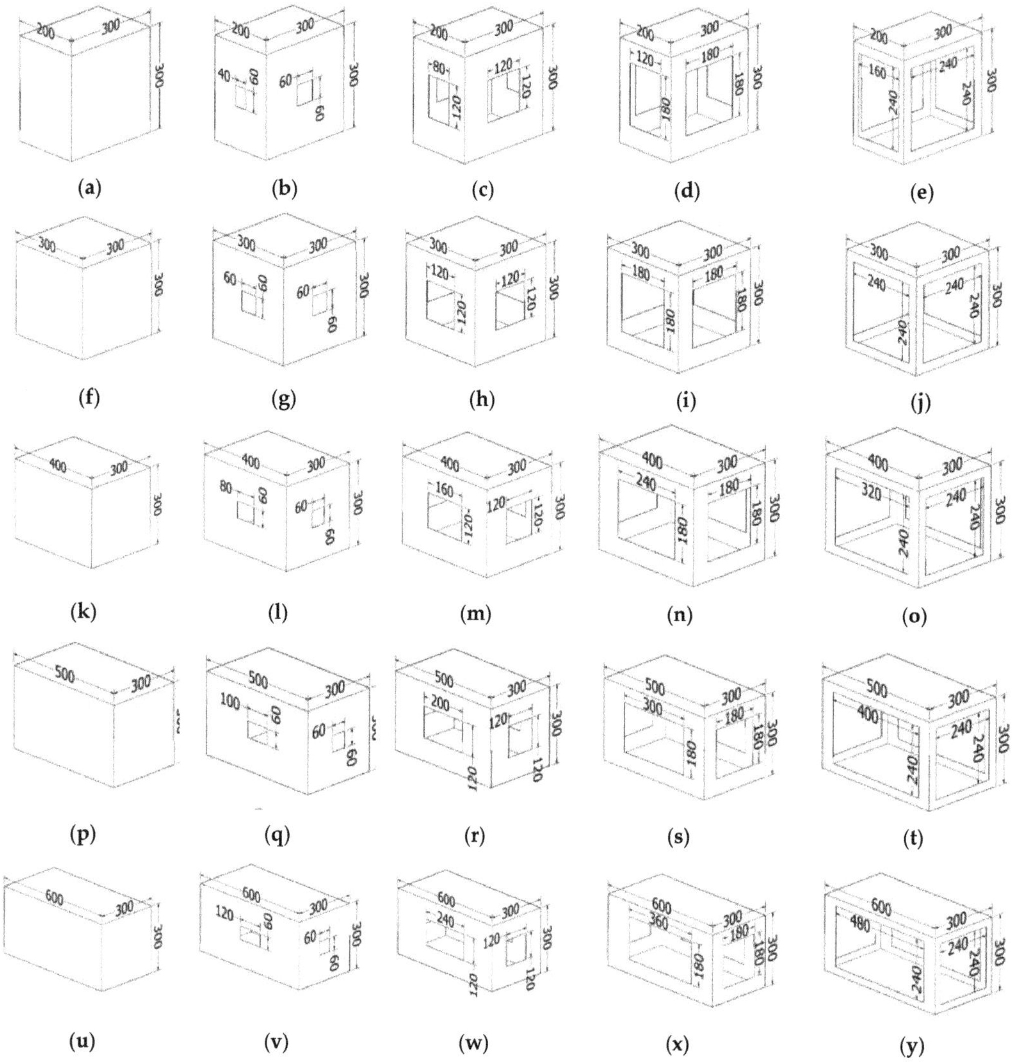

Figure 1. Cases of tested samples with different cutouts. **(a–y)** Twenty-five tested samples.

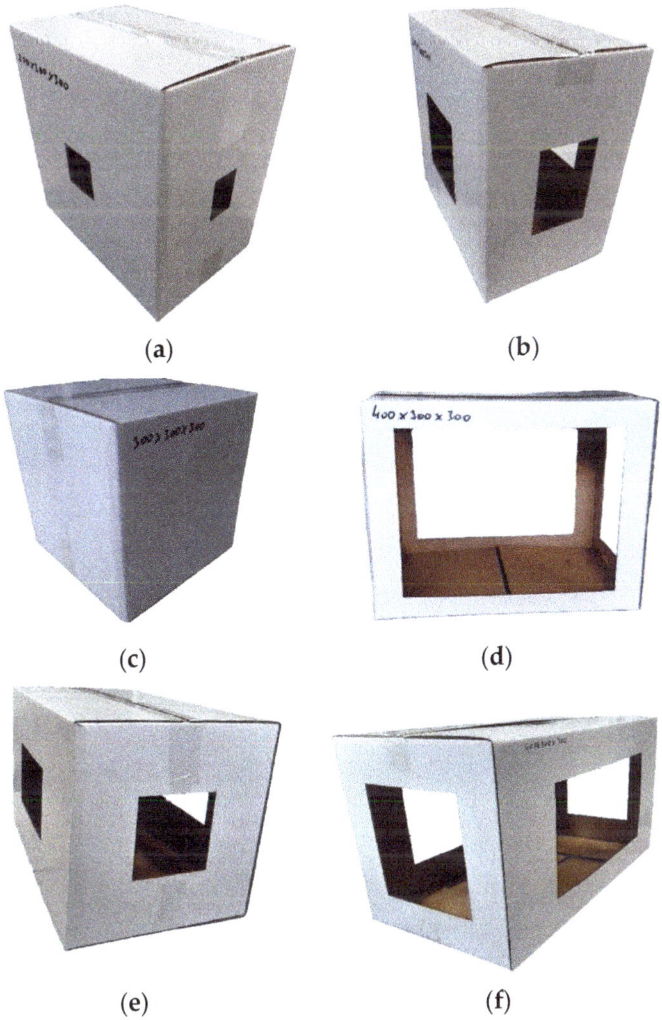

Figure 2. Examples of tested box samples: (**a**) box 200 × 300 × 300 mm with 4% cutout, (**b**) box 200 × 300 × 300 mm with 16% cutout, (**c**) box 300 × 300 × 300 mm with 0% cutout, (**d**) box 400 × 300 × 300 mm with 64% cutout, (**e**) box 500 × 300 × 300 mm with 16% cutout, (**f**) box 600 × 300 × 300 mm with 36% cutout.

2.2. Experimental Setup

The experiment consisted of a BCT test (box compression test) to measure the strength behavior of different boxes. The BCT setup is shown in Figure 3. Before the test, the boxes were preconditioned at 30 °C ± 1 °C and 20–30% RH (relative humidity) for 24 h and then conditioned at 23 °C ± 1 °C and 50 ± 2% RH for 24 h in a climate-testing chamber in accordance with the ASTM D4332 standard. Right after the conditioning process, the BCT measurement was performed according to the ASTM D642 standard. The testing speed was 12.7 mm/min ± 2.5 mm/min until failure of the box occurred. The recorded data were the compression force and the deformation, continuously during the measurement. For statistical evaluation, 10 samples were tested for each box design.

Figure 3. BCT measurement setup.

2.3. *Numerical Model of Cardboard Boxes with Different Cutouts*

Numerical calculations were performed in Abaqus Unified FEA software [37]. Twenty-five different packaging models with the dimensions shown in Figure 1 were built to compute their compressive strength. To speed up and simplify the analysis, only 1/8 of the box was modeled. The top and bottom of the packaging was also omitted because they do not affect the load-bearing capacity. Figure 4 shows an example model of 1/8 parts of the packaging for a case with a length of 200 mm and cutout rates of 36%.

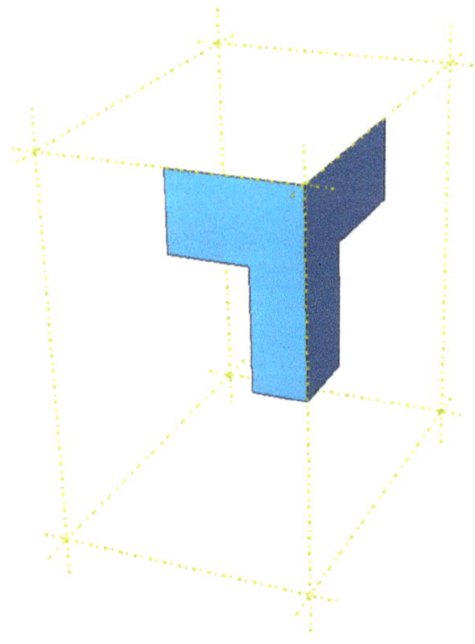

Figure 4. Scheme of the 1/8 parts of the packaging.

The proper behavior of the box under load was ensured by defining symmetry boundary conditions in the x-direction ($u_x = 0$, $\varphi_y = 0$, $\varphi_z = 0$), y-direction ($u_y = 0$, $\varphi_x = 0$, $\varphi_z = 0$), and z-direction ($u_z = 0$, $\varphi_x = 0$, $\varphi_y = 0$), where u_i is the displacement along the i-axis and φ_i is the rotation angle about i-axis (see Figure 5). The out-of-plane displacement of the top edges was blocked ($u_x = 0$, $u_z = 0$), which results from the existence of flaps in the real packaging. A vertical displacement, u_y, was also applied to the top edges, which simulates the box compression test. The analysis was carried out in two calculation stages.

In the first, a buckling analysis was performed to obtain the mode of global imperfections. The previously determined shape of imperfection was applied to the packaging in the second computational step, and it was loaded with a vertical displacement of the top edges in order to obtain the box compressive strength.

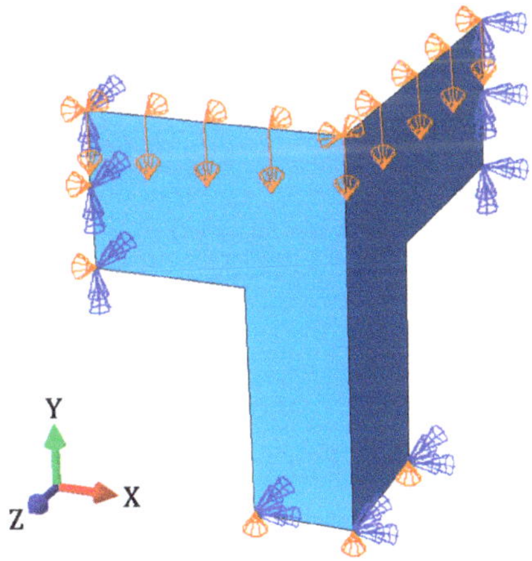

Figure 5. Boundary conditions of the box.

In the strength analyses, the linear elastic orthotropic material model was used. Additionally, Hill plasticity was used to differentiate the reference yield strength only in the machine and cross direction [38]. The packaging was made of B-flute cardboard with a grammage of 512 g/m^2; therefore, the material was marked as B-512. In Table 2, the corrugated board input data to the constitutive model are shown. The mechanical parameters of the material were analytically determined by the BSE System via FEMAT [39] based on the laboratory test data contained in Table 1. Columns 2–7 of Table 2 represent elastic orthotropic material parameters: E_1 and E_2 are the moduli of elasticity in the MD and CD, ν_{12} is the Poisson's ratio, G_{12} is the in-plane shear modulus, and G_{13} and G_{23} are the transverse shear moduli. Columns 8 and 9 contain plastic material parameters: σ_0 is the initial yield stress and R_{11} is the yield stress ratio in the MD.

Table 2. Material parameters of the B-flute corrugated board.

Grade	E_1	E_2	ν_{12}	G_{12}	G_{13}	G_{23}	σ_0	R_{11}
	(MPa)	(MPa)	(-)	(MPa)	(MPa)	(MPa)	(MPa)	(-)
B-512	2149	1474	0.36	3348	3	5	1.83	0.86

For each packaging geometry, buckling analysis and main compression calculations were performed, which gives a total of 50 numerical analyses. The 4-node quadrilaterals shell elements without integration, named S4R, were used in all computations [37]. Different global mesh sizes were assumed for different geometries. For example, for the model shown in Figures 4 and 5, the global mesh size was equal to 5.5 mm, which ultimately resulted in 856 nodes, 783 elements, and 5136 degrees of freedom.

3. Results

For each box sample, the maximum compression force values were determined both with measurement and FEM analysis. These experimental and numerical results are shown in Table 3. The results of the numerical tests (Table 3) show that the 400 × 300 × 300 with 0% cutout is the stiffest box, with a compression force of 2731 N. This shows a good match with the experimental data, in which case, the 400 × 300 × 300 box with 0% cutout also has the maximum compression force, 2651 N. The same can be seen for the weakest box because from both tests (numerical, experimental), the 200 × 300 × 300 box with a 64% cutout has the lowest compression force value.

Table 3. Maximum compression force values (N).

Type	*L*	Cutout Rates (%)				
	(mm)	0	4	16	36	64
experimental	200	2261	2218	1851	1347	615
	300	2367	2275	1981	1373	734
	400	2651	2537	2291	1656	946
	500	2402	2203	2066	1603	877
	600	2339	2189	1980	1591	862
numerical	200	2333	2202	1763	1082	712
	300	2501	2440	2163	1539	828
	400	2731	2556	2185	1625	899
	500	2494	2296	1959	1534	1005
	600	2158	1983	1711	1445	898

In Figure 6, typical BCT measuring data can be seen. Figure 6 shows the force displacement recorded data of the 400 × 300 × 300 box with 36% cutout; similar graphs were obtained in each case. There are 10 graphs in Figure 6 due to the number of tested samples being 10 in each box group. The maximum compression force was calculated using the average of the maximum values of the 10 samples.

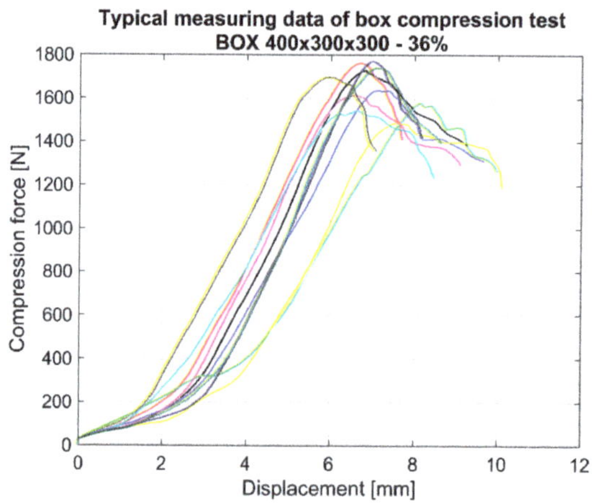

Figure 6. Typical experimental BCT plots for this study.

In Table 4, the percentage differences are shown. If the difference is positive, that means the numerical model overestimates the maximum BCT force, whereas if the value is negative, that shows an underestimation. In only six cases, the percentage difference is greater than 10%. The worst predictions occur in the 64% cutout group due to the absolute average difference being the highest in this group. The 200 × 300 × 300 box with 36% cutout has the highest percentage difference, −20%.

In Figure 7, the maximum compression forces are shown, which come from the numerical analysis and the BCT measurements. The dotted blue and continuous red lines represent the linear curves that were fitted in the numerical and the experimental data points. In Figure 7, the R^2 values (coefficient of determination) are also presented, which come from all data points for the comparison of the numerical and experimental data. These R^2 values are very high, in the 0.959–0.996 range.

Table 4. Differences between the numerical and experimental compression force results.

L (mm)	Cutout Rates (%)				
	0	4	16	36	64
	Percentage Difference				
200	3%	−1%	−5%	−20%	16%
300	6%	7%	9%	12%	13%
400	3%	1%	−5%	−2%	−5%
500	4%	4%	−5%	−4%	15%
600	−8%	-9%	−14%	−9%	4%

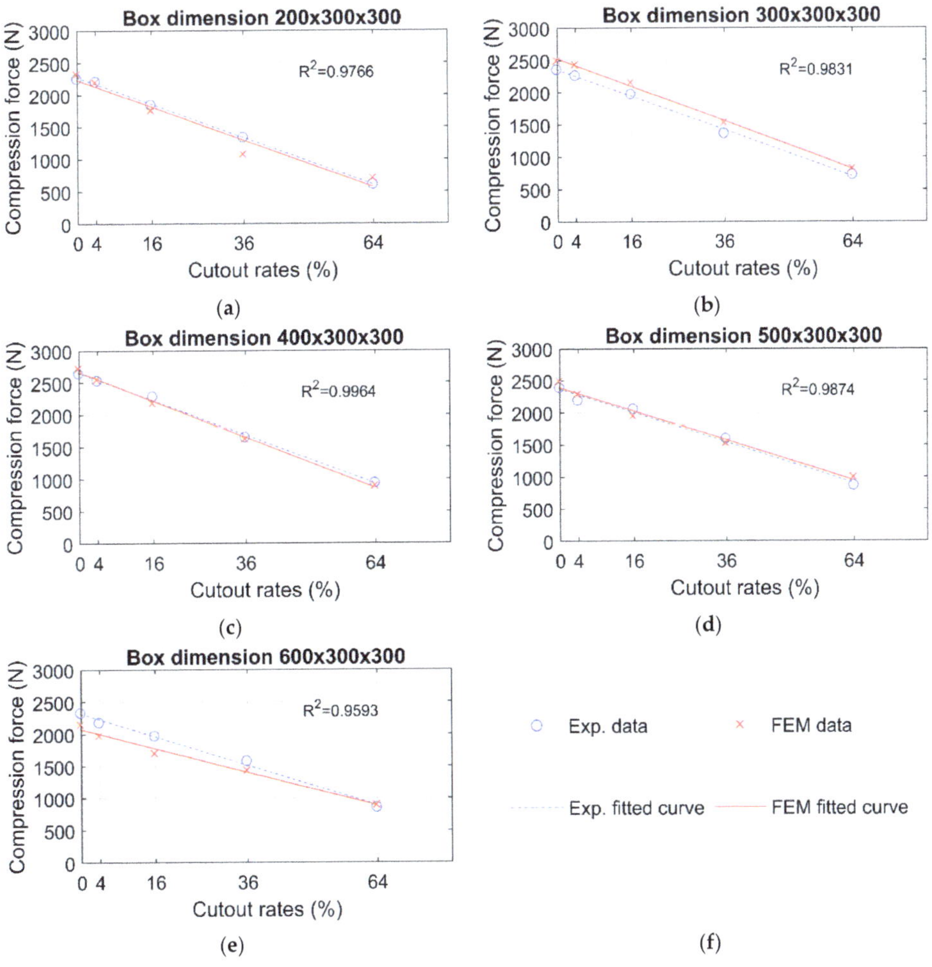

Figure 7. Numerical and experimental maximum compression forces: (**a**) 200 × 300 × 300 mm box, (**b**) 300 × 300 × 300 mm box, (**c**) 400 × 300 × 300 mm box, (**d**) 500 × 300 × 300 mm box, (**e**) 600 × 300 × 300 mm box, (**f**) legend.

A few deformation shape examples are shown in Figure 8. In Figure 8a,c,e, the deformation shapes come from experiments, and on the right side of Figure 8b,d,f, the results come from FEM analysis.

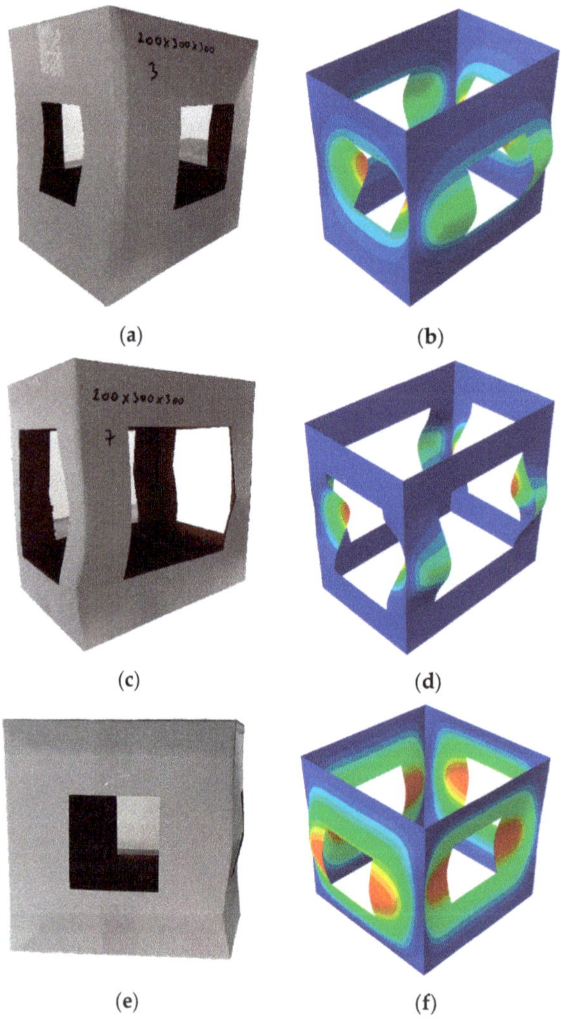

Figure 8. Deformation shapes from experiments and numerical analysis: (**a**) experimental box 200 × 300 × 300 mm with 16% cutout, (**b**) numerical box 200 × 300 × 300 mm with 16% cutout, (**c**) experimental box 200 × 300 × 300 mm with 36% cutout, (**d**) numerical box 200 × 300 × 300 mm with 36% cutout, (**e**) experimental box 300 × 300 × 300 mm with 16% cutout, (**f**) numerical box 300 × 300 × 300 mm with 16% cutout.

4. Discussion

The authors showed the evaluation of the BCT tests for the sample boxes with large cutout areas (Figure 1) in [40,41], and this study is a continuation of those. In this work, the goal was to create a numerical model using the finite element method, which gives good predictions for the BCT values. The finite element method is widely used to model mechanical properties of corrugated boxes [1,8,15,16,18,36,42–54]. Other authors have conducted similar work [1,2,43,55,56], but in those, the cutout area was not as high as 64%.

Table 3 shows the maximum compression force values from the numerical analysis and from the BCT tests. Comparing these results, it can be seen that the compression force values that come from the numerical analysis are very close to the measured ones. This is even more noticeable in Table 4, which shows the percentage differences between the numerical and experimental compression forces. Overestimation and underestimation of the numerical analysis occur in almost half of the cases, but the differences in most cases are very low. The average absolute percentage difference is 7%. This means the FEM model predicts the reality with a very good accuracy, although the accuracy of the numerical model slightly decreases with higher cutout rates.

In Figure 7, a linear regression is presented. In the previous paper of the authors, it was shown that the linear regression models described the measured data with very high accuracy [40]. This can also be concluded for the data obtained by numerical analysis. Figure 7 shows the R^2 values (coefficient of determination) that come from all data points. These very high R^2 values also show a very good fit to the obtained experimental and numerical data.

In some cases, the deformation shapes are very similar in comparison with the numerical model and the experiment. This can be seen in Figure 8a–d. In most of the cases, however, the deformation shapes are different in the same group, since the deformation shapes are highly driven by the imperfections of the boxes. This phenomenon can be seen in Figure 8e–f. These imperfections occur in each case; therefore, the deformation shapes are different if multiple same size boxes with the same cutout rates are tested with BCT. The imperfections could be caused by different things such as the raw material, inappropriate manufacturing or handling, etc.

The results show that the BCT results of the boxes from a low to high cutout rate can be predicted with a high accuracy using this numerical model. The novelty of this paper is to show the ability of FEM analysis to estimate the BCT results of corrugated cardboard boxes with very high cutout rates. In this work, a wide range of the box sizes and the cutout rates were involved, but this range is not comparable with the different type of boxes used in the industry; therefore, all findings apply only to the tested box types. Therefore, a future study should investigate to design and test a box for industry usage with a higher cutout rate.

Moreover, the presented numerical model for this study provides a cost-effective and efficient alternative way in comparison to the traditional experimental testing methods, which can be time-consuming and expensive. From a practical point of view, the presented numerical method is accurate enough for the authority of use. By using numerical simulations, the number of physical prototypes needed for testing can be significantly reduced, leading to cost savings in the design and development of corrugated boxes with cutouts. It also has to be mentioned that the numerical model can be used to perform parametric studies, where the effect of different cutout sizes on the BCT values can be evaluated without the need for additional physical testing.

Overall, the results of this study demonstrate the potential of using finite element method simulations to accurately predict the BCT values of corrugated boxes with larger cutout areas than what has been shown before. The good estimation for the numerical and experimental results, as well as the low average absolute percentage difference, indicate that the developed model is a reliable tool for predicting the performance of corrugated boxes in real-world applications. Furthermore, this study highlights the importance of considering the effect of cutout area on BCT values, as this can significantly impact the strength and durability of corrugated boxes.

5. Conclusions

This paper presents a numerical simulation and experimental verification method for the investigation of cutout problems of a single-wall corrugated board box. Although the method of choosing samples for the experimental test follows theoretically located and sized cutouts, the numerical result of analysis shows surprising accuracy in load capacity estimating for a corrugated cardboard box, the structure of which is basically made by viscoelastic material. The accuracy of the model decreases how the cutout rate increases from 0% to 64%. The results give new information for engineers to better understand the strength reduction effect of cutout holes such as carrying or ventilation holes.

Author Contributions: Conceptualization, D.M., R.P. and L.F.; methodology, P.B., D.M. and R.P.; software, D.M.; validation, D.M.; formal analysis, R.P., L.F. and D.M.; investigation, D.M., R.P., L.F. and P.B.; writing—original draft preparation, L.F. and D.M.; writing—review and editing, L.F., D.M., R.P. and P.B.; visualization, D.M., R.P. and L.F.; supervision, P.B. All authors have read and agreed to the published version of the manuscript.

Funding: This research received no external funding.

Institutional Review Board Statement: Not applicable.

Informed Consent Statement: Not applicable.

Data Availability Statement: The data published in this research are available on request from the first author and corresponding authors.

Conflicts of Interest: The authors declare no conflict of interest.

References

1. Fadiji, T.; Coetzee, C.; Opara, U.L. Compression Strength of Ventilated Corrugated Paperboard Packages: Numerical Modelling, Experimental Validation and Effects of Vent Geometric Design. *Biosyst. Eng.* **2016**, *151*, 231–247. [CrossRef]
2. Fadiji, T.; Berry, T.; Coetzee, C.; Opara, L. Investigating the Mechanical Properties of Paperboard Packaging Material for Handling Fresh Produce Under Different Environmental Conditions: Experimental Analysis and Finite Element Modelling. *J. Appl. Packag. Res.* **2017**, *9*, 20–34.
3. Opara, U.L.; Fadiji, T. Compression Damage Susceptibility of Apple Fruit Packed inside Ventilated Corrugated Paperboard Package. *Sci. Hortic.* **2018**, *227*, 154–161. [CrossRef]
4. Fadiji, T.; Coetzee, C.; Chen, L.; Chukwu, O.; Opara, U.L. Susceptibility of Apples to Bruising inside Ventilated Corrugated Paperboard Packages during Simulated Transport Damage. *Postharvest Biol. Technol.* **2016**, *118*, 111–119. [CrossRef]
5. Fadiji, T.; Coetzee, C.; Pathare, P.; Opara, U.L. Susceptibility to Impact Damage of Apples inside Ventilated Corrugated Paperboard Packages: Effects of Package Design. *Postharvest Biol. Technol.* **2016**, *111*, 286–296. [CrossRef]
6. Marsh, K.; Bugusu, B. Food Packaging—Roles, Materials, and Environmental Issues. *J. Food Sci.* **2007**, *72*, R39–R55. [CrossRef]
7. Singh, S.; Singh, J.; Paek, Y. New Pressure Sensitive Device to Measure and Predict Package Drops. *J. Appl. Packag. Res.* **2009**, *3*, 149–160.
8. Beldie, L.; Sandberg, G.; Sandberg, L. Paperboard Packages Exposed to Static Loads–Finite Element Modelling and experiments. *Packag. Technol. Sci.* **2001**, *14*, 171–178. [CrossRef]
9. Lu, F.; Ishikawa, Y.; Kitazawa, H.; Satake, T. Measurement of Impact Pressure and Bruising of Apple Fruit Using Pressure-Sensitive Film Technique. *J. Food Eng.* **2010**, *96*, 614–620. [CrossRef]
10. Fei, L.; Ishikawa, Y.; Kitazawa, H.; Satake, T. Assessment and Prediction of Repetitive Impact Damage to Apple Fruit Using Pressure-Sensitive Film Technique. *J. Food Agric. Environ.* **2012**, *10*, 156–160.
11. Pathare, P.B.; Opara, U.L. Structural Design of Corrugated Boxes for Horticultural Produce: A Review. *Biosyst. Eng.* **2014**, *125*, 128–140. [CrossRef]
12. Fadiji, T.; Berry, T.M.; Coetzee, C.J.; Opara, U.L. Mechanical Design and Performance Testing of Corrugated Paperboard Packaging for the Postharvest Handling of Horticultural Produce. *Biosyst. Eng.* **2018**, *171*, 220–244. [CrossRef]
13. Talbi, N.; Batti, A.; Ayad, R.; Guo, Y.Q. An Analytical Homogenization Model for Finite Element Modelling of Corrugated Cardboard. *Compos. Struct.* **2009**, *88*, 280–289. [CrossRef]
14. FEFCO Code Design Style Library for Corrugated Board Products. 2022. Available online: https://www.paperfirst.info/fefco-launches-the-12th-edition-of-the-fefco-code/ (accessed on 22 February 2023).
15. Garbowski, T.; Gajewski, T.; Grabski, J.K. The Role of Buckling in the Estimation of Compressive Strength of Corrugated Cardboard Boxes. *Materials* **2020**, *13*, 4578. [CrossRef]
16. Garbowski, T.; Gajewski, T.; Grabski, J.K. Role of Transverse Shear Modulus in the Performance of Corrugated Materials. *Materials* **2020**, *13*, 3791. [CrossRef]
17. Garbowski, T.; Gajewski, T.; Grabski, J.K. Torsional and Transversal Stiffness of Orthotropic Sandwich Panels. *Materials* **2020**, *13*, 5016. [CrossRef] [PubMed]
18. Biancolini, M.E.; Brutti, C. Numerical and Experimental Investigation of the Strength of Corrugated Board Packages. *Packag. Technol. Sci.* **2003**, *16*, 47–60. [CrossRef]
19. McKee, R.C.; Gander, J.W.; Wachuta, J.R. Edgewise Compression Strength of Corrugated Board. 1961. Available online: https://smartech.gatech.edu/bitstream/handle/1853/1660/1108-4_000_07111961.pdf (accessed on 19 February 2023).

20. Garbowski, T.; Knitter-Piątkowska, A.; Grabski, J.K. Estimation of the Edge Crush Resistance of Corrugated Board Using Artificial Intelligence. *Materials* **2023**, *16*, 1631. [CrossRef]
21. Kellicutt, K.; Landt, E. Development of Design Data for Corrugated Fiberboard Shipping Containers. *Tappi J.* **1952**, *35*, 398–402.
22. Yan, J.; Wang, G.; Li, Q.; Zhang, L.; Yan, J.D.; Chen, C.; Fang, Z. A Comparative Study on Damage Mechanism of Sandwich Structures with Different Core Materials under Lightning Strikes. *Energies* **2017**, *10*, 1594. [CrossRef]
23. Zhang, X.; Jin, X.; Xie, G.; Yan, H. Thermo-Fluidic Comparison between Sandwich Panels with Tetrahedral Lattice Cores Fabricated by Casting and Metal Sheet Folding. *Energies* **2017**, *10*, 906. [CrossRef]
24. Duan, Y.; Zhang, H.; Sfarra, S.; Avdelidis, N.P.; Loutas, T.H.; Sotiriadis, G.; Kostopoulos, V.; Fernandes, H.; Petrescu, F.I.; Ibarra-Castanedo, C.; et al. On the Use of Infrared Thermography and Acousto—Ultrasonics NDT Techniques for Ceramic-Coated Sandwich Structures. *Energies* **2019**, *12*, 2537. [CrossRef]
25. Garbowski, T.; Jarmuszczak, M. Homogenization of Corrugated Paperboard. Part 1. Analytical Homogenization. *Prz. Pap.* **2014**, *70*, 345–349.
26. Garbowski, T.; Jarmuszczak, M. Homogenization of Corrugated Paperboard: Part 2. Numerical Homogenization. *Prz. Pap.* **2014**, *70*, 390–394.
27. Buannic, N.; Cartraud, P.; Quesnel, T. Homogenization of Corrugated Core Sandwich Panels. *Compos. Struct.* **2003**, *59*, 299–312. [CrossRef]
28. Biancolini, M.E. Evaluation of Equivalent Stiffness Properties of Corrugated Board. *Compos. Struct.* **2005**, *69*, 322–328. [CrossRef]
29. Abbès, B.; Guo, Y.Q. Analytic Homogenization for Torsion of Orthotropic Sandwich Plates: Application to Corrugated Cardboard. *Compos. Struct.* **2010**, *92*, 699–706. [CrossRef]
30. Marek, A.; Garbowski, T. Homogenization of Sandwich Panels. *Comput. Assist. Methods Eng. Sci.* **2017**, *22*, 39–50.
31. Garbowski, T.; Marek, A. Homogenization of Corrugated Boards through Inverse Analysis. In Proceedings of the 1st International Conference on Engineering and Applied Sciences Optimization, Kos Island, Greece, 4–6 June 2014; pp. 1751–1766.
32. Hohe, J. A Direct Homogenisation Approach for Determination of the Stiffness Matrix for Microheterogeneous Plates with Application to Sandwich Panels. *Compos. Part B Eng.* **2003**, *34*, 615–626. [CrossRef]
33. Thakkar, B.K.; Gooren, L.G.J.; Peerlings, R.H.J.; Geers, M.G.D. Experimental and Numerical Investigation of Creasing in Corrugated Paperboard. *Philos. Mag.* **2008**, *88*, 3299–3310. [CrossRef]
34. Beex, L.A.A.; Peerlings, R.H.J. An Experimental and Computational Study of Laminated Paperboard Creasing and Folding. *Int. J. Solids Struct.* **2009**, *46*, 4192–4207. [CrossRef]
35. Leminen, V.; Tanninen, P.; Pesonen, A.; Varis, J. Effect of Mechanical Perforation on the Press-Forming Process of Paperboard. *Procedia Manuf.* **2019**, *38*, 1402–1408. [CrossRef]
36. Han, J.; Park, J.M. Finite Element Analysis of Vent/Hand Hole Designs for Corrugated Fibreboard Boxes. *Packag. Technol. Sci.* **2007**, *20*, 39–47. [CrossRef]
37. Abaqus Unified FEA—Mechanical Simulation. Available online: https://www.3ds.com/products-services/simulia/products/abaqus/ (accessed on 19 February 2023).
38. Hill, R.; Orowan, E. A Theory of the Yielding and Plastic Flow of Anisotropic Metals. *Proc. R. Soc. Lond. Ser. A Math. Phys. Sci.* **1997**, *193*, 281–297. [CrossRef]
39. FEMat BSE System. Available online: http://fematsystems.pl/bse-system_en/ (accessed on 19 February 2023).
40. Fehér, L.; Pidl, R.; Böröcz, P. Compression Strength Estimation of Corrugated Board Boxes for a Reduction in Sidewall Surface Cutouts—Experimental and Numerical Approaches. *Materials* **2023**, *16*, 597. [CrossRef]
41. Pidl, R.; Fehér, L.; Böröcz, P. The Effect of Side Wall Cutout Sizes on Corrugated Box Compression Strength in the Function of Length-to-Width Ratios—An Experimental Study. *Appl. Sci.* **2022**, *12*, 6939. [CrossRef]
42. Garbowski, T.; Gajewski, T.; Grabski, J.K. Estimation of the Compressive Strength of Corrugated Cardboard Boxes with Various Openings. *Energies* **2021**, *14*, 155. [CrossRef]
43. Garbowski, T.; Gajewski, T.; Grabski, J.K. Estimation of the Compressive Strength of Corrugated Cardboard Boxes with Various Perforations. *Energies* **2021**, *14*, 1095. [CrossRef]
44. Mrówczyński, D.; Garbowski, T.; Knitter-Piątkowska, A. Estimation of the Compressive Strength of Corrugated Board Boxes with Shifted Creases on the Flaps. *Materials* **2021**, *14*, 5181. [CrossRef]
45. Urbanik, T.J.; Frank, B. Box Compression Analysis of World-Wide Data Spanning 46 Years. *Wood Fiber Sci.* **2006**, 399–416.
46. Nordstrand, T.M.; Carlsson, L.A. Evaluation of Transverse Shear Stiffness of Structural Core Sandwich Plates. *Compos. Struct.* **1997**, *37*, 145–153. [CrossRef]
47. Nordstrand, T. Basic Testing and Strength Design of Corrugated Board and Containers. Ph.D. Thesis, Lund University, Lund, Sweden, 2003.
48. Avilés, F.; Carlsson, L.A.; May-Pat, A. A Shear-Corrected Formulation for the Sandwich Twist Specimen. *Exp. Mech.* **2012**, *52*, 17–23. [CrossRef]
49. Urbanik, T.J.; Saliklis, E.P. Finite Element Corroboration of Buckling Phenomena Observed in Corrugated Boxes. *Wood Fiber Sci.* **2003**, *35*, 322–333.
50. Sohrabpour, V.; Hellström, D. Models and Software for Corrugated Board and Box Design. In Proceedings of the International Conference on Engineering Design, ICED11, Kongens Lyngby, Copenhagen, Denmark, 15–18 August 2011; pp. 392–401.

51. Suarez, B.; Muneta, M.L.M.; Sanz-Bobi, J.D.; Romero, G. Application of Homogenization Approaches to the Numerical Analysis of Seating Made of Multi-Wall Corrugated Cardboard. *Compos. Struct.* **2021**, *262*, 113642. [CrossRef]
52. Garbowski, T.; Gajewski, T. Determination of Transverse Shear Stiffness of Sandwich Panels with a Corrugated Core by Numerical Homogenization. *Materials* **2021**, *14*, 1976. [CrossRef]
53. Nguyen-Minh, N.; Tran-Van, N.; Bui-Xuan, T.; Nguyen-Thoi, T. Static Analysis of Corrugated Panels Using Homogenization Models and a Cell-Based Smoothed Mindlin Plate Element (CS-MIN3). *Front. Struct. Civ. Eng.* **2019**, *13*, 251–272. [CrossRef]
54. Garbowski, T.; Knitter-Piątkowska, A. Analytical Determination of the Bending Stiffness of a Five-Layer Corrugated Cardboard with Imperfections. *Materials* **2022**, *15*, 663. [CrossRef]
55. Ambaw, A.; Mukama, M.; Fadiji, T.; Opara, U.L. Fresh Fruit Packaging Design Verification through Virtual Prototyping Technique. *Food Packag. Shelf Life* **2022**, *32*, 100858. [CrossRef]
56. Fadiji, T.; Coetzee, C.J.; Berry, T.M.; Opara, U.L. Investigating the Role of Geometrical Configurations of Ventilated Fresh Produce Packaging to Improve the Mechanical Strength – Experimental and Numerical Approaches. *Food Packag. Shelf Life* **2019**, *20*, 100312. [CrossRef]

MDPI AG
Grosspeteranlage 5
4052 Basel
Switzerland
Tel.: +41 61 683 77 34

Materials Editorial Office
E-mail: materials@mdpi.com
www.mdpi.com/journal/materials

www.ingramcontent.com/pod-product-compliance
Lightning Source LLC
Chambersburg PA
CBHW061517210926
51850CB00008B/1951
9783725820740